THE KAM STORY

A Friendly Introduction to the Content, History, and
Significance of Classical Kolmogorov–Arnold–Moser Theory

THE KAM STORY

A Friendly Introduction to the Content, History, and
Significance of Classical Kolmogorov–Arnold–Moser Theory

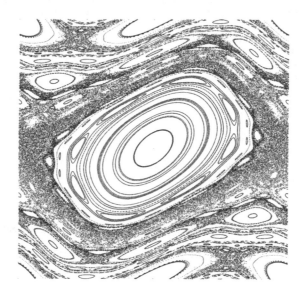

H Scott Dumas
University of Cincinnati, USA

World Scientific

NEW JERSEY · LONDON · SINGAPORE · BEIJING · SHANGHAI · HONG KONG · TAIPEI · CHENNAI

Published by

World Scientific Publishing Co. Pte. Ltd.

5 Toh Tuck Link, Singapore 596224

USA office: 27 Warren Street, Suite 401-402, Hackensack, NJ 07601

UK office: 57 Shelton Street, Covent Garden, London WC2H 9HE

Library of Congress Cataloging-in-Publication Data
Dumas, H. Scott, author.
   The KAM story : a friendly introduction to the content, history, and significance of classical Kolmogorov-Arnold-Moser theory / by H. Scott Dumas (University of Cincinnati, USA).
      pages cm
   Includes bibliographical references and index.
   ISBN 978-9814556583 (alk. paper)
   1. Kolmogorov-Arnold-Moser Theorem. 2. Science--History. I. Title.
   Q172.5.C45D86 2014
   003'.85--dc23

2014000204

British Library Cataloguing-in-Publication Data
A catalogue record for this book is available from the British Library.

Cover illustration courtesy of Ben Vaughan, showing KAM circles and chaotic orbits in the Chirikov standard map.

Figures by the author, unless otherwise indicated.

Printed in Singapore

To Warren Farnholtz

Preface

In many ways classical mechanics serves as the bedrock of physical science, yet surprisingly, it has crucial features that are not widely known. Many people know something about 'chaos theory'—how mathematical models of certain deterministic classical systems fail to predict the evolution of those systems in a practical sense. If they're interested in history, they also know that much of chaos theory was understood by mathematicians almost a century before it was popularized by way of computer models in the last third of the 20th century. But there is a deeper, more interesting story that is not well known outside a circle of experts, and the aim of this book is to tell this story to a wider audience.

The story in a nutshell is this: Right from the start, after enunciating his laws of mechanics and gravitation, Isaac Newton ran into difficulties using those laws to describe the motion of three bodies moving under mutual gravitational attraction (the so-called 'three body problem'). For the next two centuries, these difficulties resisted solution, as the best minds in mathematics and physics concentrated on solving other, increasingly complex model systems in classical mechanics (in the abstract mathematical setting, to 'solve' a system means showing that its trajectories move linearly on so-called 'invariant tori'). But toward the end of the 19th century, using his own new methods, Henri Poincaré confronted Newton's difficulties head-on and discovered an astonishing form of 'unsolvability,' or chaos, at the heart of the three body problem. This in turn led to a paradox. According to Poincaré and his followers, most classical systems should be chaotic; yet observers and experimentalists did not see this in nature, and mathematicians working with model systems could not (quite) prove it to be true either. The paradox persisted for more than a half-century, until Andrey Kolmogorov unraveled it by announcing that, against all expectation, many of the invariant tori from solvable systems remain intact in chaotic systems. These tori make most systems into hybrids—they are a strange, fractal mixture of regularity and chaos. This stunning announcement was later affirmed with rigorous mathematical proofs by Vladimir Arnold and Jürgen Moser, and the names Kolmogorov, Arnold, and Moser were combined in the acronym KAM, by which the theory has since been known. Thus the true picture of classical mechanics—often thought to have been essentially sketched in the 17th century—was not complete until the latter part of the 20th century. And although the mathematical theory is indeed mostly complete, certain applications to problems in physics (especially in celestial and statistical mechanics) have been developed only with great difficulty, and some remain controversial and uncertain even today.

To compare the practical impact of KAM theory to that of relativity or quantum theory is not realistic (to be frank, the practical impact of KAM theory has been limited). Yet in the history of ideas and the philosophy of science, it is not a stretch to rank KAM theory alongside the revolutions in modern physics. But KAM theory—and the paradox that precipitated it— also had the misfortune of playing out over roughly the same interval during which the revolutions of modern physics took place. Not surprisingly, in that period, physicists abandoned classical mechanics to the few hardy mathematicians who remained interested in it. The physicists returned with wondrous stories of their exploits in quantum mechanics, relativity, and nuclear physics. The time has come for mathematicians to tell their tales from this period in a broad setting, too.

When I asked specialists why none of them had yet written a broad overview of KAM theory, they invariably answered that, with several different 'schools' having descended from the original founders of the theory, it would be awkward for any one individual to take up that task. In other words, KAM theory is still slightly controversial, and the experts are understandably touchy about each other's contributions. Since I don't belong to any particular school, I am prepared to step into the breach, or break the ice. I hope the experts will follow me, not with pitchforks, but with first-hand accounts, corrections, and further detail.

H.S. Dumas, December 2013

Acknowledgments

Like its subject, this book has long roots, and there are many people to thank for its development. My interest in history of science began early and was fueled in my undergraduate days by contact with Albert Van Helden and Salomon Bochner. I heard of KAM theory from Asim Barut at the beginning of graduate school, and was inspired to learn more through discussions with my mentors Jim Ellison, Bill Sáenz, Tom Kyner, Vageli Coutsias, François Golse, and especially Pierre Lochak, who taught me much of what I know about the subject and pointed me toward the Russian and Italian literature. Later, I had the good fortune to meet and talk with Jürgen Moser and Vladimir Arnold, along with some of their students, which only magnified my interest.

Once I formed the idea of this book, I was strongly encouraged by Hildeberto Cabral, Jie Chen, Chuck Wells, and especially Teresa Stuchi. When I needed to find more detail about Weierstrass and Poincaré, Alain Albouy directed me to original sources; he and Alain Chenciner later helped me understand those sources in context. Many people looked at early versions of the draft and encouraged me, including Mathias Vogt, Ken Meyer, Qiu Dong Wang, Patricia Yanguas, Jesús Palacián, Klaus Heinemann, Ning Zhong, and Bing Yu Zhang. Once written, the manuscript would not have found a good publisher without the key endorsements of Reuben Hersh, Jacques Féjoz, and Steve Wiggins.

I thank the following people for looking over parts of the manuscript and suggesting references or changes, or correcting errors of fact: Dave Levermore, Pierre Lochak, Jacques Laskar, Jacques Féjoz, Jürgen Pöschel, Jean-Pierre Marco, Steve Wiggins, Bruce Pourciau, Rafael de la Llave, Florin Diacu, Jim Murdock, Lawrence C. Evans, and Luigi Chierchia, whose e-mail tutorials were particularly enlightening and helpful. I'm especially grateful to those who read through and helped edit or correct the entire manuscript, including Heinz Hanßmann, Jim Ellison, Alain Albouy, and Armelle J.F. Clark. Much thanks also to Carles Simó and Ben Vaughan for providing numerically plotted illustrations. Perhaps more than anyone else, I'm indebted to Mikhail Sevryuk for his critical and constructive reading of the book. I also thank him—along with Phil Korman and Leonid Slavin— for help with transliterations and other things Russian that were unfamiliar to me.

Finally, I thank the Charles Phelps Taft Research Center for its support, Nancy Diemler for assistance with LATEX and other practical matters, and my editors Lai Fun Kwong and Rok Ting Tan at World Scientific for their invaluable assistance.

Contents

Chapter 1

Introduction

In his 1954 plenary address to the International Congress of Mathematicians in Amsterdam, the Russian mathematician Andrey Kolmogorov announced a theorem that wowed the mathematical world. Mathematicians quickly realized that, if true as stated, the theorem resolved a paradox that had stood since Henri Poincaré's work at the end of the 19th century, and possibly also invalidated Ludwig Boltzmann's *ergodic hypothesis*[1] that lay at the foundations of statistical mechanics. Even more, if the theorem could be successfully applied to models of planetary motion based on Newtonian physics, the centuries-old goal of showing that the solar system is stable might finally be reached.

It's rare for a mathematical theorem to have such impact, and although Kolmogorov sketched a proof of the theorem that year (in a Russian mathematics journal [Kol54]), and discussed it a few years later (in the proceedings of the Amsterdam congress [Kol57]), the world still waited for definitive mathematical proof with all details spelled out. This came several years later in a series of remarkable papers by Kolmogorov's young student Vladimir Arnold and the German-American mathematician Jürgen Moser. Arnold was the first to show that Kolmogorov's proof-techniques 'worked' by using them to solve a previously intractable 'circle map' problem [Ar61]. The following year, Moser combined Kolmogorov's proof-techniques with other methods to prove a specialized (low-dimensional) version of Kolmogorov's theorem [Mos62] (with one hypothesis that was unexpectedly weak—making the theorem unexpectedly *strong*). Then in 1963, Arnold proved a version of Kolmogorov's theorem valid in all dimensions [Ar63a] (as Kolmogorov had announced in 1954), together with a closely related

---

[1] Terms appearing in *slanted text* (as opposed to *italics*) are defined or discussed in the glossary in Appendix F.

version that he applied to models of solar system stability [Ar63b], though under very restrictive conditions.

Thus was Kolmogorov-Arnold-Moser theory[2] born, and it soon became customary to use the acronym KAM[3] to refer to it. And although KAM theory has continued to evolve to the present day, passing through periods of fashionability and even mild controversy, it also unfortunately suffers undeserved obscurity among non-specialists.

1.1   What this book is, and how it came about

This book presents classical[4] KAM theory in its broadest context. It is intended for mathematicians, physicists and other interested scientists whose training in *classical mechanics* stopped at the level of, say, (one of the editions of) H. Goldstein's book [Gold59], [Gold80], [GoldPS02] but who are nevertheless curious about what lies beyond. Experts may also find certain portions interesting, and I hope that they will add to or correct parts of the story with which they're especially familiar.[5] Finally, the historical and speculative parts should also appeal to anyone interested in the history of ideas.

But let me be frank right from the start: this book will not teach you about KAM theory at a very deep mathematical level. I do not present a complete proof of a KAM theorem in these pages. Instead, the mathematical part of the story is connected by a century-long thread running from Henri Poincaré to Kolmogorov, Arnold, Moser, and beyond. I trace this thread by way of a Hamiltonian function in modern notation, using it to

---

[2]One hears 'KAM theory' more often than 'the KAM theorem.' As was evident right from the start when the founders announced several different versions, there is no 'one' theorem, but instead many variations, each reflecting choices made in the underlying hypotheses and methods of proof. Many of these variations will be detailed below. For another succinct discussion of the early results focusing more on priority, see Part D.1.1 of the reader's guide in Appendix D.

[3]The acronym KAM was coined in [IzC68] by F.M. Izrailev and B.V. Chirikov. Note that in English, one customarily pronounces the three letters separately ('K-A-M'), whereas in Russian (and French), it is a true acronym, pronounced as the one-syllable word 'kam.'

[4]By 'classical KAM theory,' I mean the theory as it was originally developed for finite-dimensional Hamiltonian systems and twist maps of the annulus. The expansion of KAM theory outside its original framework is also touched upon in this book, but is not a main emphasis.

[5]See the book's website   http://thekamstory.wordpress.com/   to read or to submit corrigenda.

show, in a simplified way, how mathematicians dealt with the problem of transforming a 'slightly nonintegrable' Hamiltonian into 'integrable form.' This approach should give the newcomer an idea of what the founders did, and a taste of the new techniques they (and others) created along the way. Since there is no shortage of rigorous mathematical treatments of KAM theory in the literature, readers who want to see complete proofs can choose from a wide selection.[6]

What does seem to be missing from the literature—and what I provide here alongside the simplified mathematics—is an overview of KAM theory, something that explains its content, history, and significance in relatively simple terms. I mean to clear up some common misunderstandings, to give a rough but understandable account of the main ideas, and to show how and why these ideas are important in mathematics, physics, and the history of science.

I can reveal one of the reasons for KAM theory's celebrity right away: Henri Poincaré famously said that understanding perturbations of *integrable Hamiltonian systems* was the 'fundamental problem of dynamics.' This innocuous sounding statement by the 'father of dynamical systems' conferred upon *Hamiltonian perturbation theory* (HPT) a fashionability that it enjoys to this day. Since KAM theory is the key result of HPT, it of course basks in the same glory; but it receives a further—very dramatic— boost from the fact that Poincaré not only did not foresee KAM theory, but hinted that he thought it could not be true. In this sense (and in others to be explained) KAM theory went against the grain of its time.

This book grew out of an informal lecture on KAM theory that I gave in a number of places during the last decade. My view of the subject was formed by many years of being an American in Paris, where in the early days I worked in an area of HPT called Nekhoroshev theory[7] which is closely related to KAM theory. Because Paris is a crossroads of European mathematics, I had a front-row seat from which to view many developments in the subject. As I looked on in amazement, over the years several odd things became evident. First of all, for a mathematical discipline, KAM theory—or HPT generally—is somewhat unsettled. Along each of several dimensions, there's a wider range of views than is ordinarily the case for a

---

[6]See Part D.1.2 of Appendix D for suggestions of where to find nice proofs of KAM theorems.

[7]So named after its developer Nikolai Nekhoroshev (1946–2008), a former student of V.I. Arnold at Moscow State University. See §6.2 for more details.

relatively mature mathematical subject. Let me run through just a few of these dimensions: physicists and mathematicians often differ markedly in their understanding of, use of, and enthusiasm for KAM theory. Researchers from different countries often seem to view and understand KAM theory differently. Occasionally, disagreement erupts over how much Kolmogorov proved in 1954 (some say his sketch-of-a-proof had such big gaps that it wasn't a proof at all; others say that it was complete enough to drop the A and M and simply call the KAM theorem 'Kolmogorov's theorem'). Still others think that C.L. Siegel's name should be attached to the theorem (cf. §4.1 below to see why). In the early days after the announcement and proofs of KAM, there was some controversy over what the theorem might mean for mathematical physics, and physics generally. Did KAM really imply that the solar system was stable (or just that a 'toy model' of it was)? Did it really invalidate the ergodic hypothesis, thus throwing statistical mechanics into a foundational crisis? Later, in the area of HPT dealing with instability, a number of published results were found to have errors, and an uncharacteristic rancor and controversy erupted. Finally, although KAM theory sits right at the heart of *chaos* theory[8] and is called by enthusiasts 'one of the high points of 20th century mathematics,' there is remarkable ignorance of it among scientific journalists and chroniclers of chaos theory, especially in the U.S. All these things—and more—are well known among experts, but experts themselves are rare.

1.2   Representative quotations and commentary

To show the reader that what I say above is not simply a way to generate interest in the subject—that KAM theory really does evoke a wide range of reactions among mathematicians and physicists—I offer here some quotes from (relatively) recent books, in chronological order.

First, from an edition of the book most often used in American universities over the last half-century to teach classical mechanics to graduate students in physics, we have this (the only mention of KAM theory that appears[9]):

---

[8]It's difficult to write the words 'chaos' or 'chaos theory' without quotation marks, as these terms are quite elastic and have never been given universally accepted meanings by mathematicians. (But see the *chaos* entry in the glossary in Appendix F.) This ambiguity also makes them very useful terms, and I won't shy away from them in the sequel.

[9]However, the latest (2002) edition of this book [GoldPS02] (now with coauthors) contains a new chapter on 'classical chaos' with a brief (2-page) section on KAM theory.

"Only in the last few decades has the [solar system] stability question been freshly illuminated, by the application of new (and highly abstract) mathematical techniques. [...] A series of investigations, associated with the names C.L. Siegel, A.N. Kolmogorov, V.I. Arnold, and J. Moser, have shown that stable, bounded motion is possible for a system of n bodies interacting through gravitational forces only. [...] The brilliance of the achievement and the power of the new methods are probably of greater significance than the specific result, for the fate of the solar system will likely be determined by dissipative and other nongravitational forces."

—H. Goldstein, *Classical Mechanics* (2nd Ed.), 1980 [Gold80] (p. 530)

Next, from a mathematics book that does include a chapter on KAM theory, with an outline of a proof:

"The KAM theorem originated in a stroke of genius by Kolmogorov [...]"

—P. Lochak & C. Meunier, *Multiphase Averaging for Classical Systems*, 1988 [LocM88] (p. 154)

From another mathematics book that provides careful and detailed treatments of many topics in perturbation theory comes a kind of apology for *not* treating KAM theory:

"... in the conservative case, the theory is very technical and deserves to be considered one of the high points of twentieth-century mathematics. It is called Kolmogorov-Arnold-Moser theory (frequently abbreviated to KAM), and is far too difficult to discuss in any detail here."

—J. Murdock, *Perturbations. Theory and Methods*, 1991 [Mur91] (p. 332)

One of the more interesting and revealing passages comes from a book intended for graduate students in physics:

"In many ways the KAM theorem possesses sociological similarities to Gödel's famous theorem in logic. (a) Both are widely known and talked about, yet many people

are rather vague on what the theorems actually state, and
very few have actually read the proofs, much less validated
them. (b) Each has been called, by different mathematicians, the most important theorem of the twentieth century. (c) Neither is very useful for practical calculations:
[...] the stable phase space estimated by the KAM theorem
is typically too conservative to be of value."

 —L. Michelotti, *Intermediate Classical Dynamics With
Applications to Beam Physics*, 1995 [Mic95] (pp. 305–306)

And finally, in a book by mathematicians popularizing the last century-
and-a-half of achievements in dynamical systems and celestial mechanics,
we have the following high praise:

"... the great edifice of KAM theory"

[And, at a later point in the book, also in reference to
KAM theory:]

"one of the more remarkable mathematical achievements of this century ..."

 —F. Diacu & P. Holmes, *Celestial Encounters*, 1996
[DiH96] (p. 146, p. 165)

In these quotations, it's interesting that authors seem compelled to pay
tribute to KAM theory, to praise it and its inventors. Physicists (and even
some mathematicians) seem also to want to avoid a direct encounter with
it, saying it's too 'abstract' or too 'hard.' But in the quotations from the
physics books by Goldstein and Michelotti, we also hear another reason for
avoiding it: it's not very useful. Once you know that many physicists think
this, you realize that much of their praise is politely dismissive.

 Mathematicians and physicists are generally civil with each other, and
no one would write a strong statement about the uselessness of KAM theory in a book. But in spoken encounters over the years, I've heard much
stronger statements and questions, such as "What's so great about KAM
theory?" [10] or "What practical result has ever come from KAM theory?",
or even "I'm tired of hearing so much hype about KAM theory." In this
book, I'll explore how remarks like these partly reflect a lack of knowledge,
and partly reflect justified frustration on physicists' part.

 Finally, following these brief quotations from books, I should also point

---

[10]This became the title of one of my talks to audiences of physicists.

out that in *dynamical systems* papers of the 1980s and 90s, it became so common to see the term 'celebrated KAM theorem'[11] that you might think the adjective 'celebrated' had been permanently attached as part of the theorem's name.[12]

1.3   Remarks on the style and organization of this book

In these pages, I'm going to tell KAM theory as a 'story,' and I'm going to use several simplifying features to try to make the narrative more readable. First, at one end of a spectrum, I imagine people I'll call advocates or 'enthusiasts' for KAM theory (the reader can think of West European or Russian mathematicians). At the other end, I imagine 'skeptics' or 'detractors' of KAM theory (think of hard-boiled American physicists). Now even if these are largely mythological characters, it will nevertheless be more fun to think of things this way as we go. We can draw on a number of stereotypes and clichés to keep us awake and make certain points—we already know that physicists and mathematicians like to tell jokes about themselves and each other that turn on these clichés. Likewise, Americans (and sometimes Britons) comfort themselves about their ignorance of continental European thought by picturing a 'fog of pointless theory' emanating from the old world, while Europeans are shocked by the crass pragmatism in America, a place whose main contribution to philosophy has been to enquire about the 'cash value of truth.'[13] From this point of view, my task on the one hand is to try to educate the reluctant skeptics by penetrating the fog of theory to find the underlying cash value of KAM theory. On the other hand, if I occasionally make fun of a few enthusiastic theorists along the way, so be it.

As mentioned earlier, I don't think KAM theory can be appreciated properly without at least some knowledge of its history. In order to tell it quickly and vividly, I'll recount many parts of the narrative using the 'great man' point of view, rather than the more painstaking process of documenting all the individuals who contributed, though I'll at least list

---

[11] Whenever I see the word 'celebrated' used this way, I can't help but think of Mark Twain's short story *The Celebrated Jumping Frog of Calaveras County.* The reader may recall that in that story, the said celebrated frog did not live up to its owner's expectations, further emphasizing my feeling that much of the praise heaped upon KAM theory is not wholly heartfelt.

[12] I stopped looking once I found a dozen papers with this locution. Unfortunately—dare I admit it?—one of the papers was my own.

[13] As gleefully propounded by the Harvard-trained scholars C.S. Peirce and W. James.

some of them. I'll also employ some terminology anachronistically; a more detailed approach would carefully follow changes in the meaning of terminology over several centuries.[14] In the places where the anachronisms are especially misleading, I'll say so. Also regarding the way language is used here, and despite the long time I spent in Europe and the European flavor of the subject, I've used something very close to ordinary American[15] vernacular to write this down. For my part, that's only natural, but it's also curiously unprecedented—KAM theory is rarely discussed 'in American.'

Finally, this book comes with a lot of scholarly apparatus which the reader may use according to his or her taste: in-text references to a single long bibliography, several appendixes that together are almost as long as the main text, and many footnotes—a few pages have more footnotes than ordinary text. I felt it was important to include the additional material as a way of pointing to deeper layers of the story. As the reader will see, the nature of KAM theory means that its story can be told on many levels.

---

[14] As Salomon Bochner puts it, "[In history of science] more than in any other history, the past discloses itself in the future." (See the interesting discussion from which this fragment is quoted on p. 60 of [Boch66].)

[15] Here 'American' is used to mean the variety of English commonly spoken in the U.S.

Chapter 2

Minimum Mathematical Background: Integrable Hamiltonian Systems

This chapter introduces the basic ideas and terminology needed to understand KAM theory. It's best to see this material first, since I'll use much of this terminology (often anachronistically) when discussing the history. The reader who prefers a wholly non-technical survey of KAM theory and its significance may skip directly to Chapter 5. Though the material here is mathematical, the presentation is conversational and informal. More precise definitions of many terms may be found in the glossary[1] (Appendix F), and suggested resources for additional background and deeper study are given in the reader's guide (Appendix D).

2.1 Dynamical systems

Roughly speaking, a *dynamical system* is a mathematical means of describing something (a 'system') that changes over time. The idea is to first find a suitable *phase space* (or *state space*), each point of which gives all relevant information about the system. For example, if the system you want to describe is the population of a herd of animals, the appropriate phase space might be $\mathbb{N}_0$ (the set of nonnegative integers[2]). A single number—the population—tells you all you want to know, so we say the phase space has *dimension* one. More realistically, you might want to know how two populations interact—say, a population of predators and a population of

---

[1] The first time a term defined in the glossary is used, it appears in *slanted text*. For emphasis, or when appearing again after an interlude, some glossary terms may appear more than once in slanted form.

[2] See the list of symbols at the beginning of the glossary (Appendix F) which gives notational conventions for recurring symbols such as $\mathbb{N}^*$, $\mathbb{N}_0$, etc.

prey—in which case you need two distinct numbers to describe the system, and so you have a phase space of dimension two. And if you're tracking a satellite, you may wish to know its three coordinates in space and the three components of its velocity. The appropriate phase space for the satellite has points described by six numbers, so the phase space has dimension 6. In fact, a phase space may have any dimension (including *infinite* dimension), but in this book, we'll limit ourselves to systems with phase spaces of finite dimension n.

In basic dynamical systems, phase points are represented by coordinates in an ordinary n-dimensional *Euclidean space* ($\mathbb{R}^n$). In more sophisticated dynamical systems, phase points are still '*locally* represented' by points in $\mathbb{R}^n$, but the overall phase space may have the global curvilinear structure of a *manifold* (see Fig. 2.1 below, where part (a) shows a dynamical system on $\mathbb{R}^2$, and part (b) shows a dynamical system on a 2-dimensional manifold).

Once the phase space is determined, the second step in constructing a dynamical system is to posit a mathematical *rule* or *evolution law* which says where the system goes next in phase space, based on where it is at present. This law of evolution can take different forms depending on whether one thinks of the time as 'discrete' or 'continuous' (i.e., jumping or flowing) and whether one thinks of the change as being *deterministic* or 'stochastic' (completely determined, or with an element of randomness). All types of evolution laws are important, but in what follows we'll focus on the first kind to be identified and understood: deterministic systems with continuous time, or so-called *continuous dynamical systems*. We'll also mostly limit our attention to so-called *autonomous* systems, meaning that the evolution law is not an explicit function of the time.

But how, in practice, do we find evolutionary laws that are good models of physical systems? One way to describe the discovery that Isaac Newton ranked among his greatest in 'natural philosophy' is as follows: Nature gives us its evolutionary laws in the form of *differential equations*.[3] In the modern view, an *ordinary differential equation* (ODE) is a *vector field* on an appropriate phase space. In other words, at every point of phase space

---

[3]To quote V.I. Arnold (from the opening paragraph of the preface to [Ar83–88]): "Newton's fundamental discovery, the one which he considered necessary to keep secret and published only in the form of an anagram, consists of the following: *Data aequatione quotcunque fluentes quantitae involvente fluxiones invenire et vice versa.* In contemporary mathematical language, this means: It is useful to solve differential equations." (The anagram—appearing in Newton's second letter to Leibniz via Oldenburg in 1677— once deciphered might be more literally translated as "Given an equation involving any number of fluent quantities, to find the fluxions, and vice versa.")

there's an 'arrow' whose direction literally points where the system goes from there, and whose magnitude tells how fast it goes. If you start at a particular place in phase space, the system dutifully follows the arrows: speeding up, slowing down, and changing direction as indicated. The resulting track through phase space is called the *phase curve* (or *orbit* or *integral curve*) of the system corresponding to the given *initial condition* (or IC—the particular place where the curve starts; see Fig. 2.1 (a) to get an idea of how this works). Sometimes you need to distinguish between the phase curve viewed as a geometric object, and the curve viewed as a *function* of time; in the latter case, it's usually called a solution (or *trajectory*) of the differential equation (but, alas, different authors use different conventions, so none has become standard).

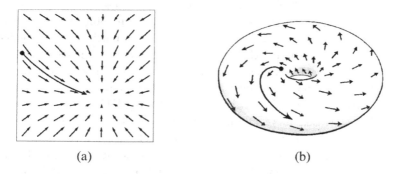

(a)   (b)

Fig. 2.1   **(a)** Pictorial representation of a dynamical system arising from a vector field in the plane ($\mathbb{R}^2$), with one integral curve sketched in, starting at the IC represented by the dot and approaching an attracting *fixed point*. **(b)** A dynamical system given by a vector field on a two-dimensional manifold, a more general sort of phase space (in this case a 'fat torus').

One of the first things mathematicians worry about in dynamical systems theory is the kind of vector fields that give rise to well-defined solutions at every point (the so-called existence problem). The next item of interest is whether phase curves of the system ever cross one another; if not, the system is said to have unique solutions (since non-unique solutions must branch or cross somewhere). Although they can be treated separately, these two items are usually bundled together as existence-uniqueness theory, and it's very satisfying to learn—and work through the proof of—the simple fact that vector fields which are *locally Lipschitz continuous* do have unique solutions at every point (this includes all vector fields that are of

smoothness[4] *class C^k with $k \geq 1$*). Finally, even when we know that solution curves exist uniquely through every IC in phase space, we can still
investigate the longest time interval on which each solution is defined—the
so-called maximal existence interval. For *nice* (e.g., locally Lipschitz) vector
fields, it turns out that there are only two ways that the maximal existence
interval of a solution may fail to be all of $\mathbb{R}$: either the solution reaches the
boundary of the phase space in finite time (in case the phase space has a
boundary), or the solution 'runs away to infinity' in finite time (in case the
phase space is unbounded).

In summary, the nicest sorts of vector fields are those giving rise to
unique solutions passing through all points of phase space, with every solution defined for all time $t \in \mathbb{R}$. But minor pathologies are common, and
many of the most interesting and useful vector fields have solutions that are
either not defined for all ICs, or not defined for all times $t \in \mathbb{R}$, or both.[5]

It's also important to understand that, even when solutions' existence
and uniqueness are assured, in most cases there is no mathematical means
of finding them explicitly; in other words, most differential equations (DEs)
cannot be 'solved' in the explicit, traditional sense one learns in elementary
courses. This difficulty arises on two levels. First (and despite what one
learns in a first course in DEs), it is quite rare to find DEs having solutions
that can be expressed in terms of *elementary functions* 'in closed form'
(i.e., without using infinite *series*).[6] One can then move to the next level
of solvability, seeking DEs that are *integrable* (one form of *integrability* is
described below), so that solutions may be found by *quadrature*. However,
even this lesser form of solvability is rare in the universe of all DEs. We'll
come back to some of these issues below; for now I just alert the reader that
they make up one of the key points of our story, and it took mathematicians
a long time to understand them and appreciate their significance.

Returning to our basic terminology for dynamical systems, once we
know that solutions of a system are uniquely defined at every point of some
region in phase space, then we can speak collectively of the solutions as
the *flow* of the system in that region. If M is the phase space and the

---

[4]Mathematicians say that "the *function* f is of *smoothness class* C^k on the *domain* D"
(abbreviated "f is of class C^k on D," "f is C^k on D," or in symbols $f \in C^k(D)$) to
indicate that f has *continuous derivatives* up to order k at every point of D. When the
domain D is understood, one simply writes $f \in C^k$, or says "f is C^k." See *smoothness*
in the glossary (Appendix F) for more information.

[5]The mathematically inclined reader will find detailed discussions of these issues in
references listed at the end of Part D.2.1 of the reader's guide (Appendix D).

[6]This has been known in one form or other since J. Liouville's 1835 paper [Liou35].

flow is global (defined at all points of M and for all times t), then in more precise terms, the flow is a one-parameter family of maps $\varphi_t : M \to M$ with parameter t. The flow generates all solutions of the system by taking any IC $x \in M$ to its new location $\varphi_t(x)$ in M at time t, satisfying the basic 'group action' relations[7] $\varphi_0(x) = x$ and $\varphi_{t+s} = \varphi_t \circ \varphi_s$.

One can bypass the whole process of constructing a dynamical system by simply positing the pair (M, φ_t) (a phase space M and a flow φ_t); one then works in reverse to get the vector field (or ODE) on M by taking time derivatives along the flow. The original forward procedure (starting from the vector field or ODE) is how dynamical systems arise naturally from physics, where evolutionary laws occur as differential equations. The reverse procedure is one of the ways dynamical systems are studied abstractly in the mathematical setting. For now, we're more interested in the traditional (forward) process.

2.2 Hamiltonian systems

A *Hamiltonian system* is a special kind of dynamical system, in many ways the apotheosis[8] of mathematical models of classical mechanics. The Hamilton in Hamiltonian stands for the Irish mathematician W.R. Hamilton (1805–1865) who first drew attention to this way of doing mechanics in 1833. His formulation is a direct descendant of Newton's laws of motion set forth in the *Principia*.[9] Nowadays it's commonplace for mathematicians to learn about Hamiltonian systems in a purely abstract way as part of *symplectic geometry*, but historically, they were derived by combining J.-L. Lagrange's formulation of mechanics—itself derived from Newton's laws—with variational methods (in turn related to 'least action' principles described by P.L.M. de Maupertuis in the early part of the 18th century).

Both the Lagrangian and the Hamiltonian formulations of classical mechanics remain important today, but physicists usually learn more about the Hamiltonian approach because of its importance in quantum mechanics.

---

[7]Even if the flow is not global, the group action relations hold for all x, s, and t at which the elements in the relations are well defined.

[8]Religious terminology is not unknown in classical mechanics: note the repeated use of the term *canonical* in reference to things Hamiltonian. (In the more culturally sensitive modern era, this term has largely been supplanted by *symplectic*; see the brief 'translator's dictionary' in the next subsection.)

[9]*Principia* is short for Newton's *Philosophiae Naturalis Principia Mathematica* [The Mathematical Principles of Natural Philosophy], written in Latin and first published in 1687, with 2nd and 3rd editions appearing in 1713 and 1726.

In fact some physicists regard classical mechanics as a closed subject, and see its study merely as a kind of warm-up exercise for related techniques used in more 'modern' subjects. This view is partly understandable, but also unfortunate, as it excludes deep, ongoing mathematical results that are not only beautiful, but have proved themselves useful in approaching unanswered questions in theoretical physics (see Chapter 7 below).

In physics, Hamiltonian systems are used, for example, to model macroscopic systems that evolve without loss of energy due to friction or other forms of dissipation (typical examples are planetary systems, and systems of charged particles orbiting in modern accelerators). In fact, Hamiltonian systems also model heat flow at the microscopic level, since the classical theory of heat involves elastic (frictionless) collisions between molecules. (There is a certain irony that heat flow—the inevitable result of any dissipative process—should be modeled by Hamiltonian mechanics.)

As was already the case in our discussion of dynamical systems, we'll concentrate mostly on *autonomous* Hamiltonian systems (those without explicit time dependence). This simplifies much of the discussion, and is not (technically) less general than the non-autonomous case.[10] The nicest feature of an autonomous Hamiltonian system is that the Hamiltonian function is constant in time (reflecting conservation of energy in physical systems). But it's important to remember that Hamiltonian systems are not limited to conservative systems only, and conversely, there are non-Hamiltonian systems with conservation laws.

Finally, in order to discuss Hamiltonian systems concretely, we must confront a quirk in the way the subject has developed. Namely, it tends to appear in one of two forms: in 'traditional form,' or in 'mathematicians' form' (though it also often appears as a mixture of the two). Recognizing this fact will not only be important in the rest of this book, but especially in the wider literature of KAM theory.

---

[10]Technically speaking, any non-autonomous Hamiltonian system can be reformulated as an autonomous Hamiltonian system by defining the time as an (additional) configuration variable. But this increases the number of *degrees of freedom* by 1 (or at least by 1/2, since many like to say that a non-autonomous system with n degrees of freedom behaves roughly like an autonomous system with '$n + 1/2$' degrees of freedom). Often, however, there are good reasons for retaining the non-autonomous formulation.

2.2.1 *Two pictures (and two languages) for Hamiltonian systems*

The traditional form—or language—of Hamiltonian systems has grown steadily, without a major break, since the work of the founders of mechanics, and is characterized by its ties to physical systems and its reliance on coordinates. On the other hand, following a long period of gestation going as far back as Lagrange,[11] the mathematicians' form became widespread only relatively recently (around the middle of the 20th century), and is characterized by its abstract geometric formulation, often without reference to mechanical systems or specific coordinates.

The mathematicians' form incorporates and supersedes the traditional form, yet the older form survives alongside the new, partly by tradition, but also because the mathematicians' form requires considerably more advanced training to understand. It was the fate of KAM theory to be established at roughly the same time that the new form began to supplant the traditional; this can be seen in the language and style of the founding articles (Kolmogorov's first article [Kol54] is essentially in traditional form, but thereafter, the language of KAM theory shifted inexorably toward the new form).

To keep this book 'friendly,' I will keep almost all discussions of Hamiltonian dynamics at the traditional level, using mostly the traditional vocabulary. However, I'll sometimes use the newer vocabulary, a few terms of which are discussed in the 'translator's dictionary' below, and also further in the glossary. This mixed use is partly by necessity, since recent work in KAM theory is written in the modern form.

So before proceeding further, it will be useful to sketch the bare outlines of the two forms separately. For simplicity and clarity, I'll limit both outlines to the case of *autonomous* Hamiltonians of natural type,[12] but most of what is described here carries over to more general cases without too much difficulty.

---

[11] In [Ig98], P. Iglesias shows that the *symplectic form*—the essential feature of *symplectic geometry* and the modern form of Hamiltonian systems—was already recognized and used (though not named) by J.-L. Lagrange as early as 1810.

[12] One says that the Hamiltonian function H is 'natural' if it can be written as the sum $H = T + V$ of kinetic energy T and potential energy V, where V depends only on the configuration variables.

The traditional picture (for autonomous, 'natural' Hamiltonians)

The traditional picture[13] of Hamiltonian systems begins with a mechanical system and is constructed by way of the system's *Lagrangian*.[14] One first identifies the *configuration space* Q of the system (the set of all physically possible positions the system may assume, subject to any constraints), and introduces so-called *generalized coordinates* $(q_1, \ldots, q_n)$ for Q. Although the choice of these coordinates is not unique, the minimum number n of independent coordinates required is unique (once any constraints are taken into account), and is called the number of *degrees of freedom* of the system. Once the generalized coordinates q_k are chosen, their time derivatives $\dot{q}_k$ are called the *generalized velocities*. One next introduces the Lagrangian $L = L(q_1, \ldots, q_n, \dot{q}_1, \ldots, \dot{q}_n)$ of the system. (Recall that the Lagrangian for a natural system is $L = T - V$, where T and V are the kinetic and potential energy of the system, V being a function of the q_k only.)

At this point, we could begin to analyze the system's dynamics by writing down the Euler-Lagrange equations of motion $\frac{d}{dt}(\partial L/\partial \dot{q}_k) = \partial L/\partial q_k$ ($k = 1, \ldots, n$), but instead we proceed to the Hamiltonian picture. For this purpose, we define the *generalized momenta* p_k by $p_k = \partial L/\partial \dot{q}_k$. For each $k = 1, \ldots, n$, members of the pair (q_k, p_k) are said to be *conjugate* to each other (i.e., given a configuration variable q_k, its conjugate momentum variable is p_k and vice-versa). Together, the q_k and p_k form a set of *canonical coordinates* for the system. These may be gathered into vectors $q = (q_1, \ldots, q_n)$ and $p = (p_1, \ldots, p_n)$, so the system of canonical coordinates $(q_1, \ldots, q_n, p_1, \ldots, p_n)$ is abbreviated (q, p). These will be the coordinates for the *phase space* of the Hamiltonian system.[15] Using the Lagrangian L, we define the Hamiltonian H by way of the Legendre transformation: $H = \sum_{k=1}^{n} \dot{q}_k p_k - L$. By inverting the earlier expression $p_k = \partial L/\partial \dot{q}_k$, we express the generalized velocities $\dot{q}_k$ as functions of the q_k and p_k; then substituting these relations into the Legendre transformation, we express the Hamiltonian in terms of the canonical coordinates, i.e., as $H = H(q, p)$. Finally, in this formulation, we find that the Euler-Lagrange equations of

---

[13]For more detail concerning the traditional approach to Hamiltonian systems, see Chapter 8 (especially §8.1) of Goldstein [Gold80]. The opening pages (§1.1) of Marsden and Ratiu [MarsR94] also give a nice compact presentation in 'mixed' (traditional + modern) language.

[14]Here 'Lagrangian' is an abbreviation of 'Lagrangian function.'

[15]The order of coordinates (q, p) used in this book is traditional in most physics books in the West, but in his influential text [Ar78–97], V.I. Arnold uses the order (p, q) instead.

motion have been transformed to *Hamilton's equations*[16]

$$\frac{dq_1}{dt} = \frac{\partial H}{\partial p_1}, \qquad \frac{dp_1}{dt} = -\frac{\partial H}{\partial q_1}$$

$$\frac{dq_2}{dt} = \frac{\partial H}{\partial p_2}, \qquad \frac{dp_2}{dt} = -\frac{\partial H}{\partial q_2}$$

$$\vdots \qquad\qquad \vdots \qquad\qquad (2.1)$$

$$\frac{dq_n}{dt} = \frac{\partial H}{\partial p_n}, \qquad \frac{dp_n}{dt} = -\frac{\partial H}{\partial q_n}.$$

This system of $2n$ scalar, first-order ordinary differential equations (with its skew symmetry indicated by the negative signs), has proven to be one of the most fruitful models in all of mathematical physics. Its applications are not limited to classical mechanics, but extend broadly (in suitably generalized form) to electromagnetism and quantum mechanics, including quantum field theory.

A last feature of the traditional picture is worth mentioning here. Among the most common procedures for solving differential equations are variable changes. These are often applied to the system (2.1), but one often also wants a variable change that preserves the special form of the system. That is, one seeks a variable change $(q, p) \mapsto (s, r)$ which transforms the original Hamiltonian H to a new Hamiltonian $K(s, r) = H(q(s, r), p(s, r))$ in such a way that Hamilton's equations retain their original form (i.e., the transformed equations agree with what one obtains by replacing H by K, q by s, and p by r in (2.1) above). In this way, the (s, r) may be viewed as (new) canonical coordinates for a (new) Hamiltonian K of the original system. This sort of variable change $(q, p) \mapsto (s, r)$ is called a *canonical transformation*, about which more will be said below.[17]

We now turn—very briefly—to the modern mathematical view of Hamiltonian systems (again limiting ourselves to autonomous, natural Hamiltonians). This view was motivated broadly by dramatic developments in differential topology and geometry beginning in the 19th century, and more specifically by the observation that the special skew-symmetric form of system (2.1) does not reside in the Hamiltonian function itself, but rather in the geometry of the phase space.

---

[16] Also called the *canonical equations of motion.*

[17] Canonical transformations are discussed in traditional form in Chapter 9 of Goldstein [Gold80].

Warning: The following few paragraphs are written in a distinctly different language (some call it 'symplectology') from most of the rest of the book. No one can be expected to fully understand this language without prior training, but this material is placed here for readers who want a glimpse of the modern picture, and also as prelude to the short discussion below of how to translate (at least a word or two of) the new language into the traditional language.

The modern mathematical picture (for autonomous, 'natural' Hamiltonians $H = T + V$)

In the modern mathematical picture, Hamiltonian systems are a part of *symplectic geometry*, a branch of *differential topology* and differential geometry centered on the study of *symplectic manifolds*. A symplectic manifold (M, ω) is, roughly speaking, a $2n$-dimensional *smooth manifold* M equipped with a so-called *symplectic form* ω. Given a symplectic manifold (M, ω), we may consider any smooth, real-valued *function* $H : M \to \mathbb{R}$ as a Hamiltonian on M. The *vector field* X on M generated by H is given by $\omega(X, \cdot) = dH$. This equation[18] may be seen as a coordinate-free version of the canonical equations of motion (2.1) above. The flow of the vector field is called the Hamiltonian flow of H on M, and the collection of all such flows (over all possible Hamiltonians H) is the Hamiltonian dynamics of (M, ω).

Thus far, we have not spoken of coordinates, or of mechanical systems. To see that this abstract picture does include the traditional picture, we proceed as follows. Given a mechanical system, we first find its configuration space Q as before (but we do not yet need to choose a coordinate system for it). Generally, Q will be a smooth manifold of dimension n. We now consider the *cotangent bundle* T^*Q of Q. It turns out that T^*Q has dimension $2n$, and has the natural structure of a symplectic manifold;[19] i.e., it has a

---

[18]To define the precise meaning of $\omega(X, \cdot) = dH$ requires the language of differential forms (from differential topology), and is not provided in this book (see, for example, Chapters 7 and 8 of [Ar78–97] or Chapter 3 of [AbM78]). However, one can begin to understand this equation by writing it in coordinates and verifying that it agrees with Hamilton's equations (2.1). One method for doing this is described in the next paragraph above; a closely related method introduces *Darboux coordinates* (q, p) and leads to Hamilton's equations in the form $(\dot{q}, \dot{p})J = DH$, where J is the *standard symplectic matrix*, and DH is the gradient of H written as a row vector. (This also shows that in Darboux coordinates, ω has the matrix form J.)

[19]For discussions and proofs of this fact, see §37B of [Ar78–97], or Theorem 3.2.10 of [AbM78].

natural symplectic form ω. By *Darboux's theorem*,[20] we may choose *local coordinates* $(q, p) = (q_1, \ldots, q_n, p_1, \ldots, p_n)$ on T^*Q (called *Darboux coordinates*) in which ω may be written[21] $\omega = dq \wedge dp = dq_1 \wedge dp_1 + \cdots + dq_n \wedge dp_n$. If we express the total energy $H = T + V$ of the mechanical system in these coordinates (i.e., as $H = H(q, p)$), then the abstract Hamiltonian vector field X generated by H agrees with Hamilton's equations (2.1) above, and we have recovered the traditional picture of the mechanical system.

A (one-term) translator's dictionary between the traditional and modern pictures

By comparing the two pictures as outlined above, we can establish a very short translator's dictionary between them. The most basic correspondence is that usually, what is called 'canonical' in the traditional picture is called 'symplectic' in the modern picture. (However, this correspondence isn't perfect: canonical coordinates may also be called Darboux coordinates in the modern terminology; and one does not speak of 'symplectic equations of motion.') A canonical transformation may loosely be called a symplectic transformation, but one usually says 'symplectomorphism' to indicate that the transformation is also a *diffeomorphism*. There are many other subtleties related to the modern geometric viewpoint (e.g., generalized velocities are tangent vectors, while generalized momenta are cotangent vectors), but this very short list should serve our purposes.

With these brief descriptions of the language(s) used for Hamiltonian systems, we next look into the solution of such systems.

2.2.2   *What does it mean to 'solve' a Hamiltonian system?*

The answer to the question posed as the title of this subsection is not as simple as might be supposed at first thought. The remarks already made above (§2.1) about the subtleties of finding solutions of differential equations certainly also apply here. One can try to 'solve' or 'integrate' Hamilton's equations (2.1) by 'reducing them to *quadrature*'; i.e., by expressing the phase coordinates as explicit functions of time (these functions may never-

---

[20]For statements and proofs of Darboux's theorem, see §43 B of [Ar78–97], or Theorem 3.2.2 in [AbM78].

[21]The symbol $\wedge$ is the 'wedge product' of differential forms; again see Chapter 7 of [Ar78–97] for background. Note however that we can already identify the conjugate variables from the traditional picture as the paired factors of wedge products when ω is expressed in Darboux coordinates.

theless contain integrals that cannot be expressed in terms of elementary functions). When it's possible, this process of integration leads in principle to explicit formulas for the flow of the system, but the process may nevertheless be very difficult to carry out in practice. When the integration or solution process can't be carried out directly (using, say, the techniques one learns in integral calculus or differential equations), the usual approach in seeking solutions is to change coordinates using a *canonical transformation* $(q, p) \mapsto (s, r)$, which is guaranteed to preserve the form of the canonical equations in the new variables (as described above in §2.2.1). Judicious choice of new variables (s, r) may lead to simpler canonical equations that can be solved by elementary means.[22] Over time, much effort has gone into finding canonical transformations that permit integration of systems, and we summarize the most important result in this direction below.

As mentioned earlier, whether or not this process is possible—i.e., the *integrability* of Hamiltonian systems—is a key question, one that already greatly vexed Newton in his attempts to study the motion of the moon,[23] and one that largely drove developments ultimately leading to the discovery of 'chaos' and KAM theory. We devote the next section to a more careful description of what is meant by an integrable Hamiltonian system, the concept that perhaps best answers the question in the title of this subsection, and a notion without which KAM theory cannot be understood even on a rudimentary level.

2.2.3   *Completely integrable Hamiltonian systems and the LMAJ theorem*

Precise notions of integrability developed slowly after Newton and Leibniz, and are still evolving; the type of integrability we describe below is the most standard modern form, often called Liouville integrability. The basic ideas of integrable systems were first roughed out by C.G.J. Jacobi [Jaco43], rediscovered and made more explicit by J. Liouville [Liou55], described and proved in more modern form by H. Mineur [Min35], [Min36], written and proved in a form suited to Hamiltonian perturbation theory by V.I. Arnold [Ar63b], [ArA68], [Ar78–97], and generalized to arbitrary *symplectic manifolds* and made still more precise by R. Jost [Jost68], by L. Markus and

---

[22]However, it may also turn out that the simplest or most useful solution of a Hamiltonian system is achieved by a change of variables that is *not* canonical.

[23]Of course, this is an anachronistic way of speaking, since Hamiltonian systems were not conceived for more than a century after Newton's death.

K. Meyer [MarkM74], and by others. When these results are combined into a theorem, Arnold calls them 'Liouville's theorem'; other authors call them the 'Liouville-Arnold theorem,' the 'Liouville-Mineur-Arnold theorem,' or the 'Liouville-Arnold-Jost theorem.' In an effort to be inclusive and international, I will say the 'Liouville-Mineur-Arnold-Jost theorem,' which I abbreviate here with the acronym LMAJ.[24] Historical notes in the book [CuB97] cite at least a dozen variations of the LMAJ theorem (with proofs), and there are many others, especially when different kinds of integrability are taken into account.

To describe integrability and the LMAJ theorem in a basic way, we need a few more definitions, starting with the *Poisson bracket* { , } which operates on pairs of functions to produce a third function. If the phase space M of a Hamiltonian system has canonical coordinates $(q, p) = (q_1, \ldots, q_n, p_1, \ldots, p_n)$, and if f and g are two smooth real-valued functions on M (i.e., $f, g : M \to \mathbb{R}$), then their Poisson bracket is the function $\{f, g\} : M \to \mathbb{R}$ given by

$$\{f, g\} = \sum_{k=1}^{n} \left(\frac{\partial f}{\partial q_k} \frac{\partial g}{\partial p_k} - \frac{\partial g}{\partial q_k} \frac{\partial f}{\partial p_k} \right). \tag{2.2}$$

Three further preliminary definitions are needed:

Two smooth functions f, g on M are in *involution* if $\{f, g\} \equiv 0$ (i.e. their Poisson bracket vanishes on M). A set of more than two functions is in involution if each pair of functions in the set is in involution.

A finite set of smooth functions $\{f_1, f_2, \ldots, f_n\}$ on M is *independent*[25] on $K \subset M$ if the corresponding set of gradient vectors $\{Df_1, Df_2, \ldots, Df_n\}$ is *linearly independent* at *almost every* point of K.

Finally, a non-constant smooth function $f : M \to \mathbb{R}$ is a *first integral* (or *constant of motion*) for the system (2.1) if it is constant when evaluated along solutions of (2.1) (i.e., if $(q(t), p(t))$ is a solution of (2.1), then $f(q(t), p(t)) = \text{const.}$ at all t for which the solution is defined[26]).

---

[24]Ken Meyer pointed out to me that Arnold's proof of this theorem in [Ar78–97] has an error, or at least a sizeable gap. Regarding the mild controversy over this, and over the name of this theorem, everyone interested in mathematics (and especially dynamical systems) should be aware of the repeated clashes between Russian and Western mathematicians over priority and the level of detail required in proofs. These points are discussed further in Appendix C.

[25]Though it's not as formal as saying "The set $\{f_1, f_2, \ldots, f_n\}$ *is* independent," one also says "The functions $f_1, f_2, \ldots, f_n$ *are* independent."

[26]A compact way of indicating that f a first integral of H is to write $Df \not\equiv 0$ and $\{f, H\} \equiv 0$ (on M). The first part says that f is non-constant on M and the second part says f is constant along solutions of (2.1).

We now give the basic definition of integrability:

Definition.   The Hamiltonian system (2.1) with n degrees of freedom is *completely integrable* if it has n constants of motion $f_1, f_2, \ldots, f_n$ which are independent and in involution.

This is a good place to point out that the requirement of only n constants of motion in the above definition reflects the special structure of Hamiltonian systems. A general system of $2n$ ordinary differential equations— without the special structure—requires $2n - 1$ constants of motion to be integrable.

Although the modern definition sounds abstract, it remains closely linked to the older meaning of integrable, which applied to systems that could be 'solved explicitly,' or at least reduced to quadrature. This is embodied in the nice result mentioned above (the Liouville-Mineur-Arnold-Jost (LMAJ) theorem) which we now describe.

Paraphrase of the LMAJ theorem. *Consider a completely integrable n-degree-of-freedom Hamiltonian system (M, H) with n independent first integrals $f_1, f_2, \ldots, f_n$ in involution. From standard results in differential topology we know that setting each of the first integrals equal to a (separate) constant produces an n-dimensional smooth surface—a* manifold—*in the 2n-dimensional phase space M. Because it is 'cut out' in this way by first integrals, this surface is clearly invariant under the flow; i.e., any solution beginning on the surface remains on it for all times thereafter. The LMAJ theorem asserts this, and says further that whenever such a surface is bounded and connected, it is an n-dimensional torus, denoted $\mathbb{T}^n$ and defined as $\mathbb{T}^n = \mathbb{R}^n / \mathbb{Z}^n$ (or as an n-fold Cartesian product of circles).*

But the LMAJ theorem says considerably more: Whenever there's a 'nice' region of phase space where these surfaces are tori, there are special canonical coordinates $(\theta, I) = (\theta_1, \ldots, \theta_n, I_1, \ldots, I_n)$ in that region, called action-angle variables, *in which the Hamiltonian H takes the simple form $H = h(I)$ (i.e., H becomes independent of the angles θ_k) with canonical equations*

$$\frac{dI_k}{dt} = 0, \qquad \frac{d\theta_k}{dt} = \omega_k, \qquad k = 1, \ldots, n \tag{2.3}$$

having the very simple solutions

$$I_k(t) = I_k^0, \qquad \theta_k(t) = \theta_k^0 + \omega_k t \ (\text{mod} 1), \qquad k = 1, \ldots, n. \tag{2.4}$$

Here I_k^0, θ_k^0 *are the ICs, and the* $\omega_k = \dfrac{\partial h}{\partial I_k}(I_1^0, \ldots, I_n^0)$ *are called the frequencies. When these quantities are collected into the vectors* $I^0 = (I_1^0, \ldots, I_n^0)$, $\theta^0 = (\theta_1^0, \ldots, \theta_n^0)$, *and* $\omega = (\omega_1, \ldots, \omega_n)$, *the solution may be written in the more compact form*

$$I(t) = I^0, \qquad \theta(t) = \theta^0 + \omega\, t, \tag{2.5}$$

where the addition in each θ *component is understood to be modulo unity.*[27]

Finally, transforming from the action-angle variables (θ, I) *back to the phase coordinates* (q, p) *originally used in the nice region requires only 'elementary' operations, meaning that the solution of the Hamiltonian system in original variables may be obtained by* quadrature. *This is the connection to the older meaning of integrability mentioned earlier.*

Several remarks (and a picture; cf. Fig. 2.2) may help in further understanding complete integrability and the LMAJ theorem.

First of all, the invariant surfaces that are tori are of course called *invariant tori* (or *Lagrangian tori* to distinguish them from other invariant tori that may be of dimension different from n), and when they occupy a region of phase space, they fit together snugly in concentric layers.[28] Mathematicians describe this (and more) by saying that the tori *foliate* the region of phase space they occupy.

Since *action* values are fixed under the flow, each *torus* may be labeled by the initial action I^0 of a trajectory on it (or to put it another way, the tori are parametrized by the actions). Once an individual torus $I = I^0$ is selected, the angles $\theta = (\theta_1, \ldots, \theta_n)$ form a coordinate system on it (each angle coordinate θ_k corresponding to an independent cyclic direction on the torus). Individual trajectories of the flow wrap around the torus with constant angular *frequency* ω_k in each angular coordinate θ_k (the $\omega_k = \omega_k(I)$ are constant on each torus because the actions $I = I^0$ are constant there). Simple flow of this type on a torus is called *conditionally periodic* flow. All trajectories on the torus $I = I^0$ have the same frequency vector $\omega = \omega(I^0)$ (so we may speak of the 'frequency vector of a torus'), and distinct phase curves never intersect. As we move from one torus to another however, the frequency vector generally changes, so that trajectories may

---

[27]To be explicit, one uses the mathematical notation $\theta_k(t) \equiv \theta_k^0 + \omega_k t$ (mod 1). See the entry *mod, modular, modulo* in the glossary.

[28]It's customary to say that they fit together like 'layers of an onion,' but this expression is by now so hackneyed that it's hard to repeat with a straight face, much like the cliché of describing chaos by the effect of a butterfly's wings.

wind around different tori in qualitatively different ways (this last point is important, and is the subject of the last subsection we present below before beginning the 'Story of KAM' in earnest).

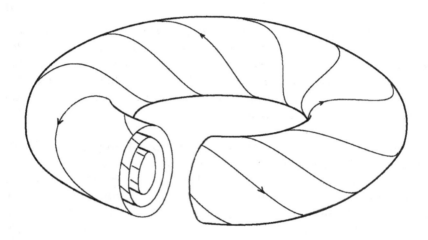

Fig. 2.2 A portion of phase space foliated by tori supporting conditionally periodic flow for a completely integrable Hamiltonian system with 2 degrees of freedom. The cut-away diagram shows a 3-dimensional slice of the 4-dimensional phase space. Note how the frequency (represented by the 'slope') of the flow changes from torus to torus. The break in the tori may be viewed as 'cut transversely by a Poincaré surface-of-section' (as represented by the square sheet in Fig 3.1, p. 50, below). Many subsequent figures in this book show integrable behavior as concentric circles formed by cutting tori with a surface-of-section in this way.

The basic picture of *bounded* motions in a completely integrable Hamiltonian system is summed up in Fig. 2.2. There we see a cut-away diagram of the 2-dimensional invariant tori carrying conditionally periodic flow for a Hamiltonian system of $n = 2$ degrees of freedom. Although the true phase space $(\theta_1, \theta_2, I_1, I_2)$ is 4-dimensional, we take a 3-dimensional slice (and so represent it in a way people can see) by the usual trick of fixing one of the action variables ahead of time (it would work just as well to fix the Hamiltonian itself; this latter method is called 'looking at the flow on a fixed energy surface').

2.2.4 *Resonant and nonresonant tori*

To explain the idea of *resonant* versus *nonresonant* tori, it's useful to cut open a 2-dimensional torus from Fig. 2.2 and lay it out as a flat square as indicated in Fig. 2.3. When we do this we see that the flow in the interior of the square is linear with direction vector $\omega = (\omega_1, \omega_2)$ (i.e., with fixed slope ω_2/ω_1). When the flow reaches an edge of the square, it continues across it, reappearing at the corresponding location on the opposite edge (much like certain projectiles in some of the earliest video games).[29]

It's not hard to see that in this simple case of linear flow on a 2-dimensional torus, there are exactly two possibilities: Either (i) the slope ω_2/ω_1 is a rational number and the flow connects back to itself where it started (i.e., the flow is periodic); or (ii) the slope ω_2/ω_1 is an irrational number and the flow never reconnects with itself, but instead fills the torus more and more densely with increasing time.

Fig. 2.3   Cutting and unwrapping a 2-dimensional torus so the flow appears linear in the unit square. The slope of the flow is ω_2/ω_1, where $\omega = (\omega_1, \omega_2)$ is the direction vector (also called the 'frequency vector' for conditionally periodic flow).

It turns out (and it's not hard to check) that this dichotomy generalizes to linear flows on n-dimensional tori in the direction $\omega = (\omega_1, \ldots, \omega_n)$ as follows: Consider the equation

$$k \cdot \omega = 0 \tag{2.6}$$

(i.e., $k_1\omega_1 + \cdots + k_n\omega_n = 0$) where $k = (k_1, \ldots, k_n)$ is a vector with integer components, hereafter abbreviated $k \in \mathbb{Z}^n$. Then either (i) Eq. (2.6)

---

[29]See also the *very nice animated illustrations* of linear flow on 2-dimensional tori by Corrado Falcolini in the online article [ChiM10].

has nonzero solutions k (i.e., it is satisfied for some $k \in \mathbb{Z}^n$ with at least one nonzero integer component); or (ii) the only solution of Eq. (2.6) is $k = (0, 0, \ldots, 0)$ (abbreviated hereafter $k = 0$). In case (i), we say that the frequency ω is *resonant*; in case (ii) we say it is *nonresonant*. These terms also apply to the flow, and to the torus where it resides. Finally, nonresonant linear flow on $\mathbb{T}^n$ is also called *quasiperiodic flow*[30]. However, it's important to note that when linear flow on $\mathbb{T}^n$ is resonant, it may still be quasiperiodic on a lower-dimensional *subtorus* of $\mathbb{T}^n$. In fact, unless the flow is periodic, it will always be quasiperiodic on some torus of dimension $m \geq 2$. On the other hand, the only linear flow (of nonzero speed) possible on $\mathbb{T}$ (the torus of dimension 1, or 'circle') is periodic flow.

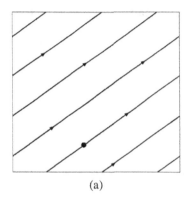
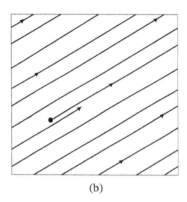

(a)   (b)

Fig. 2.4   (a) Resonant versus (b) nonresonant flow on $\mathbb{T}^2$. (Note that resonant flow in higher dimensions $n > 2$ is not necessarily periodic.)

2.2.5   *A first introduction to the idea of nondegeneracy*

There's one last related item that's important: the idea of *nondegeneracy* for completely integrable Hamiltonian systems.

Recall that in action-angle coordinates (θ, I), the Hamiltonian H of a completely integrable system takes the simple form $H = h(I)$ depending

---

[30]See the glossary for a definition of *quasiperiodic flow on $\mathbb{T}^n$* (and note that, because the flow may be quasiperiodic on a lower-dimensional subtorus, there is some subtlety in this terminology). The reader should also be aware that many authors use *conditionally periodic* synonymously with *quasiperiodic*, but in this book I distinguish between these terms.

only on the actions I. To each invariant torus $I = I^0$ we then associate the single fixed frequency vector $\omega = \frac{\partial h}{\partial I}(I^0)$. This establishes a *map* $I \mapsto \omega$ from action vectors to frequency vectors (or just 'actions to frequencies'), and it is the properties of this map that determine whether or not the system is 'degenerate' or 'nondegenerate.'

In KAM theory, there are several kinds of nondegeneracy conditions. For now, I'll just give a simple idea which can be easily described for the case of $n = 2$ degrees of freedom (other nondegeneracy conditions are given below in §4.5). For $n = 2$, instead of looking at the whole *frequency map* $(I_1, I_2) \mapsto (\omega_1, \omega_2)$, let's look at a proxy for it that can be graphed. In place of the actions, we use the radius r from the center of the concentric tori in Fig. 2.2; and in place of the frequencies, we use the slope ω_2/ω_1 from Fig. 2.3. The idea of nondegeneracy is roughly that a change in actions should induce a change in frequencies; or, in more geometric terms, the frequency map should not collapse the dimension of the action space (in this case, say, by mapping an interval to a point, as happens in Fig. 2.5b). In terms of Fig. 2.2, this means that as we move out radially from torus to torus, the slope of the flow should change; in other words our proxy map $r \mapsto \omega_2/\omega_1$ should look something like Fig. 2.5a, rather than Fig. 2.5b (where we instead show a 'degenerate' map).

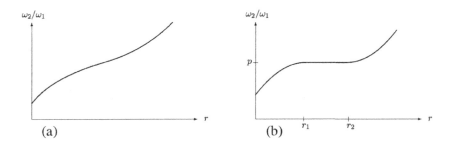

Fig. 2.5   A proxy $r \mapsto \omega_2/\omega_1$ for the frequency map $(I_1, I_2) \mapsto (\omega_1, \omega_2)$ showing **(a)** a nondegenerate case, **(b)** a degenerate case. In case (b), the interval $[r_1, r_2]$ is collapsed to the point p under the proxy map.

Finally, here's a key fact we'll return to below: For a nondegenerate system with $n = 2$, as we move out radially in r (Fig. 2.2), the resonant tori are distributed among the nonresonant tori in the same way that the rational numbers are distributed among the irrationals. In particular, the resonant tori are *dense* and the nonresonant tori have *full measure*.[31] This will be important, because one of the main messages of KAM theory is that, when a nondegenerate integrable system is *perturbed*, its resonant tori (along with some others 'close to resonant') break up chaotically, while a large subset of its nonresonant tori survive and are only slightly distorted.

We now have the bare-bones mathematical background necessary to proceed with the story of KAM theory.

---

[31]See the glossary for definitions of 'dense' and 'full measure.'

Chapter 3

Leading Up to KAM: A Sketch of the History

When we arrive later to a description of early KAM theory, the principal players will be C.L. Siegel, A.N. Kolmogorov, V.I. Arnold, and J.K. Moser. From the longer historical view leading up to KAM however, the principal players are I. Newton and H. Poincaré, with a supporting cast of many others. In this first part we try to make sense of the context of KAM theory in the largest possible historical setting.

3.1 The planets lead the way

Extreme enthusiasts intimate that the starting point for the KAM story should be the beginning of recorded history. Even without going back quite that far, at some point we have to confront the historical weight of planetary motion and its relevance to our narrative. Volumes have been written about the tantalizing allure of the 'wandering stars' and the role they've played in the development of mathematics, physics, and other intellectual endeavors, to say nothing about less intellectual but still quite imaginative endeavors such as astrology. Planetary motions beckon to be explained; historical records indicate keen human interest in the planets and their motions as far back as ancient Babylonia, and no doubt the interest goes much further back. Over the centuries the motions of the planets were observed with increasing precision, while 'theories' were set out as explanations. These theories often incorporated religious elements, and later struggled to free themselves from those elements, most notably in Europe during the period from Copernicus to Galileo (mid 16th to mid 17th century) when the animating center of the solar system was moved from the earth back to the sun (it had been located at the sun for a short time in ancient Greece, following the teachings of Aristarchus of Samos).

All of these developments were of great importance not only in the history of mathematics, mechanics, and physics, but in general human intellectual development.[1] For a gripping account of the story up to the time of Newton, the reader should look at the book [Koe59] by Arthur Koestler,[2] or at the best part of it (the biography of Kepler) published separately as a shorter book [Koe60].

3.2   Newton, Poincaré, and the most romantic view of KAM

With the appearance of Newton's *Principia* in 1687, the story takes its most momentous turn yet recorded. The *Principia* inaugurated the modern form of mathematical physics—modeling phenomena by way of differential equations—and nothing has been the same since. Right at the center of this most important book in the history of physics is the theory of gravity and its application to the motions of 'heavenly bodies.' It is truly astonishing that this theory conceived in the 17th century should be used up to the present day as the basis for calculating planetary orbits with almost arbitrary precision. The only correction to date—Einstein's general theory of relativity—applies to our solar system with practical significance only in the case of the tiny planet Mercury, which lies close enough to the Sun to suffer significant 'relativistic tides.' But if Newton was the chief architect of rational and celestial mechanics, and those in the century after him its builders, we'll see later that it was Henri Poincaré who restructured the whole edifice and extended it to new heights.[3]

---

[1] For a succinct discussion of the influence of astronomy and celestial mechanics on human intellectual development, see pp. 475–476 of P. Watson's book [Wat05]. To bring famous names into the picture, recall that Immanuel Kant likened the revolution he wrought in philosophy to the Copernican revolution; and finally, on p. 616 of D. Boorstin's book [Boo83], one finds the following quote attributed to Karl Marx in his 1841 thesis at Jena: "The glorification of the heavenly body is a cult which all Greek philosophers celebrate ... It is the intellectual solar system. Hence the Greek philosophers, in worshipping the heavenly bodies, worshipped their own mind."

[2] Some may be alarmed by Koestler's associations with 'paranormal' subjects (such as ESP), or by startling allegations in the biography of Koestler by D. Cesarani [Ces98], but considered by themselves, the books [Koe59] and [Koe60] have the kind of readability that only a gifted writer could give them.

[3] Beginning here, I follow the modern fashion of using architectural metaphor to describe the science of mechanics (the word 'edifice' in reference to mechanics is nearly as widespread as 'celebrated' in reference to KAM). For other examples, see the last quotation in §1.3 above, or the passage on p. 27 of I. Ekeland's book [Ek88] which begins "Science has striven for over a century to build an imperishable temple on the foundations Newton laid [...]."

No one reading this will be ignorant of the general outline of intellectual development leading from Aristarchus through Newton and beyond. But I recall this part (in slightly declamatory style) to point out that many of those who know the history of mathematics and physics intimately—who describe it as a 'great edifice'—give special significance to celestial mechanics and hence also to Newton and Poincaré. For some, not only did the millennia-long evolution of theories of planetary motion lay the foundation for this later edifice of mathematics and physics, but the planetary theories of Newton and Poincaré form the magnificent central tower. Now the most extravagant and romantic claim that can be made about KAM theory—one that is never voiced explicitly but can occasionally be glimpsed beneath the surface—is that it is the crowning cupola of that central tower, the design of which would startle the founding architects. If Newton were reincarnated and looked at the structure of rational mechanics to see how it had grown since he sketched its form, he would immediately pick out KAM theory as a special sort of 'crown' that he never expected to see.[4] Even Leibniz would be placated to some degree in the criticisms he brought against Newton's 'system of the world.'[5] We'll be in a better position to understand these claims later, and we'll understand how Poincaré, too, might be surprised and impressed by this crown, and might smile wryly at the fact that he did not live to see it conceived, or even to conceive it himself.

3.3 A more sober view

The foregoing is, of course, the ultimate rhapsodic giddy view, and we'll see as we go along how enthusiasts might be drawn to it in fits of ecstasy. But for now, let us try to remain more sober and simply point out that (a) theories of planetary motion were an important driving force in the development of mathematical physics; (b) these theories took on their essentially final modern form with Newton and the formulation of the 'n body problem' (see below); and (c) there is a sense in which KAM theory responds to some of the most perplexing problems to arise in the n body problem; but (d) this was not possible until Poincaré developed new insights, which, af-

---

[4]Then again, who knows—he might be too busy perusing a copy of *The Da Vinci Code* to notice. (This remark is meant to be at least slightly amusing, since Newton is notorious for expending more energy on biblical exegesis, alchemy, and other 'occult studies' than on natural philosophy itself.)

[5]*The System of the World* is the title of Book 3 of Newton's *Principia*, and is the part that drew the most interest (and criticism) among philosophically minded observers.

ter a long period of assimilation, allowed Siegel and especially Kolmogorov to have (e) the culminating insights that resulted in KAM theory, as first worked out in detail by Moser and Arnold.

In this more prosaic way of thinking, Newton retains his undisputed place in history as the greatest single contributor to celestial mechanics, and even to mathematical physics generally.[6] But however important Newton (and Poincaré) may be, I try to mostly resist here the romantic 'great man' view of history and take the view that classical and celestial mechanics were created through a large collaborative effort over centuries both before and after Newton, reaching a somewhat modern form only after efforts by the likes of Copernicus, Kepler, Galileo, Leibniz, several of the Bernoullis, Euler, Lagrange, Clairaut, Laplace, Liouville, Hamilton, Jacobi, Gauss, Kowalevski, and many others, to say nothing of what happened once Poincaré arrived on the scene.

Yet whether you prefer the 'great man' or the more collaborative picture, it was undeniably Newton who showed how to use differential equations to model the world. Afterward it was inevitable that the system of differential equations modeling planetary motions, the 'n body problem,' would be the subject of deepest mathematical investigation.

3.4   The n body problem

By the n body problem, we mean the dynamical system that describes the motions of n given masses interacting by mutual attraction according to Newton's law of gravity. By 'masses' we generally mean planets, so this is the basic mathematical model for a planetary system such as our solar system (where $n - 1$ is the number of planets, the sun or other star being the nth body), and few models in the history of science are better than this one in their practical predictive power.

Taking the specific case $n = 2$ gives the simplest nontrivial n body problem. The two body problem is readily reduced to the '$1/r^2$ central force problem' (i.e., the motion of a point mass moving in an inward central force field with force inversely proportional to the square of the distance from the center), and it was 'solved' more or less at the same time it was formulated (in geometric terms) by Newton in the *Principia*.[7] The $1/r^2$ central force

---

[6] As J.-L. Lagrange is supposed to have said, "Newton was the greatest genius who ever lived and the luckiest, for we cannot find the system of the world more than once." (See Quote **Q**[Lag] in Appendix E for the original French.)

[7] The story is often told of how in 1684 Edmond Halley asked Newton what orbit a body

problem is usually now called 'Kepler's problem,' and is completely integrable as described above in §2.2.3; one choice of three independent first integrals is: the total energy, one of the components of angular momentum, and the total angular momentum.[8] There is nevertheless a lot to be said even in this simplest case; a good overview may be found in [Al02].

The real trouble with the n body problem begins when n is three or greater, as Newton discovered to his dismay when he tried to use his newly enunciated laws to calculate motions of the earth-moon-sun system.[9] Although the n body problem itself has not changed since its creation, attempts to 'solve' it or simplify it—either in general, or for any specific $n \geq 3$—have generated an enormous and still growing corpus of literature. Solutions have been propounded (and found wanting); approximate solutions have been calculated and tested; the problem itself has been reformulated, reexamined, and generally turned upside down and inside out in attempts to solve it, find solutions in special cases, or just gain insight into its mysteries. As we'll see, a simple (and reductionistic) way of describing Poincaré's main contributions to celestial mechanics is to say that he showed definitively that the n body problem is not completely integrable, and he provided the tools for understanding it in new qualitative ways instead.

3.5 The stability problem

Even if the elusive goal of *solving* the n body problem that models our solar system is set aside, for centuries a closely related holy-grail problem for mathematicians and astronomers has been, *Is the solar system stable?* That is, even if its motions aren't periodic, or completely predictable, can we determine whether it will 'hang together,' or suffer calamity (fly apart

---

would follow if subjected to an inverse square central force, to which Newton immediately replied "An ellipse." At Halley's urging, Newton's efforts to write down the proof of this assertion eventually became the *Principia.*

[8] In fact Kepler's problem has *five* independent first integrals—two more than required for complete integrability. In this sense it's *superintegrable* (and one consequence is that all its bounded motions are periodic).

[9] The short book [ClC00] by D. Clark and S.P.H. Clark gives an interesting account of Newton's (often unscrupulous) attempts to gain access to the most accurate observations of the moon's motion, and the vexation he felt in trying to successfully describe that motion. There is also the famous remark attributed to the astronomer John Machin, according to whom Newton said that "...his head never ached but with his studies on the moon"; cf. [Wes80] p. 544. (Some readers may be interested to know that 'John Machin' sounds like 'John What's-his-name' in French.)

or suffer internal collision) over some given long interval of time (say, the next few billion years)?

This question—the problem of the stability of the solar system—was present in some form in astronomers' (and astrologers', and theologians') minds long before the mathematization of astronomy by way of Newton's laws. It was connected to ideas of the 'immutability of the heavens' and the 'purity of the celestial spheres' that were rudely contravened by the occasional appearance of what we now call meteors, comets, supernovae, and so on.[10] But the planets themselves intervened in the purity picture, since against the backdrop of the 'fixed heavens,' they appeared to 'wander.' And although these wandering motions seemed to repeat themselves at regular intervals, after closer and more accurate observations over long times, it became clear that the motions were not quite perfectly periodic.

In short, after Kepler, Galileo, and above all Newton, the ancient problem of planetary motion was recast mathematically as the n body problem, and although initial hopes of solving it in the traditional sense withered, the stability problem remained and gained importance as it too was reformulated with more mathematical precision.

At the beginning of the 18th century, Newton famously wrote that the solar system needed occasional divine intervention (presumably a nudge here and there from the hand of God) in order to remain stable.[11] This was interpreted to mean that Newton believed his mathematical model of the solar system—the n body problem—did not have stable solutions. Thus was the gauntlet laid down, and a proof of the stability of the n body problem became one of the great mathematical challenges of the age.

In the early 19th century, building on earlier work of Lagrange,[12]

---

[10]These matters of cosmic stability and purity were taken very seriously by theologians. Galileo's troubles with the Catholic Church are well known, and although much of the proceedings of Giordano Bruno's trial have been lost, it is generally assumed that when inquisitors burned him at the stake in 1600 at the Campo de' Fiori in Rome, it was partly or largely because of his heretical and unrecanted cosmological beliefs. On p. 180 of [Ab94], R.H. Abraham writes that the eternal stability of the solar system became Church dogma at the time of Bruno's death.

[11]Newton's remarks about divine intervention appear in Query 23 of the 1706 (Latin) edition of *Opticks*, which became Query 31 of the 1717 (2nd English) edition (see Quote Q[New] in Appendix E). Similar 'theological' remarks are found in scholia of the 2nd and 3rd editions of *Principia*, and in at least one of Newton's letters. In a 1715 letter to Caroline, Princess of Wales, Leibniz observed sarcastically that Newton had not only cast the Creator as a clock-maker, and a faulty one, but now as a clock-repairman (see [Klo73], Part XXXIV, pp. 54–55).

[12]Lagrange's work in celestial mechanics and stability includes his (re)discovery and treatment of the famous 'Lagrange points' (the first three of which were found earlier by

Laplace 'proved' the stability of the solar system, and although this work was an important step and is mathematically correct as far as it goes, it was soon recognized that he had not shown stability for the n body problem, but rather for an attenuated version of it in which higher-order interactions between planets are ignored. Such claims and half-successes by the mathematical giants of the ages (and many other less-than-giants) of course only added celebrity to the n body problem, and to the stability problem in particular.

For our own purposes, the most important fact about the n body problem is that, with considerable effort, certain versions of it can be recast as a 'nearly integrable Hamiltonian system' (see §3.10 and §7.1.1 below). The theory of such systems is called Hamiltonian perturbation theory, and it was the intense study of them largely started by Poincaré that eventually led to KAM theory. KAM theory in turn illuminates the n body problem—and other things—in very surprising ways.

3.6 Toward the modern era—integrability and its vulnerabilities

Up to now, we've been content to look at the history in a cursory fashion, mostly as a way to set the background stage for—and lend the appropriate air of gravitas to—what comes next. By fast-forwarding through the centuries in this way we arrive to the beginning of the 19th century, an amusing point in our story where we slow down to take a more detailed look. Several interesting things happened at that time. First, the philosopher G.F.W. Hegel wrapped up his *Habilitationsschrift*[13] at Jena, and made remarks in it that we render thus: "Mathematicians and astronomers should quit wasting their time searching for new planets. If they'd only study philosophy, they'd realize on purely metaphysical grounds that there can be

---

L. Euler): When two masses are in a circular orbit around their *barycenter*, there are five gravitational *equilibrium* points (traditionally labeled $L_1, L_2, \ldots, L_5$) carried along rigidly in the orbital plane of the system. The first three Lagrange points lie on the line through the masses' centers, while the last two lie at the vertices of the two equilateral triangles in the orbital plane with (other) vertices at the masses' centers. In the Sun-Jupiter system, one finds a number of asteroids riding along at these last two Lagrange points; they bear the collective name 'Trojan asteroids,' with the 'Greek camp' leading at L_4 and the 'Trojan camp' trailing at L_5.

[13]This is the second, longer doctoral thesis (for the *Habilitation* qualification) that still exists in Germany and many other European countries, considerably more involved than a PhD thesis.

no more than seven."[14] But as if on cue, in Palermo, Sicily, on the opening day of the 19th century (January 1, 1801), Giuseppe Piazzi discovered the asteroid now called *Ceres* and tracked its progress for several weeks until it came too close to the Sun to follow further. At that point, Carl Friedrich Gauss was called upon to predict where the new planet would later emerge, and famously calculated it with superb precision using his new method of least-squares approximation (this was in fact the event that propelled Gauss to fame).

According to this story (and the reader must admit it makes a good story), Hegel was derided,[15] and the event was seen as simply another vindication of the increasingly refined science of rational mechanics. This was after all the era of Laplace's famous remarks about the mechanistic nature of the universe,[16] and of Lagrange's (and later Hamilton's) elegant reformulations of classical mechanics. Yet when we come later to a description of KAM theory and to its place in the history of ideas, we'll see that (in a way unrelated to the number of planets) it is Hegel who has the last laugh (cf. §5.1.2).

For the moment though we remain focused on the 19th century and its growing confidence in the ascendancy of classical mechanics. New integrable systems were discovered along with systematic processes for 'integrating' them, notably C.G.J. Jacobi's methods as set out in his grand treatise on *dynamics* [Jaco43]. With hindsight, we now look back wistfully at the integrable systems (the most important of which were found before the end of the 19th century) and lament how rare they turned out to be, so few in number as to be catalogued in a relatively short list.[17] But at the time,

---

[14]This is a popular anecdote (recounted, for example in Chapter 14 of [Bel37] and also more briefly on p. 37 of [Sart52]), but friends with an interest in continental philosophy have assured me that Hegel must've written something less silly, perhaps simply adding to the discussion of the *Titius-Bode law* (on the distribution of planetary orbits) with a metaphysical bent. I hope one day to look at relevant parts of Hegel's *Habilitationsschrift* myself.

[15]Page 18 of [Guc78] cites (a loose English translation of) an excerpt from a November 1, 1844 letter from C.F. Gauss to the astronomer Schumacher in which Gauss criticizes the "muddled concepts" of philosophers, including not only Hegel, but also Plato, Schelling, and Kant.

[16]"An intellect which at a given moment knew all the forces that animate nature and the respective positions of the beings that comprise it, if vast enough to submit these data to analysis, could encompass in the same formula the movements of the greatest bodies of the universe and those of the lightest atom: for it, nothing would be uncertain, and the future as well as the past would be present before its eyes." (See Quote **Q**[Lap] in Appendix E for the original French.)

[17]A very short list of integrable systems (Hamiltonian, with finitely many degrees of

progress in finding integrals seemed to come swiftly, and it is probably only a modest exaggeration to say that until the last decade of the 19th century, most researchers thought that the majority of simple mechanical systems were integrable; the problem was just finding the means to integrate them, and this too would yield to increasingly refined methods.[18] Especially encouraging to many was the 1888 prize-winning paper [Kow88–90] by Sophie Kowalevski[19] showing the integrability of certain rotating bodies ('heavy asymmetric tops'); this was an extraordinary feat that had eluded even Euler and Lagrange.

Alas, the rosy integrability picture—with Kowalevski's top as the centerpiece—was not to last. There were some who suspected that certain systems showed more resistance to integration than they ought to, and one form of nonintegrability appeared at almost the same time as Kowalevski's paper, when H. Bruns published a proof [Bruns87] of the nonexistence of (algebraic) integrals of the three body problem beyond the ten classical integrals already then known.[20]

Not long after, more convincing analytic proofs of nonintegrability would appear, the pendulum of fashion would swing the other way, and it would become more fashionable to say that 'almost nothing is integrable' (cf. §3.12 and §3.13 below). We next examine the beginning of this turn of events more closely.

---

freedom) is: Any 1-degree-of-freedom system (trivially); the Kepler problem (Newton 1687); and more generally, any point mass moving in a central force field derived from a potential; geodesic flow on the surface of an ellipsoid (Jacobi 1839); and various special motions of rigid bodies ('tops') about a fixed point (Euler 1765, Lagrange 1788, Poinsot 1834, Kowalevski 1888). A more up-to-date and complete list would probably be no more than a dozen times as long. However, it should be pointed out that new integrable systems (KP, KdV, Toda hierarchies, etc.) have been discovered in *infinitely* many degrees of freedom (i.e., in systems of PDEs) where they have taken on increased importance since the 1960s.

[18] It should be pointed out however that the article [HasK02a] by B. Hasselblatt and A. Katok cautions against thinking that this simple view was widespread.

[19] One sees both parts of Kowalevski's name in different forms or transliterations (and she used the first name Sonya after moving from Russia to Sweden). A good transliteration of her full Russian name is Sofiya Vasilevna Kovalevskaya; but in this text, I use the form with which she signed the publications she wrote in German or French.

[20] As mentioned earlier, nonintegrability of a more basic mathematical kind had already appeared as early as 1835, when J. Liouville published a paper [Liou35] showing that a large class of otherwise 'nice' functions $f : \mathbb{R} \to \mathbb{R}$ could not be integrated by means of *elementary functions*.

3.7   Weierstrass, Poincaré, and the King Oscar prize

We now come to what is perhaps the most famous part of our story. This
has been told many times and at various levels of detail (it is described in
Chapter 7 of [Pete93] by I. Peterson, in Chapter 1 of [DiH96] by F. Diacu and P. Holmes, and it is the essential core of the book [Barro97] by
J. Barrow-Green), so I give here only the barest outline, in a present-tense,
stage-direction style.

The stage opens in Berlin in the early 1880s where we find Karl Weierstrass (1815–1897) keenly interested in the n body problem. Weierstrass is
probably the world's leading figure in mathematics, rising from humble beginnings as a *Gymnasium* teacher in Münster (1841) to full professor at the
University of Berlin (1856). This he does through the sheer brilliance of his
work in *analysis*, despite onerous teaching duties and health problems that
plague him from about 1850 on. Once in Berlin, and health permitting, he
becomes a renowned teacher, establishing new standards of *rigor* in analysis and attracting students and scholars from around the world. He has
more than 40 PhD students and teaches many others less formally, including H. Bruns, G. Cantor, G. Frobenius, L. Fuchs, E. Husserl, W. Killing,
C. Runge, and H. Schwartz. Most important in our story though, are his
protégé Gösta Mittag-Leffler and his former student Sophie Kowalevski,
both currently in Stockholm. Historians will later discover rich detail in
Weierstrass's numerous letters to Kowalevski over two decades (cf. [Koč73];
but Kowalevski's own letters were apparently burned by Weierstrass after
her early death from pneumonia in 1891).

Returning to center stage, Weierstrass calculates *formal series* solutions
(akin to so-called *Lindstedt series*) for the n body problem, suspects that
they are *convergent*, but has difficulty establishing this *rigorously*. He becomes aware of a remark[21] from 1858 that J.P.G. Lejeune Dirichlet made
only a year before his death[22] to his former student L. Kronecker, in which
he claims to have shown the convergence that Weierstrass seeks. Because

---

[21] With a little whimsy, one could say that this remark plays a role in KAM theory similar
to that played by Fermat's famous margin notations in the recent proof of 'Fermat's last
theorem': both remarks acted as strong spurs to subsequent mathematical developments,
yet the remarks' authors were almost certainly overconfident in their claims.

[22] It has always seemed odd that the three immediate successors to Gauss in his chair at
Göttingen—Dirichlet, G.F.B. Riemann, and R.F.A. Clebsch—survived in it for only 4,
7, and 4 years respectively, especially since Gauss sat there for decades. But this 'curse
of Gauss's chair' comprises only three cases of untimely death in our story; there are
others, as we shall see.

Dirichlet's reputation is high, Weierstrass attaches considerable weight to this claim, and becomes unwaveringly convinced that the Lindstedt-like series are somehow, somewhere convergent.

Meanwhile, in Stockholm, Mittag-Leffler[23] has risen to a high station in Swedish academic circles, high enough that the scientifically enlightened Oscar II, King of Norway and Sweden, asks him to suggest questions for a scientific prize competition that will honor the king's upcoming 60th birthday in 1889. Mittag-Leffler in turn consults Weierstrass and Kowalevski, who suggest the convergence problem. One of the three prize questions is therefore on the convergence of these series in the three body problem, or, as it is dramatically phrased in the competition, "on the stability of our planetary system."

A certain young French mathematician J.H. Poincaré now enters the scene. In fact both Mittag-Leffler and Weierstrass have Poincaré firmly in mind when the prize problems are announced in Mittag-Leffler's newly founded (1882) journal *Acta Mathematica*; the fact that the third member of the jury, C. Hermite, was a former member of Poincaré's thesis committee does not hurt either.[24]

To keep the story short, we cut to the end, where Poincaré is awarded the prize for his magnificent 270-page memoir [Poi90] upon which the jury heaps praise, saying that "it will change the course of astronomical dynamics forever."[25]

---

[23] As is often repeated, the persistent rumor that Alfred Nobel did not establish a Nobel prize in mathematics because Mittag-Leffler had an affair with Nobel's wife cannot be true since Nobel never married. Nevertheless, there may have been bad blood between the two (see p. 53 of [Craw84]), in which case it is somewhat odd that in the years after Nobel's death in 1896, Mittag-Leffler played a significant role in decisions about Nobel prizes.

[24] Alain Albouy pointed out to me that letters between Poincaré and Mittag-Leffler (cf. [Nab99]) make it clear that the initial aim of the prize was simply to crown some of Poincaré's recent work. But Poincaré rejected this, deciding instead to go in a very new—or as it turned out, revolutionary—direction.

[25] This story—with its king, its theme of stability of the solar system, etc.—would be famous even if it were only as outlined here. But there are ruffled feathers, intrigue, and also a 'story within the story' that give it still more celebrity. It turns out that after Poincaré was awarded the prize, and while the presses were printing his paper, the young Swedish mathematician L.E. Phragmén discovered a fundamental error in the prize paper. Poincaré (with help from Mittag-Leffler) literally had the presses stopped, snatched as many printed copies as he could find from the printers' hands, paid them (more than the sum he won from the prize) and worked furiously over the next year to correct and re-create the 'great manuscript' we know today. (And the rediscovery in the 1970s of copies of the original 156-page flawed manuscript caused a stir among the cognoscenti.) See [Barro97], [DiH96], or [Pete93] for further details.

3.8   Aftermath of the prize: the seeds of change are sown

It is certainly true that Poincaré's prize-winning paper changed the course of dynamics; but it's also true that the memoir did not 'solve' the prize problem. On the contrary, the memoir was revolutionary for going a great distance in the opposite direction, showing instead that the n body problem is insoluble in the usual classical sense. In the memoir and afterward, Poincaré also showed that the sought-after series of the prize competition could not converge for many initial conditions, and speculated—but carefully disavowed any proof—that they were *divergent* everywhere (or *almost everywhere*). We'll return to this point in §5.3.1.

If King Oscar's prize is the most well-known part of the story, the most important things to convey here come afterward. We'll examine the legend of how Poincaré's work put dynamics into a kind of slow-motion crisis which was not resolved for the better part of a century, how Poincaré laid the groundwork for much of the new kind of thinking that would be required for further progress, and how the development of KAM theory may be viewed as the single event that best encapsulates the resolution.

So just what did Poincaré discover about the three body problem that was so groundbreaking? A complete answer to that question would—and does—fill volumes (namely Poincaré's *Œuvres* [Poi15–56]). Perhaps the best intermediate-length answers available in English are the detailed book [Barro97] and the more mathematical essay [Chenc12] (Chapter 1 of [DiH96] also gives a nice non-technical overview). Here I'll try to respond in the space of a few pages, which requires a lot of compression. But before getting to the heart of the matter below in §3.11, this is a good place to provide some background on Poincaré, who abruptly entered our story above.

3.9   A quick sketch of Poincaré and his work

Readers from the English-speaking world are familiar with Newton, his revolutionary scientific accomplishments, his proclivity for 'sorcery,' and his strange, reclusive and vengeful ways as a person. But they're probably less familiar with Jules Henri Poincaré (1854–1912), the French mathematician often called the 'last universalist of mathematics' and 'father of dynamical systems,' who stands second only to Newton in revolutionizing the science of mechanics (but was of much friendlier disposition). There's no shortage of biographical information on Poincaré in scattered sources, but authoritative biographies of him are scarce.[26] (Until recently, I knew only of the preface by P. Appell in Poincaré's *Œuvres* [Poi15–56] and the biographies [Ap25] and [Belli56], all in French. But scientific biographies in English have now appeared [Ver12], [Gray13].) In this section, I'll give the briefest possible account of who he was and what he did, followed by a few special details that are relevant to our story later.

Poincaré was born to an illustrious family[27] in Nancy, in the Lorraine province of northwestern France, where he won the lion's share of top academic prizes at the *lycée*.[28] He sailed through the *École Polytechnique* as the top or nearly top student, wrote his doctoral thesis at the University of Paris, and went on to a brilliant career during which he taught at the University of Caen, the Sorbonne, and the *École Polytechnique*. He wrote innumerable papers, monographs, treatises, and books (including several popularizations of science that showcase his literary talents), and received almost as many prizes, medals, and awards, both foreign and domestic. Sadly, he lived only to the age of 58 before succumbing to an embolism, following surgery, at the height of his mathematical powers.

Although not primarily a physicist, he nevertheless studied and lectured on optics, thermodynamics, electromagnetism, potential theory, celestial mechanics, quantum mechanics, telegraphy, capillarity, elasticity, and more. He was a co-discoverer (with A. Einstein and H. Lorentz) of the special theory of relativity, but generally downplayed his contribution and encouraged the younger Einstein.

In pure mathematics, he was by all accounts a monster,[29] making fundamental advances in complex variables, *topology*, algebraic geometry, and of course differential equations. As for what most concerns us, he seismically shifted the field of ordinary differential equations so that it merged with geometry and topology and became what is now known as dynamical systems. In so doing, he found the need to invent algebraic topology (he called

---

[26] It's interesting to compare the dearth of biographies of Poincaré with the abundance of biographies of Newton. Some of this is accounted for by the difference in the subjects' scientific statures, and by the difference in the sizes of French and English language readerships. But much of it is probably because Poincaré's conduct, bearing, and personal life were utterly normal in comparison with Newton's.

[27] Henri's father Léon was professor of medicine at the University of Nancy, and his first cousin Raymond later became prime minister, then president, of France.

[28] The lycée is now the *Lycée Henri Poincaré*.

[29] In French, *monstre* is one of the complimentary ways to describe great mathematical talent. There are some who say that using the term in this way derives from a teacher's description of Poincaré as '*un monstre de mathématiques*' as recounted on p. 110 of the biography [Belli56] by A. Bellivier.

it 'analysis situs' [Poi95], a term coined earlier by Leibniz for his project of creating a 'symbolic geometry') and to produce its first important results.

Although the concept of a formal thesis student is somewhat anachronistic in France for Poincaré's time,[30] several rivers of mathematical research eventually flowed from his legacy, and his work continues to be an important source for mathematicians in the present day.[31]

In this book, we'll primarily be concerned with what Poincaré did in celestial mechanics and Hamiltonian perturbation theory (HPT), so at the very least, we need to point out his most important work in these areas. We have already described the drama surrounding the famous prize paper [Poi90] in *Acta*. After the hurried production of that memoir, Poincaré spent almost a decade afterward elaborating the ideas in it, publishing them in the three-volume *Les Méthodes nouvelles de la mécanique céleste* (or *New Methods of Celestial Mechanics*) [Poi92–99]. Volumes 1, 2, and 3 appeared in 1892, 1893, and 1899, respectively, and they're usually referred to as *Les Méthodes nouvelles*. It's these volumes even more than the *Acta* paper upon which Poincaré's fame in celestial mechanics and HPT rests, as they contain a wealth of new ideas and make free use of variational methods, integral invariants, *periodic orbits*, *asymptotic series*, the principle of *recurrence*, and more.

This last idea, recurrence, had an especially interesting career after Poincaré announced and proved it in the *Acta* memoir of 1890.[32] The statement of the principle (or theorem) is simple: If the flow of a dynamical system preserves volume and evolves in a bounded phase space, *almost*

---

[30]If we allow ourselves to be anachronistic, perhaps Poincaré's most notable student was Louis Bachelier, whose 1900 thesis *Théorie de la spéculation* partly anticipated the fundamental work of Black and Scholes in financial mathematics by more than seventy years. The thesis also contained work on continuous Brownian motion problems that preceded A. Einstein's famous 1905 paper on Brownian motion.

[31]The reader may be aware that in 2000, the Clay Mathematics Institute established its list of seven 'Millennium Prize Problems,' explaining that it was a kind of modern version of David Hilbert's list of problems from his 1900 address to the International Congress of Mathematicians. (Hilbert's list strongly influenced the direction of 20th century mathematics.) One of the problems so honored was Poincaré's 1904 conjecture in topology—that any *simply connected*, closed 3-manifold is homeomorphic to the 3-sphere. (Here 3 indicates the dimension; higher dimensional versions of this conjecture were proved earlier by S. Smale, M. Freedman, and others.) The Poincaré conjecture was the first Millennium problem to be resolved: in 2002–03, the Russian mathematician Grigory Perelman outlined a method for affirming the conjecture which has since been verified in detail. In 2006, Perelman was awarded a *Fields medal* for this work, which he declined to accept. He also declined the Millennium Prize in 2010.

[32]In fact, some form of it appears to have been present already in the original prize manuscript withdrawn in 1889.

every trajectory returns arbitrarily close to its initial condition, and does this infinitely many times.[33] When it was first formulated, some thought that the recurrence principle could be used to justify Ludwig Boltzmann's *ergodic hypothesis*, about which we'll say more shortly. Instead it was later turned *against* Boltzmann's ideas by E. Zermelo and F.W. Ostwald, who used it in the obvious way to try to invalidate Boltzmann's 'H-theorem,' a form of the *second law of thermodynamics* that necessitates the increase of *entropy* in closed systems not in *equilibrium*. Boltzmann eventually fought back, showing that recurrence times for any realistic system must be many orders of magnitude larger than the age of the universe; yet still today, discussion of these issues can be very tricky (see §7.2.1 below for further discussion).

Although Poincaré's work is vast, the now-famous quotations—both prophetic and antiprophetic—that relate to our story are drawn from just the few sources mentioned (the *Acta* paper and the *Méthodes nouvelles*). These will be pointed out later.[34] But to get the full sense of irony embodied in the quoted passages, it's important to note the clash between how they're sometimes read, and what Poincaré himself might have intended.

Poincaré had an unusual style for a mathematician. To put it simply, he wrote quickly and discursively, and he didn't look back. As befits one of the great mathematicians in history, his output often included astonishing lengths of error-free calculations of the most abstruse kind. The manuscripts show no hesitation or correction, only the quick confident script. But errors did occasionally creep in (as we already saw), and readers also often complained about the prolix style,[35] so unlike the terse polished prose that had become fashionable since Gauss. Some have complained that he did not make the effort to condense, polish, and check his work; others have expressed relief that he didn't waste time on such trivialities during the relatively brief span of his creative years.[36]

---

[33] This statement bears a remarkable similarity to F. Nietzsche's doctrine of eternal recurrence (*ewige Wiederkehr*) as he discussed it in *Die fröhliche Wissenschaft* first published in 1882. (Of course, Nietzsche was by no means the first to talk about recurrence in this way, but he gave it special significance in his philosophy.)

[34] A number of Poincaré's famous quotes are catalogued in Appendix E.

[35] This was already evident in Poincaré's thesis report, where the jurors complained of his "inability to express his ideas in a clear and simple manner" (cf. [Mill96]).

[36] Discussions of the effect of Poincaré's temperament on his science echo similar debates about Newton. There are those who lament the numerous bitter disputes in which Newton became embroiled, wondering what other discoveries he might otherwise have made; but others quite plausibly suggest that the disputes were a primary engine of his discoveries.

In any event, going back over things was apparently not to his taste, and as long as mistakes were minor, he wasn't especially concerned. He preferred to get things on paper in his own style, then move on. There's a short popular anecdote in France that encapsulates this: Poincaré sits scribbling at his desk while an assiduous assistant pores over his most recent finished page. The assistant reluctantly interrupts the master, "Sir, there's a mistake!" to which the master (without ceasing to scribble) magnanimously replies, "Well correct it my friend, correct it!"[37]

All of this contrasts sharply with later events, in several ways. Although Poincaré's work was amply recognized in his lifetime, many parts remained enigmatic long afterward. As more nuggets were dug from his *Nachlass* and it was realized how far ahead of his time he was, there emerged a tendency to read his works line-by-line, like sacred texts. These 'exegetical Poincaré studies' did (and do) occasionally lead to new insights into the master's meaning, but on the whole it's amusing to imagine what Poincaré might make of the close scrutiny given to writings that he himself took less seriously.[38]

It's also interesting to ponder the severe and mostly no-fun reaction to Poincaré's intuitionist style in the decades after his death. Something similar can be said of Ludwig Boltzmann, another intuitionist-genius relevant to our story, except that the reaction against him occurred in his lifetime. Poincaré never saw the austere anti-intuitionist *Bourbaki*[39] tradition that emerged in French mathematics in the 1930s. Boltzmann was not so lucky; some say the demons that compelled him to take his life were partly inspired by puritanical *positivists*, led by Ernst Mach.[40] The happiest part of the story, though, is that later events unfolded very much—but not entirely—

---

[37] "Monsieur, il y a une erreur!" "Alors, corrigez mon ami, corrigez!" (As told to the author by Pierre Lochak.)

[38] This tendency might be partly ascribed to certain literary practices in France: from the meticulous *explication de texte* taught in the *lycée* or earlier, to the word-by-word analysis often employed by deconstructionists and others, French academic traditions employ and encourage some of the closest textual readings in the world, sometimes with comic effect.

[39] Nicolas Bourbaki is the pseudonym first used by a group of leading French mathematicians (educated at *ENS*-Paris) who sought to place mathematics on a *rigorous* and structured foundation. They became highly successful and also controversial; for more information, see Parts D.3.2 and D.8 of Appendix D, as well as the *Bourbaki* entry in the glossary (Appendix F).

[40] Boltzmann was later defended against Mach by none other than Vladimir Lenin, who saw the positivists as 'subversive' and Boltzmann as an undeclared materialist (see pp. 13–14 of S. Brush's introduction to his English translation of Boltzmann's book [Bolt96–98]).

in favor of Poincaré and Boltzmann. We'll have more to say about these things later.

3.10   HPT: 'The fundamental problem of dynamics'

We now turn to the part of Poincaré's work most relevant to our story: Hamiltonian perturbation theory (or HPT). Using modern language and notation, we consider a *smooth* (or even *analytic*) completely integrable n-degree-of-freedom Hamiltonian system with Hamiltonian $h = h(I)$ in the action-angle variables (θ, I), valid in a nice region U of phase space. To this integrable system, we add a smooth (or analytic) function of both actions and angles having the form $\varepsilon f(\theta, I, \varepsilon)$ and called a *perturbation* of h. This gives a so-called *perturbed Hamiltonian system*

$$H(\theta, I, \varepsilon) = h(I) + \varepsilon f(\theta, I, \varepsilon) \tag{3.1}$$

or *nearly integrable Hamiltonian system.*[41] The ε is simply a parameter—a number—that can be used to vary the 'strength' of the perturbation, while $f = f(\theta, I, \varepsilon)$ represents any sufficiently smooth function.

Hamiltonian perturbation theory is the study of systems of the form (3.1). In fact, Eq. (3.1) is the usual starting point for discussions of KAM theory. If justification of its importance is needed, a single sentence (seen again and again in the literature) usually suffices: "Poincaré called it the fundamental problem of dynamics."[42] But apart from Poincaré's authority,[43] what is really so fundamental about Eq. (3.1)?

Perturbation theory—the idea of starting from a known entity (here $h = h(I)$), then 'perturbing' it slightly as a way of exploring the unknown—was introduced by Newton in the *Principia* and soon became the standard means of studying the n body problem for $n \geq 3$. In this case, a two body problem is the known entity, and smaller additional bodies comprise the perturbation. It was by varying this approach that many of the fundamental advances in planetary motion were made in the 18th and 19th centuries (and by the usual suspects: the Bernoullis, Euler, Lagrange, Laplace, . . .).

Poincaré went further, seeing that the majority Hamiltonian systems could be explored in this way. Starting from the system $h = h(I)$ which we

---

[41] Yet a third terminology (from the French) is quasi-integrable Hamiltonian system.

[42] In French: *'le problème général de la dynamique'* (see Quotation **Q**[Poi2] in Appendix E below).

[43] For a rough idea of Poincaré's continuing stature in dynamical systems, see p. 11 of [Man89a], where B.B. Mandelbrot writes "For students of chaos and fractals, Poincaré is, of course, God on Earth."

understand thoroughly, we can use it as a kind of island base from which we set sail in 'perturbation-boats' εf to chart the vast unknown sea of systems surrounding h. In this sea-faring analogy, we may think of moving out from the charted island h a distance ε in the direction f. Poincaré's intuition was later vindicated when the theory of *Birkhoff normal forms* was used to show more rigorously that 'general' Hamiltonian systems may be considered as perturbations of completely integrable systems [Bir27], [Bos85], [MarkM74]. We next look at what Poincaré himself discovered when he 'set out to sea.'

3.11   From small divisors to nonintegrability and chaos— what Poincaré did

From the point of view of our subject, Poincaré's most important discovery in Hamiltonian perturbation theory was that 'most'[44] Hamiltonian systems are not integrable in the classical sense described above in §2.3, and it is this point that we'll examine more closely now.

To understand something about what Poincaré did, let's roughly follow one of his lines of reasoning to see if there's a way to bring Eq. (3.1) back to integrable form. Again in the sea-faring analogy, this corresponds to leaving the nice, integrable island h to see what portions of the surrounding oceans might also be integrable.

The basic method of verifying integrability, developed largely by C.G.J. Jacobi in the early 19th century, is to look for a smooth canonical change of coordinates $(\theta, I) \mapsto (\phi, J)$ such that in the new action-angle variables (ϕ, J) the Hamiltonian will depend only on J. Since the perturbation εf in (3.1) is $O(\varepsilon)$, we expect the coordinate change to be $O(\varepsilon)$-close to the identity transformation.

One still-standard way of finding canonical coordinate changes is to use Jacobi's original 'mixed variable' method, as people mostly did in Poincaré's time. But there's an equivalent method created later that has the advantage of being easier to understand and use, at least on the formal level we present here. It's called the Lie series method (after Sophus Lie), and it consists of looking for a 'generating function' $\chi = \chi(\phi, J)$ which we put in the left slot of the Poisson bracket $\{\ ,\ \}$ to make a 'Liouville operator' L_χ. This operates

---

[44]Speaking in this context, modern mathematicians replace the word 'most' by *generic*, which has a precise meaning given in the glossary (Appendix F). In the sequel, I'll sometimes use the term generic as a forward anachronism; the reader can of course substitute a word like 'most' if desired.

on sufficiently smooth phase-space functions (such as the Hamiltonian H) like this: $L_\chi(H) = \{\chi, H\}$. Finally, we create a near-identity variable change by placing ε in front of L_χ and exponentiating to get the operator $e^{\varepsilon L_\chi} = \mathbb{I} + \varepsilon L_\chi + O(\varepsilon^2)$ (we've only expanded through $O(\varepsilon)$, indicating the remainder by $O(\varepsilon^2)$).

Before proceeding further, it's probably best to point out that this is the formal, 'operator style' of making a symplectic change of variables $T : D' \to D$ defined as the time-one flow of the Hamiltonian $\varepsilon\chi$. Here the new variables $(\phi, J) \in D'$ and the old variables $(\theta, I) \in D$. This may seem backwards at first glance, but in fact it's the right way to proceed: one gets the new Hamiltonian $K = K(\phi, J)$ in the new variables by the nice formula $K = e^{\varepsilon L_\chi} H = H \circ T$ in which the action of the operator $e^{\varepsilon L_\chi}$ is defined by the second equality. In a more careful discussion of this method, one describes the *domain* D' and *codomain* D and shows that T is a *diffeomorphism* between them. For more details, see [Duma93].[45]

Performing this variable change on $H = h + \varepsilon f$ (see Eq. (3.1)) gives the transformed Hamiltonian $K = K(\phi, J, \varepsilon)$ which we calculate as $K = e^{\varepsilon L_\chi} H = H + \varepsilon L_\chi(H) + O(\varepsilon^2)$, or more explicitly

$$K(\phi, J, \varepsilon) = h(J) + \varepsilon f(\phi, J) + \varepsilon\{\chi, h\}(\phi, J) + O(\varepsilon^2). \qquad (3.2)$$

(Observe that $\varepsilon L_\chi(H) = \varepsilon\{\chi, h\} + \varepsilon^2\{\chi, f\}$, so we've absorbed the second term $\varepsilon^2\{\chi, f\}$ into the $O(\varepsilon^2)$ remainder above.)

In this first approach, we aren't going to worry about what happens in the $O(\varepsilon^2)$ remainder; we're just going to try to get rid of the ϕ-dependence in the $O(\varepsilon)$ term. Looking at (3.2), we see that we need to solve

$$f + \{\chi, h\} = 0, \quad \text{or more explicitly} \quad f(\phi, J) = -\sum_{k=1}^{n} \frac{\partial\chi}{\partial\phi_k} \frac{\partial h}{\partial J_k}. \qquad (3.3)$$

(The second sum of terms in the expansion of the bracket $\{\chi, h\}$ is missing because h does not depend on the angles ϕ.) This is an example of a 'homological equation,' or fundamental linear equation of perturbation theory. In order to try to solve this simple linear partial differential equation for χ, we use the fact that f and χ have unit period in each component of ϕ to expand both of them in multiple *Fourier series*:

$$f(\phi, J) = \sum_{k \in \mathbb{Z}^n} \hat{f}_k(J)\, e^{2\pi i k \cdot \phi}, \qquad \chi(\phi, J) = \sum_{k \in \mathbb{Z}^n} \hat{\chi}_k(J)\, e^{2\pi i k \cdot \phi}.$$

---

[45] Another clear and rigorous discussion of a related method for generating symplectic transformations is given in §6.1 of [MeyHO09].

Now substituting these expressions in (3.3) and using $\omega(J) = \partial h/\partial J$, we get

$$\sum_{k \in \mathbb{Z}^n} \hat{f}_k(J)\, e^{2\pi i k \cdot \phi} = -2\pi i \sum_{k \in \mathbb{Z}^n} \hat{\chi}_k(J)\, k \cdot \omega(J)\, e^{2\pi i k \cdot \phi}. \tag{3.4}$$

We then solve for the Fourier coefficients of χ to find (at least *formally*):

$$\hat{\chi}_k(J) = \frac{-\hat{f}_k(J)}{2\pi i\, k \cdot \omega(J)}. \tag{3.5}$$

These are the Fourier coefficients of the purported solution to our problem; they are supposed to give us the generating function χ for a change of variables that will eliminate the ϕ-dependence of $K = K(\phi, J, \varepsilon)$ through $O(\varepsilon)$. But there are several problems with this innocent-looking 'solution.' In the denominator of the right-hand side of Eq. (3.5) we find the product $k \cdot \omega(J)$ which we encountered above in §2.4. We see immediately that Eq. (3.5) makes no sense for $k = 0$. However, this by itself is not a problem: Since the harmonic $f_0(J)e^{2\pi i\, 0 \cdot \phi} = f_0(J)$ is the one term in the *Fourier series* of f that never has any ϕ-dependence, we do not need to eliminate it, and we can safely (re)define $\hat{\chi}_0 \equiv 0$. But now, having dealt with $k = 0$, we examine Eq. (3.5) more closely and we begin to worry about the possibility of other 'zero divisors.' In fact, based on our earlier discussion in §2.4 and §2.5, where we saw that the resonant tori (those with $k \cdot \omega(J) = 0$) of nondegenerate systems are *dense*, we are forced to the following conclusion:

> **Small divisor problem.** The nearly integrable Hamiltonian $H = h + \varepsilon f$ with nondegenerate integrable part h will have zero divisors $k \cdot \omega(J) = 0$ (i.e. *resonances*) in any open set of phase space. For 'ordinary'[46] perturbations εf, there is therefore *no hope* of using Eq. (3.5) to construct the desired variable change in the classical sense (i.e., so that it's defined on a 'nice' domain—one that's open, or contains nonempty open sets). While it's true that the portion of phase space where the divisors vanish has *Lebesgue measure* 0, even if this portion is removed, the divisors still become arbitrarily small on the remaining part of phase space. Thus, even on this (highly spongiform) part of phase space from which all zero divisors have been removed (and which contains no nonempty open sets), we are

---

[46] We could of course find special perturbations without the problems described here (e.g., perturbations with finite Fourier series), but these are so rare that we call them *nongeneric*, while the problematic perturbations are *generic* (see Appendix F for further details concerning *genericity*).

unable to ensure the *convergence* of the Fourier series for χ using the coefficients of Eq. (3.5).

The situation just described is a prototypical example of the notorious problem first encountered by mathematical astronomers in the mid 19th century as they tried to use transformation techniques together with J. Fourier's new series to simplify the n body problem.[47] The reader wishing to follow the purely mathematical thread here to see how the small divisor problem was finally resolved can skip to the beginning of Chapter 4 (and thereby skip roughly a century of history).

It was Poincaré who first recognized that the small divisor problem made integrable Hamiltonian systems exceedingly rare and that most[48] systems are not integrable in the classical sense defined above in §2.3. The fact that no classical transformation to integrable form exists for generic systems is called 'Poincaré's non-existence theorem' in [BenGG85a]. (Using *Birkhoff normal forms* not yet available in Poincaré's time, the theorem was later sharpened and stated in more modern form by C.L. Siegel [Sie54].)

Yet Poincaré went considerably further than this. In [Poi90], he showed that the resonant tori not only produce vanishing divisors, but that the fate of these tori under generic perturbations is strange indeed. Using a device now called the *Poincaré map*[49] to simplify the flow of a 2-degree-of-freedom system in the neighborhood of a periodic orbit, he showed that even if some of the orbits on the resonant tori survive perturbation more or less intact, the majority of them are forced to follow the contortions of what he called 'a sort of trellis' and what modern dynamicists call a 'homoclinic tangle.'[50] The tangle can be seen by taking a 'slice' or 'section'[51] *transverse* to the flow of the system. Surfaces invariant under the flow appear as invariant curves in this section, and orbits of the flow appear as *sequences* of points along the curves, so that the original flow is reduced to a Poincaré map (see Fig. 3.1).

By showing that invariant curves emanating from *fixed points* of the map 'split' under perturbation, then analyzing the subsequent behavior of

---

[47]The first researcher to emphasize the difficulties surrounding the small divisor problem was C.-E. Delaunay, in his two-volume treatise on lunar motion [Delau60–67].

[48]Again, we now say *generic*.

[49]Sometimes called by the longer name 'Poincaré first return map.'

[50]In fact, dynamicists distinguish two kinds of tangles: 'homoclinic' are those formed by invariant curves emanating from the same *fixed point*; 'heteroclinic' are formed by invariant curves from different fixed points.

[51]Or, in full, a 'Poincaré *surface-of-section*.'

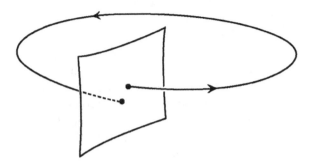

Fig. 3.1   Schematic diagram of the Poincaré map. The lower dot is the *image* of the upper dot after 'one return.' Note that many of the later figures in this book may be interpreted as invariant curves of a Poincaré map in a *surface-of-section* (represented here as the approximately square sheet transverse to the flow).

the curves using uniqueness and integral-invariance, Poincaré was led to imagine a figure of bewildering complexity. This was already evident in the (corrected) prize paper [Poi90], but there is also a very famous passage in *Les Méthodes nouvelles* ([Poi92–99], Vol. 3, p. 389) in which Poincaré comments on the tangle:[52]

> When one tries to depict the figure formed by these two curves and their infinity of intersections, each corresponding to a doubly asymptotic solution, these intersections form a sort of trellis, web, or infinitely tight mesh; neither of the two curves can ever intersect itself, but must fold back on itself in a very complex way in order to intersect all the links of the mesh infinitely many times.
>
> One is struck by the complexity of this figure that I shall not even attempt to draw.[53] Nothing is better suited to give us an idea of the complexity of the three body problem and all of the problems of dynamics in general where

---

[52]See Quote **Q**[Poi6] in Appendix E for the original French.

[53]This remark often draws smiles from readers, because Poincaré was notorious for his inability to draw. (Although he finished first in the entrance exam—and second in the exit exam—at *École Polytechnique*, he needed a 'special exemption' from the drawing portion, having received the normally disqualifying score of *zero*.) Further irony comes from the fact that one now routinely sees T-shirts with homo- or heteroclinic tangles emblazoned upon them (as in Fig. 3.2).

there is no uniform integral and Bohlin's series diverge.

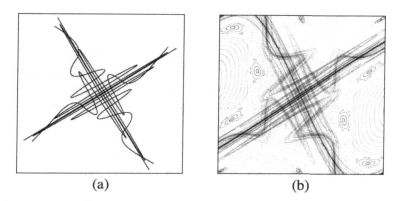

(a)   (b)

Fig. 3.2   Poincaré's 'trellis' in its modern incarnations: **(a)** sketch of a homoclinic tangle (a typical emblem of chaos); and **(b)** numerically computed orbits in a tangle of the *Chirikov standard map*. (Part (b) courtesy of Carles Simó.)

Although it's now commonplace knowledge (and approximations to tangles are routinely sketched or computed numerically) it seemed astonishing at the time to think that the vast majority of Hamiltonian systems were nonintegrable and contained homo- or heteroclinic tangles. Later it was discovered that the dynamics near tangles is even more complex than Poincaré's description suggests. In fact, the existence of tangles (often inferred by calculating a 'Melnikov integral'; cf. footnote 13, p. 132, below) entails the presence of a so-called *Smale horseshoe*, which in turn implies a number of dynamical features now seen as hallmarks of 'chaos': infinitely many periodic orbits of arbitrary periods, an invariant *Cantor set* with dense orbits, and a dynamics on this invariant set that is *conjugate* to a 'symbol shift'—meaning among other things that orbits starting very near each other can quickly end up far apart (the *sensitive dependence on initial conditions* so strongly associated with chaos, and something of which Poincaré was quite aware, as is often demonstrated by quotations[54]).

Nonintegrability, homoclinic tangles, sensitive dependence on initial conditions, recurrence: all of this was present in Poincaré's work on HPT and the three body problem. As has been pointed out repeatedly (see e.g. [AubiD02]), this embodies most of 'chaos theory' as it became fashionably

---

[54]See for example Quotes **Q**[Poi1] and **Q**[Poi7] in Appendix E.

known in the 1970s. We next briefly examine the broad effects that these discoveries had in the decades following Poincaré's death.

3.12   The post-Poincaré era

Since the development of 'chaos theory' in the 1970s and 80s and the 'discovery' that the bulk of it was already known to Poincaré, for many people a central question surrounding Poincaré's legacy has been, Why did it take almost a century for his ideas to catch on? In the context of chaos and its applications to various sciences, this is a worthy question, and I'll examine it briefly below (cf. §3.12.2). Yet in the context of mathematics itself, the questions about Poincaré's legacy are somewhat different. Poincaré had tremendous impact both during his lifetime and after, and was certainly never forgotten, even by those who—like the Bourbaki—reacted against him. But even within several areas of mathematics it could be argued that there was an unusually long period of assimilation. In other words, Poincaré was far ahead of his time in many ways, and it was not so easy to pick up where he left off.

And yet, surprisingly, in the narrower context of HPT and KAM theory, we'll see that Poincaré's ideas—blended with those of L. Boltzmann—were *too* readily assimilated, and were pushed *too far* by some of his disciples.

We begin first with a brief overview of activity in dynamical systems in the post-Poincaré era.

3.12.1   *Poincaré's legacy in dynamics*

Setting aside Poincaré's influence in algebraic geometry, complex function theory, and much of topology, let's examine what happened in the parts of mathematics pertaining to our narrative.

The foremost disciple of Poincaré—the one who immediately picked up in dynamics where Poincaré left off—was the American mathematician George David Birkhoff.[55] Within a year of Poincaré's death, Birkhoff completed the proof [Bir13] of Poincaré's 'last geometric theorem' [Poi12] and went on to generalize and enlarge Poincaré's work, publishing his 300-page memoir *Dynamical Systems* [Bir27] in 1927, then proving a strong version

---

[55] Birkhoff (1884–1944) is usually considered the third home-grown American mathematician of real historical significance. (The first two are George William Hill (1838–1914), who did important work in *celestial mechanics*, and Josiah Willard Gibbs (1839–1903), the co-developer of statistical mechanics with Boltzmann and Maxwell.)

of the famous *ergodic theorem* [Bir31] in 1931 (see §7.2.2 below). The next year, he showed [Bir32] that in a map of the annulus having at least two periodic orbits with different periods, there are complicated limit sets separating the orbits' domains of attraction.

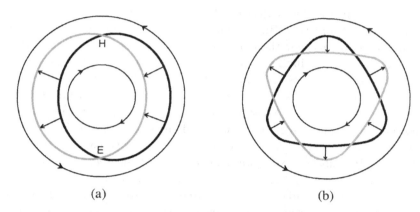

(a)   (b)

Fig. 3.3   Illustration of the Poincaré-Birkhoff theorem for *twist maps* of the annulus. (a) Suppose the area-preserving homeomorphism T of the annulus leaves the boundary circles invariant but turns them in opposite directions (here the inner boundary circle is turned CW and the outer circle CCW). Using continuity of the rotation, it can be shown that there is a closed curve C (bold curve) each of whose points is mapped radially (i.e., without rotation) to its image point, forming the closed image curve $T(C)$ (gray curve). Because T preserves area, C and $T(C)$ must intersect an even number of times (at least two times, as here). These intersections are clearly fixed points of T. Local analysis (indicated by arrows) shows that these fixed points are alternately *elliptic* (E) and *hyperbolic* (H). (b) A more elaborate example showing six fixed points (three elliptic, three hyperbolic). Birkhoff was already aware that perturbations of twist maps could produce 'zones of instability' around the hyperbolic fixed points, as partly described by Poincaré. Alas, neither Poincaré nor Birkhoff foresaw the KAM theory toward which this theorem dimly pointed.

Because of their connections with later KAM theory, some of Birkhoff's results deserve a more detailed description here. Poincaré's 'last geometric theorem' (now also often called the Poincaré-Birkhoff theorem; see Fig. 3.3 above) concerns area-preserving *homeomorphisms*[56] of an annulus (the region between two concentric boundary circles) in which the inner and outer boundary circles are turned in opposite directions.[57] In characteristic fash-

---

[56]In Poincaré's and Birkhoff's original treatments, T was an *analytic diffeomorphism*, but later versions of the theorem reduce the *regularity* requirement to simple *continuity*.
[57]This sort of mapping is now commonly called a *twist map*; in fact, the first complete proof of a KAM theorem was given for twist maps by Moser [Mos62].

ion, Poincaré distilled this mapping from the flow of the dynamics of a three body problem; his goal and conjectured theorem was to show that the mapping had a *fixed point*, from which he could infer the existence of a *periodic orbit* of the three body problem. He did succeed in proving the theorem in certain special cases, but conjectured that it was true more generally [Poi12]. It was this conjecture that Birkhoff famously proved [Bir13] and later extended further [Bir25]. The fixed points in this theorem (and in similar theorems—by now there are many) are related to KAM theory; in fact both they and the invariant curves of Moser's twist theorem are special cases of the invariant sets in Aubry-Mather theory (cf. §6.1).

Among the many achievements in Birkhoff's book *Dynamical Systems*, one finds a development of the *normal form* now called by his name.[58] These normal forms provide natural coordinates near *elliptic fixed points* of Hamiltonian systems, and are now one of the most common ways of preparing a Hamiltonian for application of a KAM theorem (cf. §7.3.1).

Concerning Poincaré's followers closer to home in France, it's natural to think of Jacques Hadamard[59] who used Poincaré's methods to study the strong *statistical properties* (the 'chaos,' we might say) of geodesic flow on surfaces of negative curvature [Had98]. (These ideas were continued in turn by Birkhoff's PhD student H.C.M. Morse and postdoctoral assistant E. Hopf, then by Morse's student G.A. Hedlund, and had a marked influence on the later development of *ergodic theory* and *symbolic dynamics*.) Also in France, A. Denjoy extended Poincaré's work on maps of the circle from *homeomorphisms* to *diffeomorphisms*. (Developments along this line of inquiry in 'low-dimensional dynamics' would later have crucial significance in refinements of KAM theory; see Appendix B.2 for a brief discussion.)

Beginning in Poincaré's lifetime, others took up the problems of celestial mechanics, even exploring the possibility that convergent series solutions to the (restricted) three body problem could be found, despite Poincaré's doubts. Incredibly, following a line of 'regularization' work beginning with T. Levi-Civita and P. Painlevé, in the first decade of the 20th century the Finnish astronomer Karl Sundman did obtain convergent series solutions that affirmed the original King Oscar prize question in the strictest sense[60]

---

[58]The Birkhoff normal form is also sometimes called the Birkhoff-Gustavson normal form, following F.G. Gustavson's further development and use of them in *celestial mechanics* [Gus66].

[59]Hadamard (1865–1963) is best known for his 1896 proof of the prime number theorem, one of the foremost achievements in number theory, and in 19th century mathematics.

[60]It could be argued, however, that the spirit of the prize question (originating as it did mainly with Weierstrass) was not to show the convergence of just any series solutions,

[Su07], [Su09], [Su12]. There was initially some question as to whether these solutions could be consistent with Poincaré's results on the nonexistence of integrals (they are consistent). But even once Sundman's series were validated, they did not arouse as much interest as might be expected, mainly because their convergence was so slow as to make them nearly useless in practical applications. (This feature of very slowly converging series had been pointed out by Poincaré, and he contrasted it with the practical usefulness of asymptotic series.) See F. Diacu's article [Di96] for an interesting discussion of Sundman's work and subsequent developments.

All of these efforts buttressed and extended Poincaré's work, and have proved to be important in retrospect (especially Birkhoff's, as we'll see). But although students did carry the work forward to some degree, before the 1960s nowhere in Western Europe or the U.S. saw the emergence of a truly vigorous school of dynamical systems in Poincaré's qualitative style. Instead, the place where Poincaré's ideas took strongest root was Russia and then the Soviet Union. This was not apparent at first, but by the middle of the 20th century it would be plain to anyone who cared to look. It would also be a source of some tension between Western and Russian or Soviet mathematicians once the West began to catch up.[61]

The Russian school of dynamics began around 1890 with Aleksandr Lyapunov's thesis in Moscow on the general problem of stability of motion. Lyapunov soon recognized his work's affinity with that of Poincaré, and initiated a correspondence with Poincaré which led to cross-fertilization, and later to Lyapunov being recognized in the West. His quantitative *estimates* of the rate of divergence of nearby solutions in dynamical systems (*Lyapunov exponents*) are now a standard tool for measuring chaos. Starting in the late 1920s, research in the Soviet Union continued with the work of L.I. Mandelshtam, A.A. Andronov and L.S. Pontryagin at the so-called Gorki school, where Poincaré's and Lyapunov's ideas were combined to study nonlinear oscillations, bifurcations, dissipative systems, control theory, and what is now called *structural stability*. (S. Lefschetz would later import ideas from the Gorki school to the U.S. in his post at Princeton.) In the 1930s, another important Soviet school arose in Kiev, Ukraine, under the leadership of N.M. Krylov and N.N. Bogoliubov, who studied nonlinear mechanics using quantitative, analytic methods inspired mainly by Volume 2 of Poincaré's *Méthodes nouvelles*. It was this Ukrainian school that most influenced Kolmogorov's later work in dynamics.

---

but rather the convergence of the *Lindstedt* (or Lindstedt-like) *series*.
[61]See Appendix C.

Before Kolmogorov, no single individual in the Soviet school of dynamics outshone Birkhoff as Poincaré's successor. Nevertheless by the mid 20th century, the Soviet school as a whole was unquestionably the world's strongest in dynamics, and had the advantage of building and sustaining itself until the arrival of Kolmogorov, who would finally see past the limitations imposed by Poincaré.

3.12.2   The chaos debate

It has often been claimed that chaos theory was discovered in the 1960s (E. Lorenz's paper [Lor63] is usually cited as the starting point), then developed into a 'revolution' in the 1970s and 80s. Without a doubt, the book—or perhaps misreadings of the book—most responsible for this view in the U.S. was the best-seller [Glei87] by James Gleick, from which I quote (pp. 5–6):

> The most passionate advocates of the new science [of chaos] go so far as to say that twentieth-century science will be remembered for just three things: relativity, quantum mechanics, and chaos. Chaos, they contend, has become the century's third great revolution in the physical sciences.[62]

As mentioned several times already, it has just as often been pointed out that the lion's share of chaos theory was discovered by Poincaré in the 19th century, to which the 'revolutionaries' reply "Yes, but most of that was forgotten until the 1970s." The debate continues by saying that mathematicians never forgot and continued working diligently (as amply demonstrated in the previous subsection); that physicists were not interested because of the more important revolutions in quantum and relativity theory (where relatively 'low-hanging fruit' was available for easy picking); and even that the two world wars disrupted dynamics[63] in a way that was not true for other parts of mathematics or physics (certainly not nuclear physics). But the most compelling reason for the sudden growth of chaos theory around 1970 is simply the arrival of easily programmable personal computers, with which scientists of all sorts—many of whom would not dream of reading Poincaré—could experiment with simple mathematical models of nonlinear

---

[62] In the otherwise enlightening and highly detailed article [AubiD02], a quote overlapping this one is cited and abridged in a way that makes it sound as though Gleick is himself one of the 'advocates.'

[63] As discussed by M. Mashaal in the book [Mas02], egalitarian impulses in France ("everyone must fight") led to decimation in the ranks of French mathematicians during WWI, something that was not true to nearly the same degree among other belligerents.

systems. To a large extent, they rediscovered by experiment what Poincaré and followers had discovered by thinking.

These debates and explanations are interesting in their own right, and the reader may consult a large body of literature on the subject (see Part D.3.2 of Appendix D for a summary of one 'chaos debate'; see also pp. 49–50 of Ruelle's book [Rue91] for a succinct and cogent discussion). But concerning our main subject of HPT and KAM theory, my point of view lies entirely outside this debate: I claim that, not only were Poincaré's ideas not forgotten, but leading mathematical physicists embraced them so zealously that they pushed them too far, thereby delaying clarification of the 'fundamental problem of dynamics.' In other words, if the early to middle 19th century was a time of over-indulgence in the idea that mechanical systems are integrable, then the post-Poincaré era of the early to mid 20th century saw the pendulum of indulgence swing too far in the opposite direction. This idea will be developed in the next three subsections.

3.12.3 *Ergodic theory*

Here we introduce and briefly discuss the crucial notion of *ergodicity*. Although this is now a basic mathematical concept in dynamical systems, I won't define it precisely here.[64] For now, I'll simply say that if nonintegrability is a weak form of *chaos* possessed by some dynamical systems, then ergodicity is a slightly stronger form (i.e., ergodicity is 'more chaotic,' and any system that is ergodic is necessarily nonintegrable, but not vice-versa). An ergodic system has trajectories that do not confine themselves to separate portions of phase space; instead, trajectories pass through *almost all* points in phase space that are energetically accessible.

Ergodicity was introduced[65] by Ludwig Boltzmann in 1871 [Bolt71] by way of the *Ergodensatz* (or *ergodic hypothesis*), which assumes roughly that a dynamical system has enough 'chaos' to permit the identification of certain time and phase space averages in statistical mechanics. As one of the founders of statistical mechanics (with Maxwell and Gibbs), Boltzmann was primarily a physicist, but he also had mathematical intuition of the

---

[64]See the glossary for definitions of *ergodicity* and related terms; see also the longer discussion in §7.2.2.

[65]The origins of ergodic theory are slightly obscure, and there is a tendency to find them at ever-earlier dates. I once saw a nice talk by Vitali Bergelson in which he played on this by claiming that the earliest mention of ergodicity occurs in the Book of Ecclesiastes 1:6: "The wind goeth toward the south, and turneth about unto the north; it whirleth about continually, and the wind returneth again according to his circuits."

highest order. Although he never formulated the notion of ergodicity in very precise mathematical terms, he certainly understood that it must entail nonintegrability. Much of the excess enthusiasm I'll describe here consists of researchers thinking or hoping for the converse: that nonintegrability entails ergodicity.

The truly mathematical formulation of ergodic theory did not arrive until the 1930s, as it had to await the invention and development of the modern (Lebesgue-Borel) theory of *measure* and integration. Early mathematical justifications of identifying the desired averages were called 'ergodic theorems,' and were proved in turn by John von Neumann, G.D. Birkhoff, and Birkhoff's PhD student B.O. Koopman, at about the same time in 1931. Unfortunately, it was not clear that the hypotheses of these theorems were satisfied by the systems of interest in statistical mechanics. Today ergodic theory is a large and thriving subdiscipline within dynamical systems, but the support that all of this mathematics gives to the foundations of statistical mechanics remains a matter of debate. (See §7.2 below.)

3.12.4   *Over-indulgence in chaos*

By the end of the 19th or beginning of the 20th century, the keen interest that Boltzmann and followers had generated in ergodic systems merged with the interest in Poincaré's (and others') proofs of the nonintegrability of generic Hamiltonian systems. Not only were many of the leading mathematicians and mathematical physicists aware of and enthusiastic about nonintegrability, but they pushed further in their quest to justify the ergodic hypothesis. In fact, we now know that they pushed too far, and that they conjectured and in some cases 'proved' statements that were too strong. In particular, in a series of papers [Fer23a], [Fer23b], [Fer23c], [Fer24] written before he reached the age of 24, Enrico Fermi 'proved' that 'general' (read *generic*) mechanical systems are 'quasi-ergodic' (the term for 'ergodic' in those days). Birkhoff also seems to have believed this.[66] We

---

[66] In fact, I found a reference saying that Birkhoff had conjectured in [Bir42] that Hamiltonian systems were generically ergodic, but when I looked there, I found only a mild statement made in passing which would need to be stretched substantially to constitute a conjecture. (Of course, it may be that he conjectured this elsewhere.) There is also the following statement on p. 261 of S. Smale's article [Sma80]: "This hypothesis played an important role in Birkhoff's later work. He not only believed it, but part of his work is written assuming the truth of the Hypothesis of Metric Transitivity." [*Metric transitivity* was Birkhoff's term for ergodicity.]

now know that these claims are not true,[67] but nothing is better evidence of the zeal with which some of the best researchers of the time embraced the idea of chaos than this demonstration of their belief in the ergodicity of generic Hamiltonian systems.[68]

If further evidence is needed, I can point to 1953 (the year before Kolmogorov announced KAM), where we find Fermi (again), this time together with J. Pasta and S.M. Ulam, gathered around the MANIAC I computer at Los Alamos, running numerical experiments that they hoped would show the 'rapid *thermalization*' (i.e., the *ergodicity*) of a model of a system of weakly coupled oscillators. Unfortunately, Fermi died the following year;[69] his coworkers published the 'disappointing' results of their experiment as a technical report [FerPU55], and the so-called Fermi-Pasta-Ulam problem continues to present researchers with conundrums up to the present (see §7.2.3 below).

Yet it's perhaps not surprising that leading researchers believed generic Hamiltonian systems were ergodic, and continued to believe it right up until it was shown to be false.[70] After all, once one accepts the nonintegrability of generic Hamiltonian systems, the non-existence of integrals 'ought' to mean the non-existence of invariant surfaces in phase space, so that barriers to ergodicity are no longer present. Together with the 'need' or 'desire' for systems to be ergodic in statistical mechanics, this apparent lack of barriers was convincing. But we shall see that the lack of barriers was only apparent.

From the post-Poincaré era and right up until the first announcement of KAM (and beyond), it was widely believed by some of the best researchers in the business that generic Hamiltonian systems were not only nonintegrable, but also ergodic. The pendulum of indulgence had swung to the

---

[67]Nevertheless, in the paper [BenGG85a], the authors make the case that Fermi's results from 1923–24 are still quite useful.

[68]Nevertheless, one remarkable result from this era still stands: In 1941, J.C. Oxtoby and S.M. Ulam [OxU41] showed that, in a *bounded phase space*, (there is a natural *topology* such that) the *continuous ergodic transformations* form a G_δ set in the space of *continuous measure-preserving transformations*. In other words, in a precise mathematical sense, ergodic transformations are *generic* among continuous transformations. Unfortunately this does not say anything about *smooth Hamiltonian systems* or mechanical systems.

[69]Fermi was only 53 years old when he died from stomach cancer in November 1954. His cancer was very likely the result of exposure to ionizing radiation.

[70]On p. 5 of [Conto02], G. Contopoulos writes "[...] it was generally assumed that generic dynamical systems are either integrable or ergodic," and he references Landau and Lifshitz [LanL75]. A few pages later (p. 17 of [Conto02]), he attributes a similar statement to [LanL76]. (The publication dates (1975, 1976) of these books are surprisingly late, but I wasn't able to locate the referenced statements in them myself.)

other side.

3.12.5 *Paradox and a long crisis in mechanics*

Historians agree for the most part that fundamental progress in science comes when there are problems, especially disagreements between experiment and theory. These are the crisis conditions that can lead to breakthroughs—even revolutions, as either theorists or experimentalists (usually theorists) are forced to rethink their approach. Some of the most dramatic examples in the history of physics occurred in Poincaré's time: the crisis surrounding the Michelson-Morley experiments helped precipitate the special theory of relativity, while the ultraviolet catastrophe and the photoelectric effect did much the same for the early version of quantum mechanics.

Mathematics is somewhat different, as the 'experiments' are not so concrete (although they have come to seem that way with the use of ever-more-powerful computers). After describing how some of Poincaré's and Boltzmann's ideas were pushed too far, it's very tempting to look back and find a crisis or paradox that was later resolved by Kolmogorov. Indeed, many writers see events in this way.

What is more certain is this: The paradox or crisis would have been perceived easily with high-power computers and high-resolution graphical representations of orbits of Hamiltonian systems. It's therefore even more astonishing that Kolmogorov resolved the paradox before this happened; in this case we can say that theoretical prediction preceded experimental verification.

We'll now look at this notion of paradox in more detail. It's fairly easy to describe if we confine ourselves to HPT—to a discussion of the dynamics of the system (3.1). On the one hand, in the post-Poincaré era, many theorists worked under the assumption that for generic functions $f = f(\theta, I, \varepsilon)$, as soon as the ε in the system $H(\theta, I, \varepsilon) = h(I) + \varepsilon f(\theta, I, \varepsilon)$ is nonzero, the integrability of the system is destroyed and its behavior becomes quite chaotic, even ergodic.

But those who worked with solutions of such systems in practice—the numerical astronomers and other 'calculators' of the day—saw things differently. In their experience, turning on the ε did not have such dramatic effects. They observed that, for small values of ε, most of their approximate solutions of the perturbed system $H = h + \varepsilon f$ stayed fairly close to solutions of the integrable system h for long intervals of time, and that both the closeness and the time interval could be controlled by varying ε. In other words, while theoreticians were saying that everything changes as soon as ε is switched on in Eq. (3.1), astronomers and others went ahead using perturbation theory successfully to calculate astronomical orbit approximations, just as Newton and others had done since the end of the 17th century. Experience taught them that there was some kind of continuity to integrable behavior, despite the immediate breakdown claimed by theoreticians.[71]

Poincaré had partly explained this paradox by his theory of *asymptotic expansions*, first set out in the early paper [Poi86] (see also Quotes **Q**[Poi1] and **Q**[Poi3] in Appendix E). Here he examined—and developed a general theory for—series that *diverge*, but which nevertheless may be useful if only some of the first terms are retained. (The best known example is perhaps Stirling's series for the factorial function $n!$) In fact, Poincaré was able to show that most of the series used in celestial mechanics were divergent asymptotic expansions. These series were associated with the names of the astronomers who conceived them: Delaunay, Bohlin, Newcomb, Gyldén, and of course A. Lindstedt, the originator of the type of series that Dirichlet referred to in his remarks to Kronecker, and the series that became the subject of the King Oscar prize. As mentioned before, Weierstrass always believed that these Lindstedt series converged somehow, while Poincaré doubted it after his breakthroughs in the prize paper [Poi90].

No doubt having Weierstrass in mind, Poincaré returned to the question of the Lindstedt series' convergence in *Les Méthodes nouvelles*. After showing that the series are plagued by small divisor problems (as discussed above §3.11) and therefore cannot be everywhere *uniformly convergent*, he considers the question of whether the series might converge for special initial conditions chosen in advance. There are two now-famous passages in *Les Méthodes nouvelles*[72] in which Poincaré asserts his belief that, even in these special cases, the Lindstedt series diverge. But in each case, he qualifies the statement by saying that his arguments do not allow him to assert this divergence with absolute rigor.

---

[71]Theoreticians should also have known better (and some undoubtedly did), since it is not especially difficult to use perturbation methods developed as far back as the 18th century to show that solutions of perturbed systems remain close to solutions of their unperturbed counterparts for long times. Curiously however, the *rigorous* development of perturbation theory did not start until the 20th century, and was not widely known until the appearance of the book [BogM61] by Bogoliubov and Mitropolsky. See Appendix A of [SanVM07] for historical details.

[72]See Quotes **Q**[Poi4] and **Q**[Poi5] in Appendix E below.

Most of the astronomers' series were indeed divergent, as Poincaré showed, but he also stressed that this did not diminish their utility in most cases, *unless* one tried to use them to establish stability on infinite timescales. Poincaré was also right to be cautious about his own belief that the Lindstedt series diverged everywhere (or perhaps *almost everywhere*, in modern parlance).

We'll see that others should also have been cautious in their expectations of chaotic behavior. Beyond Poincaré's disclaimers and astronomers' doubts, there were perhaps other warning signs that persistent integrable-like behavior lurked in generic Hamiltonian systems. Yet when Fermi, Pasta, and Ulam fed their model of coupled oscillators into MANIAC I, they were quite surprised that it behaved more integrably than nonintegrably. They were chaos enthusiasts and, unknowingly, KAM theory skeptics. They were about to be re-educated in a dramatic way.

Chapter 4

KAM Theory

In this chapter, I outline the breakthroughs that resulted in KAM theory and resolved the slow crisis described above. It goes without saying that the mathematicians involved—Siegel, Kolmogorov, Arnold, Moser—were among the leaders of their respective generations, and that KAM theory was only one aspect of their work, though no doubt a memorable one. Kolmogorov in particular seems to have had extraordinary mathematical vision, transforming each of the several fields (probability, turbulence, HPT, information and complexity theory) in which he took an interest.

4.1 C.L. Siegel and A.N. Kolmogorov: Small divisors overcome

The first meaningful breakthrough only appears as such in hindsight; it wasn't clear just how crucial it was at the time. It came in 1942 during World War II, while the German number theorist Carl Ludwig Siegel (1896–1981) was working at the newly established Institute for Advanced Study in Princeton, New Jersey, having fled the Nazis in 1940 on a journey that took him from his professorship in Göttingen to Denmark, Norway, and finally the United States. Although Siegel wasn't Jewish, he disliked what he saw in Germany enough to leave, and lived in what he called 'self-imposed exile in America' throughout the war.

At the time, Siegel was not working in Hamiltonian perturbation theory. Instead he was working on a stubborn part of a 'linearization problem' that was left over from research begun by E. Schröder and H. Poincaré in the 1870s. The loose ends had all been resolved save one that involved a small divisor problem. It was this that Siegel solved [Sie42], and in so doing he became the first researcher to successfully circumvent small divisors. The problem is now known as the 'Siegel center problem,' or simply 'Siegel's problem.'

Siegel's problem has a very interesting history of its own which contacts our main narrative in several places, and which I briefly describe in Appendix B. In this section, we pick up the mathematical thread where we left it at the discussion of the small divisor problem after Eq. (3.5). Instead of showing what Siegel did in his 1942 paper [Sie42], I'll give a rough idea of how Kolmogorov (and later Arnold and Moser) adapted Siegel's techniques for the purposes of KAM theory.[1] Although what I present here *does not* closely follow any of the early KAM proofs, it is related to Arnold's proof [Ar63a], and draws its inspiration from Kolmogorov.

We thus return to where we were in §3.11, and briefly recapitulate the argument leading to the small divisor problem. We begin with the simple homological equation $f + \{\chi, h\} = 0$ (cf. Eq. (3.3)) for the generating function χ of the transformation T that seeks to eliminate the $O(\varepsilon)$-term in our original nearly integrable Hamiltonian $H = h + \varepsilon f$. Recall that $f + \{\chi, h\} = 0$ first becomes $f(\phi, J) = -\sum_{k=1}^{n} \frac{\partial \chi}{\partial \phi_k} \frac{\partial h}{\partial J_k}$ and then becomes $\sum_{k \in \mathbb{Z}^n} \hat{f}_k(J) \, e^{2\pi i k \cdot \phi} = -2\pi i \sum_{k \in \mathbb{Z}^n} \hat{\chi}_k(J) \, k \cdot \omega(J) \, e^{2\pi i k \cdot \phi}$ upon using $\frac{\partial h}{\partial J} = \omega(J)$ and the Fourier representations of f and χ (cf. Eqs. (3.3)–(3.4)). This immediately gives the *formal* solutions for the Fourier coefficients $\hat{\chi}_k$ of χ; these solutions are simply $\hat{\chi}_0(J) = 0$ and $\hat{\chi}_k(J) = -\hat{f}_k(J)/(2\pi i \, k \cdot \omega(J))$ (cf. Eq. (3.5)). Following Poincaré, we saw that the presence of a dense set of resonant tori meant that small divisors in these solutions prevent χ— hence also T or $e^{\varepsilon L_\chi}$—from being defined classically. This is still true. But, perhaps inspired by Siegel, Kolmogorov would take a different approach. He would ask, What if we don't care about a classical transformation defined on a *nice* set (i.e., one with *interior*)? What if we simply look for some—any—nonempty set on which the transformation can be defined?

As an analytic number theorist, Siegel was familiar with techniques that provide the key for doing this. He knew, for example, that the Fourier coefficients $\hat{f}_k = \hat{f}_k(J)$ of a smooth or analytic function f decay rapidly

---

[1]In describing the first solutions of small divisor problems, many references say something like "Kolmogorov adapted Siegel's techniques," as I do here. However, in the sequel I'll qualify this with 'perhaps,' because, while there is no doubt that Siegel's work on small divisors preceded Kolmogorov's by a dozen years, there does not seem to be direct evidence that Kolmogorov knew about Siegel's work. In this context, on p. 5 of (the English translation of) [Ar97], Arnold writes "As far as I understand, [Kolmogorov] was aware of neither Siegel's works nor J.E. Littlewood's works on the exponential slowness of an increase in perturbations."

with the size of their index k, according to, say,[2]

$$\|\hat{f}_k\|_B \leq \frac{C}{|k|^b} \qquad (4.1)$$

where $C > 0$, and where the exponent $b > 0$ can be made as large as desired by taking f to be sufficiently smooth (and where the subscript B indicates that we use the *uniform norm* to measure $\hat{f}_k$; i.e., if B is a *closed, bounded* set of J-values—such as a *closed ball*—then $\|\hat{f}_k\|_B = \max_{J\in B}\|\hat{f}_k(J)\|$). Siegel also knew that the special set of *frequency vectors*

$$\mathcal{D}^\omega(\gamma,\tau) = \left\{\omega \in \mathbb{R}^n \mid \text{ for each nonzero } k \in \mathbb{Z}^n, \ |k \cdot \omega| \geq \frac{\gamma}{|k|^\tau}\right\} \qquad (4.2)$$

and the corresponding set of J-values defined as

$$\mathcal{D}^J(\gamma,\tau) = \left\{J \in \mathbb{R}^n \mid \text{ for each nonzero } k \in \mathbb{Z}^n, \ |k \cdot \omega(J)| \geq \frac{\gamma}{|k|^\tau}\right\} \qquad (4.3)$$

are nonempty, provided $\gamma > 0$ is small enough, with $\tau > n - 1$, and provided the unperturbed Hamiltonian $h = h(J)$ is *nondegenerate* so that the *frequency map* $\omega(J) = \partial h/\partial J$ preserves the basic structure of the ω's when mapped backward to the J's. Now the set $\mathcal{D}^\omega(\gamma,\tau)$ is a strange and remarkable thing, about which much can be said.[3] Here we limit ourselves to just a few remarks (see the *Diophantine* entry in the glossary for more information). First, the inequalities that define $\mathcal{D}^\omega(\gamma,\tau)$ are called *Diophantine conditions*, and have long been known to number theorists. The same terminology applies to the ω that satisfy the inequalities (they are Diophantine vectors, or Diophantine frequencies). Collectively these frequencies form a strange sort of set closely related to a *Cantor set*, which is a type of *fractal*. And crucially, because the frequency map is nondegenerate, the *preimage* $\mathcal{D}^J(\gamma,\tau)$ in J-space is also a Cantor-like set.

For us, the most important strange property of these Cantor-like sets is that they have empty interior (i.e., they contain no nonempty open sets). For this reason, it's hard for mathematicians to think of them as the domain of a *nice* function, especially a function that's going to be used to

---

[2]In fact, the Fourier coefficients $\hat{g}_k = \hat{g}_k(J)$ of an analytic function g drop off even more rapidly than shown in Eq. (4.1), obeying instead an exponential decay law $\|\hat{g}_k\|_B \leq Ce^{-\rho|k|}$ for suitable $C, \rho > 0$.

[3]The set $\mathcal{D}^\omega(\gamma,\tau)$ can be visualized in low dimensions ($n = 2$ or 3). For $n = 2$, $\mathcal{D}^\omega(\gamma,\tau)$ consists of the plane $\mathbb{R}^2$ with linear 'strips' removed. Each inequality $|k \cdot \omega| \geq \gamma|k|^{-\tau}$ in Eq. (4.2) deletes a strip of width $2\gamma|k|^{-\tau-1}$ between the two parallel lines $k\cdot\omega = \pm\gamma|k|^{-\tau}$ lying on opposite sides of (and equidistant from) the origin. When all strips have been removed, what remains is a union of closed half-lines pointing toward the origin. (See Fig. 1 of [Broe10] for an illustration.) The structure of $\mathcal{D}^\omega(\gamma,\tau)$ for $n = 3$ is similar, but rather than removing strips between lines, the inequalities remove 'slabs' between parallel planes $k \cdot \omega = \pm\gamma|k|^{-\tau}$ in $\mathbb{R}^3$.

define a change of variables. Siegel did not face this issue in his use of Diophantine conditions, since he worked on a problem where the frequencies were fixed. But Kolmogorov would face it a dozen years later, and would conclude that despite what Poincaré and Siegel proved about the impossibility of constructing the generating function χ in a classical way, one could nevertheless adapt Siegel's methods to get around this problem on the Cantor-like set $\mathcal{D}^J(\gamma, \tau)$, or at least to begin with, at single points of this set. Much later, Kolmogorov's ideas would lead to methods that define the generating function in a non-classical way on large portions of $\mathcal{D}^J(\gamma, \tau)$ (see §4.8 (c), below).

To get an idea of how this works, we consider J-values in the subset $S = \mathcal{D}^J(\gamma, \tau) \cap B$, on which the Fourier coefficients of χ satisfy (for $k \neq 0$)

$$\|\hat{\chi}_k\|_S = \left\| \frac{\hat{f}_k(J)}{2\pi i \; k \cdot \omega(J)} \right\|_S \leq \frac{C}{2\pi\gamma |k|^{b-\tau}}. \tag{4.4}$$

(Here we use (3.5) in the equality, and (4.1) and (4.3) in the inequality.) If we now choose f smooth enough to ensure that the exponent $b - \tau > n$, it follows that on the set S, the Fourier coefficients $\hat{\chi}_k(J)$ decay rapidly enough to get *absolute convergence* of the Fourier series $\sum_{0 \neq k \in \mathbb{Z}^n} \hat{\chi}_k(J) \, \mathrm{e}^{2\pi i k \cdot \theta}$. That is,

$$\sum_{k \in \mathbb{Z}^n \setminus \{0\}} \|\hat{\chi}_k\|_S \leq \frac{C}{2\pi\gamma} \sum_{k \in \mathbb{Z}^n \setminus \{0\}} |k|^{\tau-b} < \infty. \tag{4.5}$$

(The finiteness of the last sum is a basic fact about summability over $\mathbb{Z}^n$ for sufficiently negative exponents $\tau - b < -n$; it can easily be shown using an 'integral comparison test.') In this way we define the generating function χ, and hence also the desired transformation $\mathrm{e}^{\varepsilon L_\chi}$. This shows that the transformation is more than merely *formal*, since we specify the domain S on which its generating function converges. In doing this, we violate nothing that Poincaré proved—the transformation is not classical because S has no interior.

The idea of using Diophantine conditions to get convergence of a Fourier series like that for χ—despite the presence of small divisors—was the first essential breakthrough on the path to KAM theory, and Siegel was the first to see it. Yet even if the transformation $\mathrm{e}^{\varepsilon L_\chi}$ can be made to work in this way, we recall that it serves only to make (3.3) true; it only gets rid of the ϕ-dependence in the $O(\varepsilon)$ term of Eq. (3.2). In other words, our purported transformation is literally only the first step in a succession of infinitely many steps required to eliminate the ϕ-dependence entirely. It was Kolmogorov who pointed out how to take the remaining steps.

4.2　Kolmogorov's discovery of persistent invariant tori

As mentioned above, Andrey Nikolaevich Kolmogorov (1903–1987) was one of the leading innovators in mathematics of the 20th century. Raised in and around Moscow by his mother's sister (his mother died in childbirth), Kolmogorov persisted in his studies through tumultuous times in Russia, becoming a student of N. Luzin at Moscow and obtaining his first famous result in mathematics while still a teenager (he constructed a Lebesgue integrable function whose Fourier series diverges almost everywhere). From there he pursued research in integration theory, set theory, 'intuitionist logic,' topology, geometry, and other areas before reworking the fundamentals of probability theory and giving that subject its first true measure-theoretic foundations in the 1930s. In the 1940s, he reworked the theory of turbulence in fluid dynamics, giving it new impetus after a long period of stagnation. So by the time we arrive to Kolmogorov's discovery of invariant tori, we find that he was already among the leading mathematicians and mathematical physicists of his time. Here, I'll continue the ongoing discussion of the 'fundamental problem of dynamics' (cf. §3.10) to give an idea of what Kolmogorov did when he turned his attention to HPT.

Despite what Poincaré proved about the nonintegrability of the system (3.1) under generic perturbations (including the break-up of resonant tori on a dense subset of phase space), and despite the widespread belief (and possible conjecture by Birkhoff) that such systems were ergodic, Kolmogorov somehow conceived of a large set of invariant tori that survive perturbation and continue to support quasiperiodic flow, thus blocking ergodicity and continuing integrable behavior in a conceptually new, geometric way. In his 1954 ICM lecture, and also in the article [Kol54] published the same year, he asserted these ideas as a theorem and sketched the outlines of a proof; in fact, some of the technical details of his method were already present in his paper from the previous year [Kol53]. Vladimir Arnold has said that Kolmogorov was in fact looking for *mixing* tori[4] in perturbed systems, rather than for the abundance of quasiperiodic tori that he found, and has likened this to Columbus discovering America while searching for a short-cut to India (cf. Arnold's response to the second question in the interview [Lui96]).

Setting aside for now the question of how Kolmogorov thought to look for these tori, let's examine the technique he outlined for proving their

---

[4]Although the mixing tori have never been found, O. Knill vindicated Kolmogorov's search to some extent by finding *weakly mixing* tori in perturbed systems; cf. [Kni99].

existence. There are two basic steps. The first step is to solve a homological equation like (3.3) at first order by confining ω to the strange set $\mathcal{D}^\omega(\gamma, \tau)$; Kolmogorov (may have) adapted this step from Siegel's work, as described above. The second step is to set up an iterative procedure that allows the homological equation to be solved not just at first order, but at *every* order. Here I will only give an impressionistic idea of the difficulties involved, and how Kolmogorov suggested they could be overcome. (Again, the reader may also simply look at what Kolmogorov himself said by turning to Appendix A in this text.) In the next sections I'll examine more closely how Arnold and Moser implemented Kolmogorov's strategy.

If we set up a regular iterative perturbation procedure which, at the kth step, solves a homological equation at order ε^k, we may naïvely expect that after n steps, the original Hamiltonian (3.1) will have been transformed to integrable form through $O(\varepsilon^n)$, i.e., to

$$K(\phi, J, \varepsilon) = h_n(J, \varepsilon) + \varepsilon^{n+1} R(\phi, J, \varepsilon), \qquad (4.6)$$

where the $R = R(\phi, J, \varepsilon)$ in the remainder is $O(1)$ (i.e., bounded as $\varepsilon \to 0$). We want to continue this process 'to infinity' so that the remainder is wholly eliminated on the set where the iterated transformation is defined.

But a technical problem arises: If we look at the transformed actions J at the nth step, we find that their distance $|I - J|$ from the old actions may be up to $O(\varepsilon^{n+1})$. In turn, the new frequency $\omega = \omega(J)$ will be similarly displaced from the old. It turns out that this 'wiggle room' is too much; the ω can move too close to *resonance*, and they need to be located with more precision if they're going to continue to satisfy the Diophantine conditions (4.3) with enough precision that the next step in the iteration can proceed.

This was not an issue in the small divisor problem that Siegel treated in [Sie42], since there (the analog of) ω was fixed. Siegel was able to use something like the above approach together with a refined version of the majorization method (of which the *Weierstrass M-test* is a simple variant) to get the convergence he needed.

Anyone who tried naïvely to adapt Siegel's procedure to the perturbed Hamiltonian (3.1) would be stopped by the problem just described. But Kolmogorov saw past this. He saw that, by keeping ω fixed[5] as in Siegel's problem and using a modified Newton's method in an appropriate function space, the method could be made to converge. He cited the work

---

[5] As L. Chierchia explains in Remark 3, Section 3 of [Chi09b], when carrying out Kolmogorov's proof in detail, keeping ω fixed is rather subtle, requiring a 'two-component' transformation at each step (one component in the 'J-direction,' and another in the 'ϕ-direction').

of L.V. Kantorovich [Kan48] where Newton's method had recently been generalized to function spaces. In stark contrast to the scheme above in Eq. (4.6), after n steps Kolmogorov's procedure would transform the original Hamiltonian to something like

$$K(\phi, J, \varepsilon) = h_n(J, \varepsilon) + \varepsilon^{\alpha^n} R(\phi, J, \varepsilon), \qquad (4.7)$$

with $\alpha > 1$. Note that here the $O(\varepsilon^{\alpha^n})$ remainder shrinks much more quickly than the $O(\varepsilon^{n+1})$ remainder in (4.6). This is characteristic of Newton-like methods, starting with Newton's original tangent-line scheme for finding zeros of functions of a single variable.[6] This rapid convergence combined with Diophantine conditions is enough to overcome the small divisors at all orders. In the end, this permits a transformation of the perturbed Hamiltonian to the integrable form[7]

$$K(\phi, J, \varepsilon) = h_\infty(J, \varepsilon) \qquad (4.8)$$

for any given J in the Cantor-like set $\mathcal{D}^J(\gamma, \tau)$, thus establishing the persistence for such J of an invariant torus in the original perturbed system. This discovery marks the beginning of KAM theory.

It's worth stressing again that Poincaré showed that such a transformation cannot take place on a nice set having interior. The set $\mathcal{D}^J(\gamma, \tau)$ does not have interior, but for small γ it is still large in the sense of Lebesgue measure, and it can be shown to generate *uncountably* many invariant tori for the original system. And, as Moser later showed [Mos67], the existence of these tori (and the quasiperiodic solutions they contain) settles the question of convergence of the Lindstedt series, and can thus be seen as a long-delayed answer to the question posed in the King Oscar Prize competition. Weierstrass was right after all: the series converge for a large (and strange) set of initial conditions, and Poincaré was right to be cautious of his claims of showing that they did not converge.

4.3   A closer look at the convergence scheme

In this section, I give a glimpse of how the convergence scheme proposed by Kolmogorov was later carried out in detail by Arnold and published in 1961

---

[6]This kind of convergence is sometimes called 'quadratic,' because in Newton's original root-finding method, $\alpha = 2$.

[7]In fact, in his theorem, Kolmogorov did not use the integrable form shown here; this form came later in Arnold's work. Instead Kolmogorov used a special (very clever and in some ways simpler) *normal form* specially designed to show the existence of a single invariant torus. The normal form appears in Equation (5) of Kolmogorov's paper, reprinted in translation below in Appendix A.

for circle diffeomorphisms [Ar61] and in 1963 for the KAM theorem [Ar63a], [Ar63b]. Of course, the first complete proof of a KAM theorem (for twist maps) was given in 1962 by Moser [Mos62], who showed convergence using not only Kolmogorov's ideas but also those of J. Nash, which allows the method to work in the case of finite differentiability (see §4.3.2 (f) below). Here we first stick to the analytic case, as Kolmogorov and Arnold did.

4.3.1   *An overview of the scheme*

The reader should keep in mind that what's presented here is not intended to be an historically accurate rendering of what the founders did. Instead, it's intended to give an idea of the basic features shared by the first proofs of KAM theorems. And, depending on his or her mathematical tastes, at some point the reader may have had enough of 'descriptions' of proof-methods and simply want to read a real proof (a number of suggestions are given in Appendix D, Parts D.1.1 and D1.2.).

(a) The basic set up, and the first step

Kolmogorov's basic idea is to take the transformation T used to overcome the small divisor problem at first order (cf. §3.11 and §4.1), and iterate it so as to eliminate the perturbation at all orders. For technical reasons already mentioned, this iteration process must converge rapidly if it is to succeed. In this preliminary discussion, I'll use some simplifications that don't reflect the full nature of the problem, but which may help the reader get the idea on a first read-through. Some of these simplifications are then discussed in the next subsection.

Take an open ball $B \subset \mathbb{R}^n$ of positive radius in action space and consider a real-analytic, nearly integrable Hamiltonian $H = h(I) + \varepsilon f(\theta, I, \varepsilon)$ defined on $\mathbb{T}^n \times B$, with small parameter $\varepsilon > 0$ controlling the size of the perturbation εf. Assume that $\frac{\partial^2 h}{\partial I^2}$ is invertible for all $I \in B$ (this is the nondegeneracy assumption).

Choose an action value $I_0 \in B$ so that the corresponding frequency $\omega = \frac{\partial h}{\partial I}(I_0)$ is Diophantine (i.e., belongs to $\mathcal{D}^\omega(\gamma, \tau)$ defined in Eq. (4.2)). We may think of $\mathbb{T}^n \times \{I_0\}$ as an *embedded* torus in the phase space $\mathbb{T}^n \times B$ which is invariant under the flow of the integrable unperturbed system $H = h$. Our goal is to show that for small enough $\varepsilon > 0$, this torus persists (though slightly distorted) as an invariant torus of the perturbed system.

Now consider an appropriate complex neighborhood D_0 of $\mathbb{T}^n \times \{I_0\}$. We want to find a slightly smaller complex domain $D_1 \subset D_0$ and a symplectic near-identity transformation $T_1 : D_1 \to D_0$ which transforms H to $H_1 = h_1(J, \varepsilon) + \varepsilon^2 R_1(\phi, J, \varepsilon)$; we also want to be sure that the *preimage*[8] $I_1 = T_1^{-1}(I_0)$ lies inside D_1.

The reader may verify that the transformation T_1 sought here—with subscript 1 indicating that it is the 'first step'—is essentially the one described above (§3.11) with generating function $\chi(\phi, J)$ having Fourier coefficients $\hat{\chi}_k(J)$ given by Eq. (3.5). The convergence of the Fourier series is affirmed by the inequality (4.5), which holds because of the *estimate* (4.4), which in turn is true because ω is Diophantine (i.e., belongs to $\mathcal{D}^\omega(\gamma, \tau)$).

(b) Iteration and quadratic convergence

Once we confirm that the transformation T_1 works as stated, the idea is of course to iterate it. In other words, we repeat the process (now with $H_1 = h_1 + \varepsilon^2 R_1$ in place of $H = h + \varepsilon f$) to create a transformation $T_2 : D_2 \to D_1$ transforming H_1 to $H_2 = h_2 + (\varepsilon^2)^2 R_2$, while maintaining $I_2 = T_2^{-1}(I_1) \in D_2$ as needed. Note the appearance of the 'quadratically shrinking remainder' of order $(\varepsilon^2)^2 = \varepsilon^4$. At the next step, this remainder will again be squared to produce a new remainder of order $(\varepsilon^4)^2 = \varepsilon^8$.

It's not hard to see that this process generalizes to a sequence of transformations $T_k : D_k \to D_{k-1}$ on nested domains $D_k \subset D_{k-1}$ on which T_k transforms H_{k-1} to $H_k = h_k + \varepsilon^{2^k} R_k$ with $I_k = T_k^{-1}(I_{k-1}) \in D_k$. Again, note the quadratically convergent factor ε^{2^k} in the remainder.

(c) The limit transformation and existence of an invariant torus

One way to view this is to look at the composite maps $U_m : D_m \to D_0$ formed by $U_m = T_1 \circ T_2 \circ \cdots \circ T_m$. Each of these is well defined since the intermediate domains D_k are nested.

Application of U_m to H transforms it to $H_m = h_m + \varepsilon^{2^m} R_m$ which rapidly gets closer to integrable form on the shrinking domains D_m. In the limit, $\lim_{m \to \infty} H_m = h_\infty$, where $h_\infty = h_\infty(J, \varepsilon)$ is completely integrable on the complex domain[9] D_∞ which contains $\mathbb{T}^n \times \{I_\infty\}$. This last object is clearly an invariant torus of h_∞ supporting *quasiperiodic flow* with frequency ω.

If everything proceeds as needed, the limit transformation U_∞ is near the identity, so the image $U_\infty(\mathbb{T}^n \times \{I_\infty\})$ (in D_0) of the 'round' invariant

---

[8]Of course the equation $I_1 = T_1^{-1}(I_0)$ abuses the notation for T, which acts on both θ and I, but the meaning is clear enough for present purposes.

[9]This D_∞ has empty interior, and thus is not a 'classical' domain in the sense one usually means when speaking of variable transformations.

torus $\mathbb{T}^n \times \{I_\infty\}$ (in D_∞) is a slightly distorted invariant torus[10] of the flow of the original H. Furthermore, the flow on this distorted torus is also quasiperiodic with frequency ω, since U_∞ *conjugates* it to the flow on $\mathbb{T}^n \times \{I_\infty\}$.

The foregoing sketches Arnold's method of approach to KAM as implemented in [Ar63a]. Although it shows the existence of a single invariant torus, it may be generalized to show the existence of a whole family of tori parametrized over a Cantor-like set of actions (see §4.8 (c) below for a brief discussion and references).

4.3.2 *Technical issues*

Here we look more closely at certain details of the iterative process just described, and we point out some differences among the first KAM theorems.

(a) Complexified domains for the analytic case

In the case where H is real-analytic (assumed by Kolmogorov and Arnold), it is a standard result that H can be extended to an analytic function on a complex neighborhood D_0 of $\mathbb{T}^n \times \{I_0\}$. Thereafter, each of the domains D_k is a complex neighborhood of $\mathbb{T}^n \times \{I_k\}$, and $D_k \subset D_{k-1}$ as above. What is the purpose of the complex neighborhoods D_k? To put it simply, these help a great deal in the proof, because they permit the use of so-called Cauchy estimates. If the *uniform norm* of an analytic function over a given domain is known, a Cauchy estimate provides a *bound* on the uniform norm of the partial derivatives of the function over a smaller subdomain. So each time one uses a Cauchy estimate, one loses a portion of the domain. This is part of the explanation for why the domains D_k must shrink (another reason is discussed below). This loss of domain may also be interpreted as a loss of analyticity.

The Cauchy estimates were of course not available in the finitely differentiable case treated by Moser; see (g) below for a brief discussion.

(b) Where does nondegeneracy come in?

The nondegeneracy is used as usual to ensure that the structure of frequency space—in particular the structure of the resonances—is preserved in the action space. Without it, resonances might be blown up to unmanageable size in action space.

---

[10]Such a torus is often called a *KAM torus*.

(c) The ultraviolet cutoff

In order to avoid certain technical problems with limits, Arnold used a technique called the ultraviolet cutoff,[11] which we now describe in the context of KAM-type proofs by referring back to §3.11.

When solving a homological equation like Eq. (3.3) to a given order in ε, it is not really necessary to use all of f. Instead, by writing f as its Fourier series, one can break it into the sum of two parts $f = f^\le + f^>$ as follows:

$$f(\phi, J) = \sum_{k \in \mathbb{Z}^n} \hat{f}_k(J)\, e^{2\pi i k \cdot \phi} = \sum_{|k| \le N} \hat{f}_k(J)\, e^{2\pi i k \cdot \phi} + \sum_{|k| > N} \hat{f}_k(J)\, e^{2\pi i k \cdot \phi}$$

where the number $N > 0$ is the ultraviolet cutoff; i.e., the norm of the indices of Fourier modes at which the first sum $f^\le$ is cut off, and modes with indices of larger norm are put into the second sum $f^>$ (the 'tail'). The first sum $f^\le$ is then a trigonometric polynomial (and thus simpler than an infinite sum; no questions can arise about its convergence), while the second sum $f^>$ may be made as small in norm as desired by taking N sufficiently large.

The usefulness of the ultraviolet cutoff in proofs of KAM comes about as follows: we may choose $N = N(\varepsilon)$ so large[12] that $\|f^>\| < \varepsilon$, in which case $f^>$ ceases to be a part of the $O(\varepsilon)$ perturbation εf and it is instead pushed into the next order of the perturbation, here $O(\varepsilon^2)$. (As the iteration proceeds, the cutoff N must grow at each step in order to move the cut-off remainder $R^>$ into the next higher-order remainder term.) In addition, the remaining $O(\varepsilon)$ perturbation $\varepsilon f^\le$ contains only finitely many modes, so when one solves the homological equation (3.3), the frequency ω is only required to satisfy finitely many Diophantine conditions. In other words, ω no longer needs to belong to $\mathcal{D}^\omega(\gamma, \tau)$, but can instead belong to the larger, nicer, cut-off set $\mathcal{D}_N^\omega(\gamma, \tau)$ defined by $\mathcal{D}_N^\omega(\gamma, \tau) = \left\{\omega \in \mathbb{R}^n \,\middle|\, \text{for each } k \in \right.$ $\mathbb{Z}^n$ with $0 < |k| \le N$, $|k \cdot \omega| \ge \frac{\gamma}{|k|^\tau}\left.\right\}$. Although this set is an approximating superset of $\mathcal{D}^\omega$ (and shrinks down to it as $N \to \infty$), for fixed N it is qualitatively quite different. $\mathcal{D}_N^\omega$ is not a Cantor-like set, rather it consists of

---

[11] There are many references attributing the introduction of the ultraviolet cutoff to Arnold. However, in Appendix 34 of the book [ArA68] by Arnold and A. Avez, there is a footnote (footnote 5, p. 255 of the English edition) stating that the technique was used in (the 1962 French edition of) [BogM61], the well-known book by Bogoliubov and Mitropolsky. (Russian editions of this last book go back to the 1950s.)

[12] Roughly speaking, for an analytic function f with Fourier coefficients decreasing as $\|\hat{f}_k\|_B \le C e^{-\rho|k|}$, one can take N proportional to $\log(1/\varepsilon)$ to ensure that the norm of the tail $\|f^>\|_B$ is $O(\varepsilon)$; thus the cutoff N grows very slowly as $\varepsilon \to 0^+$.

finitely many connected components, each with nonempty interior. In this sense it has 'wiggle room' inside that $\mathcal{D}^\omega$ does not, and the corresponding domains D_k (cf. (a) above) may be viewed as 'classical' domains.

In his proof, Arnold uses these classical domains to ensure that the intermediate transformations $T_k : D_k \to D_{k-1}$ and $U_m : D_m \to D_0$ are also classical; it is only the limiting domain $D_\infty = \lim_{m\to\infty} D_m$ and limiting transformation $U_\infty : D_\infty \to D_0$ that are non-classical.

In Kolmogorov's scheme, the iteration can proceed without the ultraviolet cutoff, because ω is not viewed as a function of J; this is a particularly elegant and economical way of showing the existence of a single torus. Although Arnold's method also shows the existence of a single torus with fixed frequency, by viewing ω as a function of J, it pointed the way toward methods that would show the existence of whole families of tori simultaneously (cf. §4.8 (c) below).

(d) Diophantine conditions dependent on ε, and less rapid convergence

Although it is not necessary in showing the existence of a single invariant torus, one ultimately wants to know the size of the set $\mathcal{T}$ of invariant tori preserved under perturbations of a given strength ε. A first approach to this problem uses Diophantine conditions depending on ε. For example, one can replace γ in Eq. (4.2) by $\gamma_0\varepsilon^b$ and seek the optimal (largest) $b > 0$ for which the method converges. Of course, this changes the scheme slightly and makes the convergence more delicate as one seeks to push it to its limits. Estimates of the generating function (cf. Eq. (4.5)) are directly affected by this replacement, and one quickly finds that the quadratic convergence of the scheme (the remainder $\varepsilon^{2^m} R$) is reduced to something less rapid (a remainder of the form $\varepsilon^{\alpha^m} R$, with $\alpha < 2$). However, as long as $\alpha > 1$, we still get 'superconvergence' so that the method works. It can be shown that for generic nondegenerate h, the optimal value of b is $b = 1/2$, and with more work, one sees that the Lebesgue measure of the complement of the set of persistent invariant tori is $O(\varepsilon^{1/2})$ as $\varepsilon \to 0$, and that the size of 'distortion' of perturbed tori (from their unperturbed 'round' shape) is likewise $O(\varepsilon^{1/2})$. See the main references for these results in §4.6 (d) below.

Although the first KAM theorems did not include the results just described, Kolmogorov could already see that the measure of the set of initial conditions leading to invariant tori becomes full as $\varepsilon \to 0$ (cf. Theorem 2 at the end of Kolmogorov's paper [Kol54] in Appendix A).

(e) Where do the notorious smallness conditions on ε come in?

One theme that will soon become apparent in discussing various KAM theorems will be the very small values of ε required. Looking at the proofs, one finds a number of places where the smallness conditions arise, but the most essential are probably those ensuring that the transformations T_k are close to the identity and the domains D_k are not distorted so much that their mismatch causes the method to break down.

(f) The case of finite smoothness treated by Moser

Although the invariant curve theorem for twist maps proved by Moser differs significantly from Arnold's and Kolmogorov's theorems, its proof shares the basic iteration scheme described above. Here I'll give an idea of how Moser's proof is different, referring to the things already described that are the closest analogs to what one confronts when reformulating everything for maps.

First, because the (perturbed) twist map is not analytic but rather only of finite differentiability, analytic *extensions* to complex domains are unavailable,[13] so the Cauchy estimates mentioned above cannot be used. Proceeding naïvely in the most obvious way without Cauchy estimates, one iterates the transformation process described above by estimating, at each step, the norm of the generating function χ directly in terms of its Fourier series. Now if the (analog of the) function f is finitely differentiable (say of *class* C^p) then its Fourier coefficients $\hat{f}_k$ have norms that decay[14] as $\|\hat{f}_k\|_B \sim |k|^{-p}$. Using Diophantine conditions (such as $|k \cdot \omega| \geq \gamma |k|^{-q}$) to control the small divisors appearing in Eq. (3.5), we find that χ has Fourier coefficients with norms that decay according to $\|\hat{\chi}_k\|_B \sim |k|^{-(p-q)}$. In other words, starting with f of smoothness class C^p, we find after one step in the iteration that our transformation (hence also the remainder to be used in the next step) is only C^{p-q}-smooth, roughly speaking. We have lost q derivatives in one step of the iteration. Clearly, the iteration process cannot continue indefinitely; it will fail to produce a convergent generating function after approximately p/q steps.

To overcome this fundamental obstacle, Moser borrowed a mollification

---

[13] Although functions of finite differentiability cannot be extended to analytic functions, they may nevertheless be approximated by analytic functions, and it was this latter approach that Moser took when he proved more refined versions of the twist theorem, as in [Mos73].

[14] The converse is also true roughly speaking. That is, if f has Fourier coefficients that decay at a given rate, then it is of a corresponding smoothness class.

technique used earlier by J.F. Nash[15] and changed it into what is now called the Nash-Moser method. In essence, he restores lost derivatives at each step of the process by smoothing the remainders using a convolution operator with a smooth, localized kernel (see Section 3, pp. 9–12 of [Mos62] for more detail). The fast convergence of the Newton method is compatible with the smoothing in the sense that the iteration can be made to converge (as in the analytic case) using a perturbed twist map with only finitely many continuous derivatives (Moser showed explicitly that class C^{333} is sufficient).

Later, Moser and others developed much more refined methods for dealing with both maps and flows of finite differentiability, eventually leading to results that are optimal in many cases (see §4.7 (a) below).

4.4 Chronology of Arnold's and Moser's work

Although I haven't yet fully described the theorem that Kolmogorov announced (but I'll do better shortly; see the next section below), I'll call it Kolmogorov's theorem for now. Kolmogorov himself never completely filled in the details of the proof he outlined in [Kol54].[16] That task fell simultaneously to Kolmogorov's precocious student Vladimir Arnold, and to Siegel's young associate Jürgen Moser.

4.4.1 *Arnold's chronology*

Like Kolmogorov, Vladimir Igorevich Arnold (1937–2010) achieved fame in Moscow while still a teenager by helping Kolmogorov substantially in his

---

[15]Yes, John Forbes Nash, the subject of S. Nasar's biography [Nas98] which inspired the Hollywood film *A Beautiful Mind*. Nash had used the original method in his proof of the 'C^k embedding theorem' [Nash56], showing—to many people's surprise—that a large class of Riemannian manifolds can be globally *embedded* in $\mathbb{R}^n$. The proof is a *tour de force*, but the curious feature is that, like Kolmogorov's scheme for KAM, it too relies on a generalized Newton's method for accelerated convergence (but now with 'post-conditioning,' or smoothing via convolution). It would be interesting to see if Nash's inspiration could be traced to Kolmogorov or to Kantorovich. It's also worth pointing out that J.T. Schwartz had already refined Nash's method somewhat [Schw60] when Moser began to work with it.

[16]However, on p. 353 of [Ya02], B. Yandell reports that Kolmogorov's former student Yakov G. Sinai said that he saw Kolmogorov give a complete proof during classroom lectures in 1957.

solution of the 13th Hilbert problem[17] for which Arnold later received his candidate's degree[18] in 1961. For his doctor of science degree[19] he took on a detailed proof of Kolmogorov's theorem and its application to stability problems. Beginning in 1961, both for its own intrinsic interest and also as a kind of warm-up exercise for Kolmogorov's theorem, Arnold used Siegel's and especially Kolmogorov's ideas on a problem concerning conjugacy of analytic circle diffeomorphisms [Ar61] (this is one of the problems in the line of research described in Appendix B, part B.2, below). At the 1962 ICM in Stockholm, he announced his application of Kolmogorov's ideas to the problem of stability in (certain versions of) the n body problem, and also published these announcements [Ar62]. Then, in 1963 he published a complete proof of a slightly modified version of Kolmogorov's theorem [Ar63a] (dedicating the paper to Kolmogorov on his 60th birthday), and followed it with a second paper [Ar63b] applying the theory to stability questions for both the n body problem and the motion of charged particles in magnetic fields. Finally, in 1964 he published a short paper [Ar64] sketching a mechanism (now known as *Arnold diffusion*) by which a kind of instability may occur in parts of phase space not occupied by invariant tori, at least for systems with more than two degrees of freedom.

Arnold describes his early days in a richly detailed expository article [Ar97] which was recently translated into English. This article tells of his time as Kolmogorov's student, his work on Hilbert's 13th problem, and the work leading up to and including his proof of Kolmogorov's theorem. It also gives vivid portrayals of other mathematicians at work in those days.

Although Arnold did not work much in HPT himself after the 1960s (becoming interested in other areas, such as singularity theory), he continued to write about it in textbooks and monographs and to direct students in HPT, and his influence is still strongly felt, particularly in the area of instability as a consequence of the paper [Ar64]. In fact, efforts to resolve the issues raised in this paper have given rise to a whole sub-industry in HPT which is discussed further in §6.3.2. Arnold's early proof of a version of Kolmogorov's theorem that could be applied to the n body problem has also been very influential. This entailed a reworking of Kolmogorov's nondegeneracy hy-

---

[17]Hilbert's 13th problem asks for solutions of 7th-degree polynomial equations using functions of two parameters. Kolmogorov and Arnold showed that no such solutions are possible in the universe of continuous functions. Some say that Hilbert intended solutions to be multivalued algebraic functions; the problem interpreted in this way remains open.
[18]The candidate of science degree (*Kandidat nauk*) is comparable to an American PhD.
[19]The Russian doctor of science degree (*Doktor nauk*) is roughly equivalent to the *Habilitation* in Western Europe (some say it requires even more work than the Habilitation).

pothesis in the theorem, while its application required overcoming stubborn degeneracies in the n body problem; see §7.1.1 and §7.1.2 below. Finally, several of Arnold's many former students—especially N.N. Nekhoroshev and M.B. Sevryuk—have continued to do important work in HPT.

4.4.2   *Moser's chronology*

At age 16, while still a *Gymnasium* student, Jürgen Moser (1928–1999) was pressed into service in the German *Volkssturm*[20] against Russian armor in the late World War II battle for his home town of Königsberg in East Prussia. Moser's older brother and many of his classmates were killed, but before the city was overrun by Russian forces (and renamed Kaliningrad), Moser and his parents were evacuated by boat to the British zone. Following adventures that included escaping from a Nazi prison, Moser enrolled at the University of Göttingen in 1947 and received his doctorate there in 1952 under Franz Rellich. In the meantime, when C.L. Siegel returned to Göttingen from Princeton in 1950, he and Moser began a decades-long collaboration on problems of celestial mechanics which continued after Moser came to the U.S. in 1953 (working first at New York University, then at MIT near Boston, then back to NYU for two decades 1960–1980). This collaboration resulted in Siegel's classic treatise [Sie56] based on Moser's course notes, later reworked and translated into English with Moser as coauthor [SieM71].

The story has often been told of how Moser came to work on Kolmogorov's theorem. He was asked by the editors of *Mathematical Reviews* to review Kolmogorov's published version [Kol57] of his 1954 ICM address, but after a struggle during which he consulted the earlier papers [Kol53], [Kol54], he could not convince himself that Kolmogorov's iteration procedure converged. He thereafter became very interested in providing a complete proof of the theorem, and was encouraged by Siegel. The result he eventually announced at the 1962 ICM in Stockholm (and published with proofs the same year [Mos62]) was both less and more than what Kolmogorov had claimed. It was stated as a theorem about the persistence of invariant curves for twist maps of the plane; in that sense, it implied the existence of invariant tori only for perturbed Hamiltonian systems with $n = 2$ degrees of freedom, rather than for arbitrary n as in Kolmogorov's theorem.

---

[20]This German word was described to me as meaning "every male who could carry a gun and wasn't already drafted." It consisted mostly of boys 16 to 18 years old, and men between 50 and 60.

But, as mentioned earlier, Moser's theorem had a hypothesis that surprised even Kolmogorov: he replaced the assumption that the Hamiltonian was analytic with an assumption that it was only finitely differentiable (he required 333 derivatives). This drastic reduction in the smoothness hypothesis greatly strengthened the theorem,[21] and opened up a line of research into the minimal smoothness required for the persistence of invariant tori (see below §4.7 (a)).

Although Moser had many other mathematical interests (including subjects in geometry, partial differential equations, spectral theory, ergodic theory, etc.), he continued his work in HPT and its applications throughout his career. He first applied KAM theory to the stability of particle beams in accelerators, and to problems in celestial mechanics, and continued to work on related problems after he moved in 1980 to the ETH in Zürich (where he soon became director of the Institute for Mathematical Research). Toward the end of his career he was also involved in the variational methods that emerged with Aubry-Mather theory (discussed below in §6.1). Among Moser's many PhD students, several have done important work in HPT or small divisor theory; in particular, A. Celletti, L.H. Eliasson, M. Levi, and J. Pöschel have greatly extended KAM theory and its applications.

4.5   A prototype KAM theorem

Having set out the basic chronology of events surrounding the announcement and proof(s) of Kolmogorov's theorem, I'll now call it KAM theory, incarnated in specific instances of 'the KAM theorem.' In order to look more closely at such theorems, I'll loosely state a prototypical version of the theorem below, then discuss early variations and later refinements.

Prototype KAM theorem.   *Let $(\theta, I) = (\theta_1, \ldots, \theta_n, I_1, \ldots, I_n)$ be action-angle variables for the <u>smooth</u> completely integrable Hamiltonian $h : M \to \mathbb{R}$ with $n \geq 2$ <u>degrees of freedom</u>. Assume h is <u>nondegenerate</u>, and let $\varepsilon \in \mathbb{R}$ be a small parameter. Then for any <u>smooth</u> perturbation $\varepsilon f(\theta, I, \varepsilon)$ there is a <u>threshold</u> $\varepsilon_0 > 0$ such that whenever $\|\varepsilon f\|_M < \varepsilon_0$, the perturbed Hamiltonian $H(\theta, I, \varepsilon) = h(I) + \varepsilon f(\theta, I, \varepsilon)$ has a <u>nonempty set</u> $\mathcal{T}$ of invariant n-dimensional (Lagrangian, or KAM) tori in its phase space. On each*

---

[21] In [MatMNR00] it's reported that Moser was initially disappointed at being unable to prove Kolmogorov's theorem as originally announced (for analytic Hamiltonians), and was surprised when Kolmogorov and others hailed his proof as a breakthrough.

invariant torus of $\mathcal{T}$, *the flow of H is quasiperiodic with* <u>*highly nonresonant*</u> *frequency. Furthermore, the set* $\mathcal{T}$ *is* <u>*large*</u> *in the sense that its measure becomes full as* $\varepsilon \to 0$.

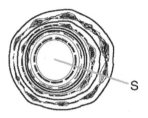

Fig. 4.1   Cross-sectional schematic diagram (for $n = 2$) of the set $\mathcal{T}$ (Kolmogorov set) of invariant KAM tori, the existence of which is guaranteed by KAM theory. ('Chaotic behavior' is shown in some of the gaps of $\mathcal{T}$.) The Kolmogorov set is a 'Cantor family' in the sense that a radial slice S intersects it in a Cantor-like set.

4.6   Early versions of the KAM theorem

Although the above statement is not very precise mathematically speaking, it should allow us to discuss variants of the KAM theorem in a rough but understandable way. I'll do this by discussing the underlined portions one-by-one, and I'll begin with the early theorems—those stated by Kolmogorov, Arnold, and Moser.

(a) <u>Regularity</u>

Here I simply reiterate that Kolmogorov's original 1954 statement [Kol54] was for *analytic* h and f, as was Arnold's much more detailed treatment [Ar63a] with proofs in 1963. But already Moser's 1962 proof [Mos62] (for 2-dimensional 'twist maps') showed that only *finitely* many derivatives are required. In his first proof, Moser made no special effort to find the minimum number of derivatives required, and in [Mos99a] he recounts how the actual number of derivatives he used (333) was a 'silly' figure designed to simplify the exposition. The question of the minimum smoothness under which KAM holds quickly became serious however; see §4.7 (a) below.

(b) Number of degrees of freedom n; two types of stability

We already saw that Moser's invariant curve theorem may be interpreted as a KAM theorem for the special case of Hamiltonians with $n = 2$ degrees of freedom, while Kolmogorov's and Arnold's versions hold for any $n \geq 2$.

There are other reasons for distinguishing the high and low ranges of n, and especially the lowest value $n = 2$, where KAM theorems have a certain 'elegance' (as Moser's theorem does), and where the invariant tori and accompanying dynamics may be visualized in three dimensions. Perhaps most fundamental is the qualitative difference in the type of stability implied by KAM theory at the lowest value $n = 2$ as compared with all higher values. This is quite simply because for $n = 2$, the phase space has dimension 4, so energy surfaces $H = E$ have dimension 3, and thus the 2-dimensional invariant tori partition each energy surface (since the difference in dimensions is precisely one). In other words, each torus has an inside and an outside, and it makes sense to talk about one torus being contained inside another, and about the narrow compartment between them where trajectories—even 'chaotic' ones—remain trapped for all time. This sort of strong stability is called topological stability.

On the other hand, for $n > 2$, the energy surfaces (of dimension $2n - 1$) are no longer partitioned by the n-dimensional invariant tori because now the dimensional difference $(2n-1)-n$ exceeds one. (Think of concentric circles [1-dimensional tori] which partition a 2-dimensional plane into annular regions, but which do not partition 3-dimensional space.) In this situation, the invariant tori no longer have insides or outsides, and topological stability is lost. What remains is a kind of 'metric' or 'probabilistic' stability: if a trajectory happens to begin on an invariant torus, by definition it remains on it thereafter and so enjoys perpetual stability. But whether or not it 'happens to begin on an invariant torus' is something that, in practical terms, can only be determined probabilistically, because the set $\mathcal{T}$ of tori is a Cantor-like set. This is discussed further in §4.6 (d) below, while the topological instability (so-called *Arnold diffusion*) that occurs for $n > 2$ is discussed in §6.3.2.

(c) Nondegeneracy conditions

As mentioned earlier, the nondegeneracy condition applies to the frequency map $I \mapsto \omega$ given by $\omega(I) = \partial h/\partial I$, hence it's a condition on the unperturbed part $h = h(I)$ of the Hamiltonian. The idea is to ensure that the structure of the Diophantine set (4.3) is preserved in action space, so that the abundance of Diophantine frequencies is reflected by an abundance of invariant tori.

A straightforward way of doing this is by using Kolmogorov's nondegeneracy condition from his 1954 announcement: $\det[\partial^2 h/\partial I^2] \neq 0$ (i.e., the *Jacobian determinant* of $I \mapsto \partial h/\partial I = \omega(I)$ is nonvanishing); this assures that $I \mapsto \omega$ is a *local diffeomorphism.*

Kolmogorov's condition is a fairly natural one for KAM theory, and is satisfied by 'many' unperturbed Hamiltonians h. But when Arnold looked into the n body problem, he found degeneracies that prevented its immediate application. He was led to introduce the so-called 'isoenergetic' nondegeneracy condition, or Arnold's condition[22]

$$\det \begin{bmatrix} \partial^2 h/\partial I^2 & \partial h/\partial I \\ \partial h/\partial I^T & 0 \end{bmatrix} \neq 0 \tag{4.9}$$

which applies on fixed energy levels of the Hamiltonian and is better suited to many of the problems in celestial mechanics. It's interesting to note that Kolmogorov's and Arnold's conditions are independent; in other words, neither implies the other. (In most applications however, the two are usually satisfied or fail simultaneously, and in [BroeHu91], H.W. Broer and G.B. Huitema use a transversality argument to prove a version of Arnold's isoenergetic KAM theorem from the 'ordinary' KAM theorem.)

Arnold also reformulated the KAM theorem for 'properly degenerate' systems (which don't depend on all the action variables) and applied it to stability questions for certain versions of the n body problem. (Unfortunately, a mistake in this application was later discovered, then repaired, first by M. Herman and J. Féjoz, then by L. Chierchia and co-workers F. Pusateri and G. Pinzari; see §7.1.1 for more details.)

Because he worked with 2-dimensional annulus maps, Moser's nondegeneracy condition is simpler: he just assumes that the rate of rotation of the unperturbed map changes monotonically (either increasing or decreasing) as one moves radially outward from circle to circle; see Fig. 4.2 below. A mapping satisfying this 'twist condition' is called a 'twist map.' (Note that the twist condition is essentially what we described in §2.5 for the corresponding Hamiltonian systems with $n = 2$ degrees of freedom; cf. also Fig. 2.5 in that section.)

---

[22]Arnold introduces this condition in [Ar63a] (in a footnote at the beginning of §2). Mikhail Sevryuk pointed out to me that the determinant in Arnold's condition appears in Poincaré's *Les Méthodes nouvelles* [Poi92–99] (Vol. 1, end of §74), where he calls it 'le hessien bordé' ('the bordered Hessian') and uses it in determining the 'characteristic exponents' for the three body problem.

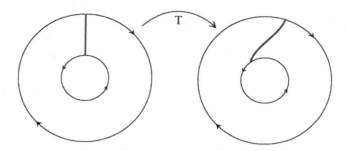

Fig. 4.2 Illustration of Moser's 'twist condition.' Here T is a continuous mapping of the annulus (region between concentric circles) to itself. T leaves each circle invariant as a set, but rotates points in each circle by an angle that varies monotonically with the circle's radius. In the case illustrated here, T rotates the inner circle CCW, then the rotation decreases with the radius until it is CW on the outer circle. The action of T on the vertical line is illustrated by its tilted image in the annulus on the right; because of this, a twist map is often said to "tilt the vertical."

And in an aside related to nondegeneracy, Moser was able to show that his twist theorem holds for a slightly larger class of annulus maps than those that are area-preserving or derived from Hamiltonian systems (it holds for those with graphs having only the so-called self-intersection property: each closed curve encircling the inner boundary circle of the annulus once intersects its image under the twist map).

(d) The (large) set of invariant tori $\mathcal{T}$

One of the key features of KAM theory is that, not only is the set $\mathcal{T}$ of invariant tori nonempty for small ε, but its measure becomes full (that is, it fills phase space in the measure sense) as $\varepsilon \to 0$. The set $\mathcal{T}$ is often called the 'Kolmogorov set,' and occasionally the 'Kolmogorov-Siegel set.'

It's perhaps worth stressing again that $\mathcal{T}$ inherits both its large size and its Cantor-like structure from the set of Diophantine frequencies; and again it's the nondegeneracy of the frequency map that ensures that these two sets correspond closely.

The early KAM theorems suggested but did not show directly that $\mathcal{T}$ is large. The first theorems show the existence of a single invariant torus or curve; one can then check afterwards that the set of initial conditions which generate such tori is a Cantor-like set of large measure. Proofs that $\mathcal{T}$ is also a Cantor-like set of large (relative) measure $1 - O(\varepsilon^{1/2})$ came later, first in dimension 2 in a paper of V.F. Lazutkin [Laz74], then in arbitrary dimensions in the work of J. Pöschel [Pös80], [Pös82], A. Neishtadt [Nei81], and L. Chierchia and G. Gallavotti [ChiG82]. These latter papers also go further and view the persistence of tori as the continuation of integrability in a mathematically precise sense (cf. §4.8 (c) below for further discussion).

Although by the early 1980s, KAM theorems rigorously estimated the size of $\mathcal{T}$ as $\varepsilon \to 0$, the theorems did not capture all the tori, and this was especially evident as ε approached the threshold ε_0 from below; see §4.7 (c) for more detail.

4.7 More recent results that are optimal (or nearly so)

Discovery may be the most exhilarating activity in mathematics, but once an important discovery is made, interest shifts to 'optimizing' it and exploring its limits. In the case of KAM theory, this means finding the weakest possible smoothness and nondegeneracy conditions, the largest possible threshold ε_0, and a precise characterization of the set $\mathcal{T}$ of invariant tori. In this section I'll describe the optimal (in some cases just the best) results that are currently available, and I should point out my indebtedness to M.B. Sevryuk's engaging article [Sev03] from which much of this material is adapted.[23] It should also be kept in mind that these results are not entirely independent of each other: like a balloon toy that bulges out in one place when squeezed in another, the hypotheses of KAM theory are highly interdependent.

(a) Optimal smoothness conditions for arbitrary n, and for twist maps

The optimal (i.e., *weakest*) smoothness conditions under which KAM results hold are now known in broad generality.

At first, only the smoothness of the perturbation was reduced: in 1969 Moser [Mos70] and later J. Pöschel [Pös80] showed that KAM holds for perturbations f of class C^r for all $r > 2n$ (in the *Hölder* sense), provided the unperturbed part h is analytic. Later, D.A. Salamon showed [Sa04] that the smoothness of h can also be relaxed to C^r, $r > 2n$, and he showed that these conditions are optimal when $n = 2$ (an early version of Salamon's results first circulated as an ETH preprint in 1986). Much more recently, C.-Q. Cheng and coauthor L. Wang have shown the optimality of these conditions in all dimensions by giving examples of the failure of KAM in any smoothness class C^r with $r < 2n$ for arbitrary $n \geq 2$ [Cheng11],

---

[23]Dr. Sevryuk also indicated to me a number of developments in KAM theory since the appearance of his article.

[ChengW13].

For an integrable *twist map* of the planar annulus to have smooth invariant curves under sufficiently small perturbations, the map must be of class C^3. In rough terms, this condition was shown to be both necessary and sufficient by M.R. Herman. But a closer look shows that the situation is more subtle. In Chapters II and III of [Herm83] (based partly on his earlier work [Herm79]), Herman indeed demonstrates the necessity of C^3 regularity by giving counterexamples of smoothness class $C^{3-\epsilon}$ for arbitrary $\epsilon > 0$. But the sufficiency of C^3 regularity comes with a qualification. In Chapter V of [Herm86], Herman shows the persistence of invariant curves under area-preserving twist maps of class C^3 which are sufficiently close to integrable twist maps in the C^3 topology. (A similar result for twist maps of class $C^{3+\epsilon}$, $\epsilon > 0$, was already shown in Chapter IV of [Herm83].) However, the rotation numbers of the maps in these results are required to be 'of constant type';[24] these numbers are known to form a set of Lebesgue measure zero (but *Hausdorff dimension* one) in the unit interval.[25] This is in distinction to usual KAM theorems for twist maps, where the rotation numbers are only required to be Diophantine (such numbers make up a set of large relative Lebesgue measure).

(b) Optimal nondegeneracy condition for the analytic case

In the analytic case (i.e., analytic h and f), the optimal nondegeneracy condition is known: KAM results hold precisely when the *image* of the frequency map $I \mapsto \omega$ does not lie *locally* in any *hyperplane* through the origin in $\mathbb{R}^n$.

The history of this condition is rather complicated (even 'dramatic,' as M. Sevryuk says in [Sev03]). It is often called Rüssmann's condition, since H. Rüssmann announced it in [Rüss89], and gave a detailed proof a dozen years later in [Rüss01]. But the first published proof [ChengS94] (of the condition's sufficiency, by C.-Q Cheng and Y.S. Sun) appeared between Rüssmann's announcement and proof, and other independent proofs have appeared [Sev96] or are known (M.R. Herman discussed his own proof in several seminars, but did not publish it before his death in 2000). Proof of the condition's optimality (i.e., its necessity) was first given by Sevryuk in [Sev95]. Although (weak) analogs of this condition are known for Hamiltonians that are not analytic, the optimal condition is presently known only

---

[24] An irrational number $\alpha \in (0, 1)$ is said to be 'of constant type' if the quotients a_k in the *continued fraction* expansion $[a_1, a_2, \ldots]$ of α are bounded; i.e., if $\sup_{k \geq 1} a_k < \infty$.

[25] See, e.g., p. 163 of [HasK02b].

in the analytic case.

And although Rüssmann appears to have been the first to use this condition explicitly in the hypotheses of KAM theorems with proofs, it has earlier origins in related contexts among Russian mathematicians. Arnold is the condition's likely originator, having used it as a hypothesis in an averaging theorem as early as 1966. Arnold's student A.S. Pyartli then studied and used it extensively in his work on Diophantine approximation [Pya69], and it was discussed around the same time in seminars by G.A. Margulis and V.G. Sprindzhuk in Moscow. Finally, in the papers [Par82] and [Par84b], I.O. Parasyuk discusses the condition's applicability to KAM theory in *reversible* systems, and in finding *coisotropic* invariant tori.

(c) The threshold ε_0

We've seen that KAM invariant tori fill up phase space in the measure sense as $\varepsilon \to 0^+$. But what about the high end of ε-values? What happens as ε approaches the largest (threshold) value ε_0 for which at least one invariant torus is guaranteed to exist? Ideally, one hopes that the threshold ε_0 is *sharp*; in other words, as ε increases toward ε_0, tori disintegrate until there is only one torus (or one family of tori) remaining—the notorious 'last surviving torus'—and this too breaks up as ε passes beyond ε_0.

In the early days of KAM theory, the thresholds established in mathematical proofs were absurdly small, so small that there were jokes about them (cf. §5.2.2, discussion of C3) and they were a source of skepticism about KAM theory's applicability to physical problems. ("What does it matter if ε_0 is positive if an electron microscope can't distinguish it from zero?") But once good numerics for nearly integrable systems became available, it was clear that invariant tori survived for much larger perturbations than indicated in theorems. Mathematicians therefore sought to increase the rigorous thresholds to 'realistic' values, especially in theorems applied to models of physical problems, such as magnetic confinement of charged particles, beam stability in particle accelerators, and of course the n body problem and its variants.

It was in the context of the so-called *Chirikov standard map* (studied most intensively by B.V. Chirikov) that this problem was first seriously attacked. This isn't surprising, as the standard map is one of the simplest systems to which KAM theory applies. It was suspected that the last surviving invariant curves would be the 'noble' or 'golden' curves, those with the most irrational or most Diophantine *rotation num-*

bers $\alpha = (\sqrt{5}\pm 1)/2$. First, numerical evidence—then more rigorous 'renormalization' techniques—indicated that indeed, golden curves survive in the standard map until the perturbation parameter ε reaches a (relatively large) threshold value near 1.

The story of how this threshold value was reached in mathematically rigorous proofs in the mid to late 1980s is a complex tale involving a number of researchers. The scope of the task is perhaps best appreciated by recalling that in the earliest KAM theorems, the thresholds were incredibly small; magnitudes of $\varepsilon_0 \approx 10^{-N}$ where N was a two- or even three-digit number were common. Since such numbers are smaller than the smallest known (nondimensional) physical quantities, physicists were understandably skeptical of the applicability of early KAM theory. Unfortunately some of this skepticism remains today, even now that proofs have mostly caught up with numerics in low dimensions.

The methods used to get good rigorous threshold values themselves involve computers. Typically, one uses symbolic computation to bring the problem into a *normal form* up to some appropriate finite order, then applies a KAM scheme to eliminate the remainder, using rigorous 'interval arithmetic' to check the convergence. Techniques of this type now yield fair agreement between numerics and theory; see §6.3.1 below for a short review of research into the threshold for the special case of the Chirikov standard map.

4.8 Further approaches and results

After highlighting some of the optimal (or nearly optimal) results that are currently available, I want to give the broader outlines of research as it has developed during the six decades since Kolmogorov's announcement. I'll sketch this in the space of a few paragraphs, giving references that may serve as entryways for the interested reader. I'll start by distinguishing the lines of research that descend from the three founders K, A, and M, then continue by condensing some of the extensive material reported in R. de la Llave's tutorial [delaL01].

(a) Validation of Kolmogorov's proof scheme

Here I simply point out that, whatever controversy surrounded the question of whether or not Kolmogorov proved a KAM theorem, the specific strategy he outlined was validated in detail, first in the paper [BenGGS84] (although the canonical transformations used there involve the Lie series method, as introduced above in §3.11, rather than the Jacobi mixed-variable generating functions originally indicated by Kolmogorov), then again in [Chi08], this time with precisely Kolmogorov's methods.

Kolmogorov's idea of using a modified Newton's method was also reformulated in more abstract fashion, as discussed next.

(b) Implicit (and inverse) function theorems

Moser's proof of his twist theorem blends Kolmogorov's ideas (based on Kantorovich's method [Kan48]) with similar approaches due to Nash [Nash56] and Schwartz [Schw60]. From this approach comes a large body of research into the convergence techniques themselves. These techniques usually go by names such as 'abstract *implicit function theorems*,' or 'hard implicit function theorems in *Banach* or *Fréchet spaces*,' and have turned out to be useful not only in KAM theory, but in the wider context of PDE and geometry. The main references are due to Moser [Mos61], F. Sergeraert [Ser72], E. Zehnder [Ze74], [Ze75–76], [Ze76], R.S. Hamilton [Ham82], L. Hörmander [Hör85], [Hör90], and Moser and Zehnder [MosZ05]. A very readable introduction to the Nash-Moser method as it's used in the proofs of implicit function theorems may be found in §6.4 of S.G. Krantz and H.R. Parks' book [KranP02]. More recently, I. Ekeland developed an *inverse function theorem* in Fréchet spaces [Ek11] which includes the Nash-Moser result as a special case and applies to functions of low regularity, but whose proof does not use a generalized Newton's method (instead it relies on variational principles and the dominated convergence theorem).

(c) Integrability on Cantor sets

As mentioned earlier, Arnold's proof-strategy may be adapted to find, not a single torus, but the set $\mathcal{T}$ of invariant tori (or at least a sizeable portion of it). This has led to more refined results which view $\mathcal{T}$ as showing the 'integrability (of the perturbed system) over a Cantor set.'

This conceptually suggestive approach was pioneered in the early 1970s when V.F. Lazutkin discovered that the invariant circles in Moser-like KAM theorems form a differentiable family or *foliation* in the sense of *Whitney* [Laz72], [Laz73]. At the beginning of the 1980s, the theory was more fully developed and generalized to the Kolmogorov set of invariant tori $\mathcal{T}$ in arbitrary dimensions by J. Pöschel [Pös80], [Pös82], and by L. Chierchia and G. Gallavotti [ChiG82]. In fact, examining the differentiable structure of the foliation formed by $\mathcal{T}$ shows it to have 'anisotropic smoothness':

it is more differentiable in directions tangent to tori than in directions *transverse* to them. Looking still more closely, on $\mathcal{T}$ one also finds smooth independent functions in involution—in other words, *integrals of motion*—and these permit solutions of the system to be reduced to *quadrature* (i.e. 'solved') on $\mathcal{T}$.

In this way, even though classical integrability of the system (cf. §2.3) is lost under perturbation, it continues in a generalized, geometrical sense as 'integrability over a Cantor set.' To me, this is the most conceptually powerful formulation of KAM theory, and the one that says most about its place in the framework of dynamical systems.

(d) KAM for reversible systems

Many physical Hamiltonian systems are *reversible* (roughly speaking, their solutions may be run forward or backward in time), but there are also important reversible systems outside the Hamiltonian framework. KAM theory was first formulated for reversible systems by Moser [Mos65], and also independently a few years later by Yu.N. Bibikov and V.A. Pliss [BibP67]. The theory evolved further with the work of other researchers, including V.I. Arnold, J. Pöschel [Pös82], J. Scheurle [Sche79], I.O. Parasyuk [Par82], M.B. Sevryuk [Sev91], [Sev98], and H.W. Broer and coauthors Sevryuk, B.L.J. Braaksma, and G.B. Huitema [BraaBH90], [BroeHu95], [BroeHuS96]. Many further references appear in the work last cited.

(e) Non-Lagrangian KAM tori

In classical KAM theory for Hamiltonian systems with n degrees of freedom, the 'standard' KAM tori are Lagrangian (i.e., they are *Lagrangian submanifolds* of the $2n$-dimensional ambient phase space, and thus have dimension n). However, one can also investigate the existence of lower- and higher-dimensional *non*-Lagrangian KAM tori (invariant tori carrying quasiperiodic orbits). These occur as three basic types: (i) Lower-dimensional *isotropic*[26] tori; (ii) higher-dimensional *co-isotropic* tori; and (iii) 'atropic' tori of fixed dimension which may be less than, greater than, or equal to n.

Invariant tori of type (i) (of dimension $< n$) were first constructed by V.K. Melnikov [Meln65] and also later by J.K. Moser, and have become increasingly important in attempts to understand the phenomenon of *Arnold diffusion* (cf. §6.3.2 below). Tori of type (ii) (dimension $> n$) were first

---

[26]See the glossary for definitions of isotropic and co-isotropic.

studied by I.O. Parasyuk [Par84a], while 'atropic' tori (neither isotropic nor co-isotropic) were first observed by Q. Huang, F. Cong, and Y. Li [HuaCL00]. It should be noted, however, that the tori of type (ii) and (iii) occur only rarely, when the *symplectic form* ω is not 'exact' (i.e., when ω cannot be written as the differential of a '1-form'; see [BroeS10] for further details).

(f) Direct methods (convergence of Lindstedt series)

Certain developments in the 1980s have a special link with our earlier story of the King Oscar Prize (§3.7). As already mentioned, KAM theory affirms the convergence of the Lindstedt series of nearly integrable Hamiltonian systems for special initial conditions. But this affirmation is indirect (it's a consequence of uniqueness), so a natural question is, Can the convergence be shown directly? The answer is yes, and because the Lindstedt series must converge *conditionally* rather than *absolutely*, the challenge of directly proving convergence lies in exhibiting massive yet delicate cancellations among the constituents of each term in the series. Such 'direct methods' have now evolved to include an array of combinatorial tools, including 'counterterms,' 'tree expansions,' 'renormalization of resonances,' and 'resummation families.'

The first results in this direction appeared in the paper [Eli96] by L.H. Eliasson (which gained notoriety by circulating as a manuscript for nearly a decade before publication). Further work by L. Chierchia and C. Falcolini [ChiF94], [ChiF96], G. Gallavotti and G. Gentile [GallG95] and others has extended and clarified these techniques; a detailed review (with emphasis on the isochronous case) may be found in the paper by M. Bartuccelli and G. Gentile [BartG02].

It's interesting to wonder what Poincaré and especially Weierstrass might have thought of these results, as they seem to respond most directly to the King Oscar prize question as formulated by Weierstrass. Though it seems very unlikely, one also can't help but wonder whether Dirichlet had something similar in mind in 1858 when he made the remarks to Kronecker mentioned in §3.7.

(g) Other methods and results

After sketching KAM research as it has evolved from the founders' methods and adding a few remarks on direct methods, I'll simply list some other areas of subsequent activity (citing only one or a few representative references). They are: KAM theory in configuration space, based on Lagrangian (rather than Hamiltonian) methods [SaZ89]; 'exclusion of parameters' methods, giving the existence of non-Lagrangian (lower-dimensional) invariant tori both for Hamiltonian and more general (e.g. volume-preserving) systems [BroeHuS96]; group renormalization methods, leading to both existence and non-existence of invariant tori [KhS86]; variational methods, also leading to existence and non-existence of invariant curves and other geometric structures (so-called Aubry-Mather sets; more on this topic below in §6.1); methods relying neither on a change of variables nor on the existence of action-angle variables [Rüss76], [CelC88], [GonJLV05], [delaLGJV05]; and a 'slowly convergent' method [Rüss10], [Pös11] that defies Kolmogorov's (and others') seemingly crucial use of rapid convergence in early KAM theory. In some cases, these new methods use computer-assisted proofs, which is itself an active area of KAM research.

Finally, I should at least mention one of the most active current areas of KAM research, namely KAM theory for infinite-dimensional systems (e.g. PDEs). I've already declared that I'll stick closely to the finite-dimensional case in this book. But infinite-dimensional KAM theory may yet prove to be the more significant branch of research, if it can be developed as thoroughly as hoped. See Appendix D, Part D.7.2, for ideas of where to find references in this developing field.

Chapter 5

KAM in Context: Questions, Consequences, Significance

In this chapter, I give a mostly nontechnical overview of how KAM theory fits into its broader setting: what it says in simple terms, why it's important in mathematics, physics, and the history of ideas, and what enthusiasts and detractors say about it.

5.1 A quick overview of KAM theory in prose and pictures

When a specialist reads a KAM theorem, he or she is aware of what makes it significant and where it sits in the bigger picture. In this section I summarize and describe KAM theory in a way that provides some of this background and setting for the newcomer or non-specialist. I begin with a description in simple pictures I call cartoons.

5.1.1 *A cartoon summary of KAM theory*

Let's begin by summarizing what KAM did for Hamiltonian perturbation theory (HPT), which, we recall, is the study of Hamiltonian systems of the form $H = h(I) + \varepsilon f(\theta, I, \varepsilon)$ with $(\theta, I, \varepsilon) \in \mathbb{T}^n \times \mathbb{R}^n \times \mathbb{R}$. For $\varepsilon = 0$, such a system is integrable, and the basic problem has always been to understand what happens when ε becomes positive; in other words, what happens when the perturbation is turned on. The state of affairs in HPT before KAM can be summed up by the cartoon in Fig. 5.1.

In the most basic sense, KAM theory resolves the half-century old paradox described in §3.12.5 by showing that positive ε does not disrupt integrability as much as previously thought. Yes, positive ε immediately breaks integrability in the classical 19th century sense. But KAM shows that in another sense—a sense beautifully captured by 20th century mathematics—

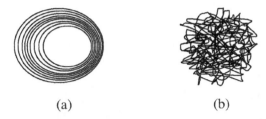

(a)   (b)

Fig. 5.1   Cartoon of the pre-KAM paradox in Hamiltonian perturbation theory (HPT).
For $\varepsilon = 0$ **(a)**, integrable systems were understood; but as soon as ε became positive
(b), systems were a mess—poorly understood, and conjectured to be ergodic.

there is a gradual continuity in the disruption of integrability as ε moves
away from zero.

The full elucidation of the content of KAM theory requires geometric
and analytic concepts that were not available until the modernization of
mathematics that began at the end of the 19th century, concepts such as
Cantor sets, Lebesgue measure, and so on. Yet the thrust of the theory can
be grasped by comparing Fig. 5.1 (the pre-KAM paradox in HPT) with
Fig. 5.2. (post-KAM HPT).

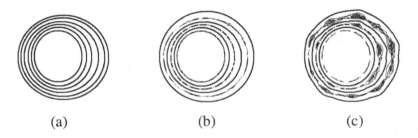

(a)   (b)   (c)

Fig. 5.2   Cartoon of post-KAM understanding of HPT, showing continuity of behavior
with increasing values of ε (left to right). In **(a)**, ε is either 0 or so close to 0 that the
system appears integrable. In **(b)**, ε is large enough that the effects of nonintegrability
begin to be apparent on a macroscopic scale. Finally in **(c)**, ε is large enough that the
system appears 'mixed,' manifesting nonintegrability on a large scale, yet still retaining
KAM tori. [Compare this cartoon with the 'real' phase portraits in Fig. 6.2.]

5.1.2 *Hegel's last laugh*

Before leaving cartoons behind, there is one more that is strongly suggested by the first two above, and may be instructive or at least entertaining to present here.

Since the 1960s, it has been fashionable to describe sudden, significant scientific advances in terms of Thomas Kuhn's theory of *'paradigm shift'* [Kuh62]. But in the case of KAM theory, this seems to be a forced fit. Much better in this writer's opinion is an older, less fashionable notion championed by our maligned philosopher of §3.6. This is Hegel's idea[1] of the dialectic, in which development or progress (in any domain, really) takes place in a multistage process of thesis, antithesis, and synthesis. The thesis appears as an idea, giving rise by way of natural reaction[2] to its negation or antithesis, thus producing a tension between the two which is eventually resolved in a synthesis of the thesis and antithesis, after which the dialectical process may begin again.

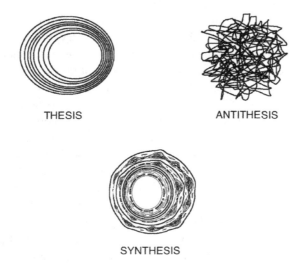

THESIS   ANTITHESIS

SYNTHESIS

Fig. 5.3   The Hegelian interpretation of KAM theory, or Hegel's last laugh.

---

[1] In fact specialists don't attribute this form of dialectical evolution directly to Hegel, saying instead that it was Hegel's interpretation of Kant, later made explicit by the Hegelian scholar H.M. Chalybäus and popularized by J.G. Fichte.

[2] The process in which the thesis-antithesis pair appears is sometimes called 'diremption' by specialists.

It does not take much imagination to reinterpret Figs. 5.1 and 5.2 along Hegelian lines as in Fig. 5.3, with 'integrability' as the 19th century thesis, nonintegrability or 'chaos' as its late-19th or early 20th century antithesis, and KAM theory as the mid 20th century synthesis of the two.

I leave for the reader to decide whether this cartoon should be taken seriously. I note however, that it seems as well-founded to me as many theories of progress (scientific or otherwise) that are approached with high-minded seriousness. Yet if we accept it as only one link in a dialectical chain, it does lead us to naturally ask what the next step is, or to put it another way, if KAM is the (current) thesis, what is the antithesis?

5.1.3   *The big historical picture*

Here I briefly recapitulate the historical setting of KAM theory, occasionally using the Hegelian picture above as a story-telling prop. Chapters 3 and 4 tell parts of this story in greater detail, but this exercise in brevity is useful for highlighting some of its features. Parenthetical references to appropriate parts of the text are included for the reader's convenience.

We start before the beginning of recorded history, when our forebears first looked skyward and wondered about the quasi-regular motions of the points and disks of light they saw there. No doubt the astronomies and astrologies that arose in the earliest Mesopotamian civilizations—with their high priests and careful records—reflect the seriousness with which these things were viewed. After a few millennia, the basic geometry of the solar system was thought to be understood in its essentials, and in the Hellenistic world a relatively sophisticated mathematical model of it was developed following Plato's dictum that celestial motions must be 'circular.' This Ptolemaic system was lost to European civilization during the dark ages, but later recovered, having been preserved by Arabic scholars. It stood for a time as the equilibrium point between the ancient and modern. Then slowly, in Europe there emerged a new way of understanding physical phenomena by observing more carefully and linking observation more directly with mathematics. This revolution traditionally proceeds through Copernicus, Kepler, Galileo, and finally Newton, at which point the mathematical n body problem is revealed and awaits solution (§3.4). The closely related (and by this point quite old) problem of the solar system's stability takes on new meaning (§3.5).

It is anachronistic to think that Newton thought about the stability problem as we do; nevertheless, his reservations about stability undoubtedly had an effect; the gauntlet thus laid down was taken up in the 18th and 19th centuries by Laplace, Lagrange, Poisson, Dirichlet, S. Haretu, and others who claimed to have shown the stability of some form of the n body problem by analyzing series expansions of solutions (§3.5). At about the same time, researchers gained more and more confidence in their ability to 'integrate' systems (§3.6), and when Weierstrass's jury announced the competition for the King Oscar prize, it was all but certain that the award would go to Poincaré for showing stability of the solar system. In the Hegelian picture of KAM theory (preceding section) this may be viewed as the high point of the 'integrability thesis.'

Poincaré tried valiantly to show stability, but the problem itself had other ideas, and Poincaré was astute enough to see them. Thus finally did the 'antithesis' (nonintegrability) emerge in his remarkable prize-winning paper [Poi90], which showed among other things that the n body problem is not integrable in the classical sense.

Almost immediately, the tension between the thesis of integrability and the antithesis of nonintegrability (or 'chaos') began to manifest itself, producing the crisis in classical mechanics described above (§3.12.5). But the balance of opinion would soon shift to the chaos side (§3.12.4). The last major voice to speak on the side of integrability seems to have been Weierstrass, who recognized Poincaré's achievement and heaped praise on it, but remained skeptical of Poincaré's further claims that the Lindstedt series were 'probably not convergent.'

In the 20th century, after Weierstrass and Poincaré, the chaos antithesis gains ascendancy, as enthusiasts for Boltzmann's ergodic hypothesis conjecture that most systems are not only nonintegrable but are furthermore ergodic (§3.12.4). Fermi (c. 1923–24) claims to have proved such a conjecture, while Birkhoff perhaps formulates his own version (but does not prove it). Eberhard Hopf also expresses support for ergodicity. In 1941, Oxtoby and Ulam show that ergodic transformations are generic in the class of continuous measure-preserving maps. Belief in the predominance of unbridled chaos continues right down to the eve of the 1954 ICM in Amsterdam.

In the meantime, ideas of an eventual synthesis are beginning to stir. Safe from the Nazis in his Princeton sanctuary, in 1942 Siegel gives the first solution of a small divisor problem, thus resolving an old conundrum related to Poincaré's 1879 thesis and to the convergence problems of the Lindstedt series (§4.1, and Appendix B.1). Half a dozen years later, Kantorovich shows how Newton's method for solving nonlinear equations can be generalized to more abstract problems in function spaces (§4.1).

Finally, a dozen years after Siegel, Kolmogorov sees how to combine Siegel's and Kantorovich's ideas to execute a synthesis that resolves the crisis in mechanics (a crisis which many researchers have perhaps forgotten in the tumult of 20th century wars and revolutions in physics). Kolmogorov's theorem—the first KAM theorem, though perhaps not (yet) really proved—reveals the first clear outlines of the picture of classical mechanics that we know today. Neither chaos nor integrability predominate in typical systems; instead they coexist in a complex *fractal*-like structure. (And later, in 1974, L. Markus and K.R. Meyer write the paper [MarkM74] whose title expresses the Hegelian synthesis: 'Generic Hamiltonian dynamical systems are neither integrable nor ergodic.')

In the early 1960s, Kolmogorov's vision is affirmed (was there ever doubt?) in the detailed proofs of Moser and Arnold, after which the acronym KAM is christened, and the theory it names attracts increasing attention. Arnold immediately shows the power of the theory by using it to prove the stability of (certain) n body problems—a mathematical holy grail is thereby reached (§4.4.1, §7.1.1). Physicists will joke about the subatomic masses of the planets in the systems to which Arnold's result applies, but the masses are positive, and the result stands (but just barely, as problems are later discovered which weaken it substantially; only after strenuous efforts, first by M. Herman and J. Féjoz, then by L. Chierchia, F. Pusateri, and G. Pinzari, is the result restored to its original strength (§7.1.1)). Alas, dynamical astronomers will show in the meantime that the real physical solar system has other ideas of its own (§7.1.3).

A slow process of assimilation and improvement now begins, interrupted (perhaps 'overtaken' is a better description) by a curious and boisterous phenomenon: chaos mania. At about the time of Moser's and Arnold's proofs of KAM, scientists working with hitherto unavailable computing power begin to notice *sensitive dependence on initial conditions* in deterministic mathematical models of physical systems. During the 1970s, 80s, and 90s, an unprecedented public interest in dynamical systems spurs developments good and bad. Money flows freely from research agencies, but many will lament the quality of what is bought this way. Mathematicians especially grit their teeth as Poincaré's, Lyapunov's, Birkhoff's, and many others' ideas are repackaged (often in debased form) and sold as new and revolutionary (§3.12.2, and Appendix D, Part D.3).

By the turn of the millennium, chaos mania will subside (along with the funding, having found new outlets in cryptography and especially mathematical biology). One happy consequence is the possibility of a more sober assessment of chaos theory and one of its central pillars, KAM theory, the intricacies of which managed to pass mostly unnoticed among non-experts during the decades of euphoria. We begin this assessment in the next section.

5.2 Pros and cons, the myths of detractors and enthusiasts

Almost as soon as Kolmogorov announced the first KAM theorem (and before its thorough validation), researchers began to take its measure. Enthusiasts were first to voice their approval, pointing to the synthesizing breakthrough described above, and to consequences throughout classical mechanics. Later there were dissenting voices, as some began to feel that the enthusiasm went a little too far, or that applications of the theory failed to live up to expectations.

In this section I record some of the things that were said in a simple list of pros and cons,[3] then discuss them briefly in the light of present understanding.

5.2.1 *A list of pros and cons*

Pros (P1–P4):

- P1. KAM theory represents a (mostly under-appreciated) revolution[4] in classical mechanics, comparable in some ways to the highly publicized revolutions in relativity and quantum mechanics.
- P2. Despite the relative obscurity of KAM theory presently (outside circles of specialists), it would have been of great interest to past masters of mechanics and mathematical physics. From Newton to Birkhoff, Jacobi to Boltzmann, all would've been keenly interested. Perhaps most of all, Poincaré would have been surprised (and Weierstrass pleased).
- P3. In celestial mechanics, KAM theory applies to the n body problem to show that, under certain (strict!) conditions, it has eternally stable, bounded solutions for $n \geq 3$. Apart from achieving a centuries-old goal of mathematical physics, this in turn goes some way toward suggesting that our own solar system is stable.

---

[3]The expression 'pros & cons' is again something that sounds quite funny when spoken in French, coming across roughly as 'professionals and idiots.'

[4]Or a *paradigm shift*, if you prefer.

- P4. In classical mechanics generally, KAM theory redresses earlier researchers' eagerness to conclude that generic Hamiltonian systems are *ergodic*. It could also eventually invalidate Boltzmann's *ergodic hypothesis* (or its modern descendant), thus forcing a reassessment of the foundations of statistical mechanics.

Cons (C1–C4):

- C1. KAM theory is quite technical—the proofs require a high level of expertise in analysis, making it difficult for non-specialists to fully understand or apply.
- C2. Hamiltonian systems are not good mathematical models of physical systems.
- C3. The KAM 'threshold of validity' (i.e., the $\varepsilon_0 > 0$ giving the maximum size of the perturbation strength ε for which KAM tori are guaranteed to exist) is absurdly small in applications to physical systems.
- C4. KAM theory is overrated, over-sold, and over-romanticized.

5.2.2 *Discussion*

Pros P1 and P2 are closely related and can be discussed together. That KAM theory represents a revolution of some kind is already established (cf. §5.1). And it certainly seems to be under-appreciated, judging from modern texts on classical mechanics or popularizations of chaos theory that fail to mention it (this is often attributed to con C1). Similarly, it seems almost ignored by historians of science. On the other hand, it is much more difficult to argue that the 'KAM revolution' is in any way comparable to those in relativity and quantum mechanics, which have undeniable historical dimensions, even outside of science. Perhaps a partial key to this lies in P2, which is a way of saying that even if the importance of KAM theory has not yet trickled down to the masses, it should (or may yet) do so, since it would so obviously be of interest to great minds of the past. This is in sharp contrast to the 20th century revolutions in physics, which have drawn much popular interest both by way of their practical applications (e.g. atomic energy, modern electronics) and by their impact on the imagination (time travel, quantum mechanical paradoxes, etc.). Apparently, no one needs to be told to think that atomic energy is important or interesting because Newton would think so. This theme—that KAM theory is of chiefly philosophical or intellectual importance—is one that is often repeated.

We now turn to the more specific pros P3 and P4, concerning the consequences of KAM in applications. It is now generally accepted that KAM theory succeeds in showing the existence of eternally stable, bounded solutions of a certain class of n body problems. However, the story of that success is more involved than is generally supposed, since Arnold's original paper on the subject [Ar63b] was found to have errors. In fact, Arnold's results were not fully established until 2004, by Herman and Féjoz, then more directly by Chierchia and coworkers (see §7.1.1 below for a brief discussion). However, even before the errors in [Ar63b] were discovered, a number of researchers complained that the result was of little value, because the class of n body problems to which it applied had thresholds of validity that were absurdly small. This objection has been partly removed by now, and is addressed below in the discussion of con C3. However, it must be stressed (because the contrary is often wrongly asserted) that KAM theory does not apply to our own physical solar system, and this failure is not simply because the available theory is not yet strong enough; rather, it seems that our solar system does not enjoy KAM stability (see §7.1.3 below). Here again, we see that a major point of division between enthusiasts and detractors arises from the 'philosophical' (as opposed to practical) nature of the results obtained by application of KAM theory.

Concerning the application of KAM theory to the *statistical properties* of dynamical systems, there is one sense in which the consequences are clear. The ergodicity of systems to which KAM theory applies is quite literally blocked by the presence of invariant tori, which form an invariant set of positive measure (and this set itself comprises many *ergodic components*). Since KAM theory applies to a generic set of Hamiltonian systems (in the sense of *Baire category*, using usual *topologies*), it follows that generic Hamiltonian systems are not ergodic (again one thinks of the great title of the article [MarkM74] that announced this result).

On the other hand, the notion that KAM theory might unlock some of the mysteries surrounding the ergodic hypothesis and the foundations of statistical mechanics has not (yet) been borne out. However, the eventual significance of KAM (or closely related Nekhoroshev theory; cf. §6.2) has not been ruled out in that regard, either. It seems that the problems at the foundations of statistical mechanics are simply much harder and subtler than supposed at the outset. (See §7.2 below.)

As for the cons (C1–C4) listed above, discussion of the first two is again rather brief. It is certainly true (as in C1) that KAM theory is technical and requires a certain expertise in analysis to understand in detail. But at the same time, one of this book's central themes is that the basic ideas behind it are accessible to almost anyone who's interested (see §5.1). It is conceivable that this more basic level of understanding would be enough for many scientists working on problems where KAM is relevant (e.g. for those scientists whose job it is to numerically model the behavior of particle beams in high-energy accelerators), or enough for historians of science interested in classical physics, or for philosophers of science working on problems related to mechanistic or deterministic models.

The contention (C2) that Hamiltonian systems are not good models of physical systems is based on the fact that essentially all (macroscopic) physical processes involve some form of dissipation, most commonly through heat loss. (Hamiltonian systems, it will be recalled, do not allow for dissipation of energy.) However, there are a few near-exceptions to this general rule, and these are among the applications briefly considered in the next chapter, namely celestial mechanics, the classical motions of charged particles in electric or magnetic fields, and certain regimes of statistical mechanics. (One of the ironies of heat loss rendering macroscopic systems non-Hamiltonian is that the theory of heat at the microscopic level is itself Hamiltonian.) It is, not surprisingly, precisely in these areas that KAM theory has proved most useful.

But there is another argument against con C2, more basic than simply pointing to the exceptional systems that are well-modeled by Hamiltonian systems. It is simply that Hamiltonian systems—the descendants of *Newton's second law*—are the basic pillar of classical mechanics, and so ought to be understood as completely as possible. After all, isn't there a sense in which classical mechanics is the most fundamental part of all of science? How can dissipative systems be understood until their idealized, non-dissipative counterparts are understood? (Again, this is an argument verging on the philosophical.)

Con C3—complaints about the small thresholds of validity in KAM theory—has an amusing history. Just a few years after the appearance of Arnold's paper [Ar63b] using KAM theory to show the stability of solutions of certain n body problems, the mathematical astronomer Michel Hénon (1931–2013) decided to check the criteria of applicability of Arnold's result. In a now-famous passage in the final section of the paper [Hén66] (pp. 64–65), Hénon applies Arnold's result to the restricted three body problem in celestial mechanics (see §7.1.1 below for a brief description of this problem). Using simple restrictions from Arnold's paper, he finds (Eq. (14), p. 64) that the threshold of validity is on the order of $\varepsilon_0 \approx 10^{-333}$. A similar estimate (Eq. (15), p. 65) based on the restrictions in Moser's paper [Mos62] gives $\varepsilon_0 \approx 10^{-48}$. He then continues in a very straight-faced way with the oft-quoted sentence that I translate as follows ([Hén66], p. 65): "Thus, these theorems, although of a very great theoretical interest, do not seem applicable in their present state to practical problems, where the perturbations are always much larger than the thresholds [above]."

Professor Hénon was a serious, prolific, and highly accomplished dynamicist,[5] but the passage cited above has often been the trigger for much laughter among physicists, if not in print, then certainly in their offices and labs (I call it 'knee-slapping' in §7.1.2 below). To belabor the obvious, the thresholds of validity obtained by Hénon are fantastically small by physical standards—especially the value 10^{-333}. Taking the perturbation ε to represent roughly the ratio of the two nonzero masses in the restricted three body problem, we cannot conceive of two such physical objects; even taking the more massive one to be the known mass of the universe, and the smaller to be the approximate mass of an electron gives a ratio 'only' on the order of 10^{-100} or so. On the other hand, the threshold of 10^{-48} from Moser's theorem is at least in the physical realm; this is the ratio between masses of, say, a single chlorine atom and the (earth's) moon.

The reaction to Hénon's remarks (and similar remarks that soon followed) is of course understandable, but many who laughed did not bother to read a few lines further in Hénon's paper ([Hén66], p. 65), where he says (again my rough translation): "The numerical results we present here, and those obtained for other problems, indicate however that the [invariant] curves continue to exist for very strong perturbations, of the same order of magnitude as the leading term."

In this passage and further on, Hénon voices concerns about KAM theory—especially invariant curve theory—that were to occupy mathematicians for several decades following. But even without indicating that the modern thresholds of validity are no longer always a source of laughter, we can sympathize with the efforts of the first proofs. After all, Arnold and Moser were not concerned with this issue in the early 1960s. Their task, as they saw it, was to open the mathematical door, so to speak, by showing that the thresholds are positive. They accomplished this by the simplest

---

[5]The Hénon map and (half of) the Hénon-Heiles system are named for him; both are important models of chaotic dynamics.

means then available, namely, long chains of *estimates* (inequalities) strung together in which each estimate is chosen for its simplicity rather than *sharpness*. If we have a chain of n inequalities, each of which fails to be *sharp* by at least a factor of α, then the end of the chain will fail to be sharp by at least a factor of α^n with respect to the beginning. It's not surprising that such submicroscopic thresholds should arise in these circumstances. Therefore, to many mathematicians, it also came as no surprise that numerical experiments indicated the existence of invariant curves (or tori, in higher dimensions) for much larger perturbations, and it became a new challenge to bring the theory into agreement with this 'experimental' fact. We can happily report that most of what Hénon wished to understand— and much more, in some respects—is now understood. A short account of some of the efforts that brought this about (at least in low dimensions) can be found below in §6.3.1.

Finally, regarding con C4—that KAM theory is overrated, over-sold, over-romanticized—well, perhaps this book itself is guilty of these sins (but in a self-conscious way). But if there is much general truth in this, it seems to be felt by those who hoped to see more 'practical' consequences of the theory, or who felt that claims of practical applications were made by KAM enthusiasts which later crumbled under closer scrutiny. This is partly understandable in light of what happened with application to the n body problem (see §7.1.1 below), or applications to statistical mechanics (see §7.2). Yet in the end, KAM theory *does* decide certain important issues in these domains; but it turns out that things are not as simple or tidy as was once hoped or supposed.

As for over-romanticizing, that is of course a more subjective issue. It's not often that anyone speaks poetically about the hard sciences or their practitioners,[6] so it can be amusing when it occurs. I've already pointed out (§3.2) how enthusiasts sometimes tend to a romantic view of KAM theory or classical mechanics. By this point, the reader should understand these tendencies better; the reader's sympathy with those tendencies is of course a matter of taste.

---

[6] One thinks reflexively of W. Wordsworth's lines about Isaac Newton, ending with "a mind for ever Voyaging through strange seas of Thought, alone." (See *The Prelude, or Growth of a Poet's Mind*, Book Third: Residence at Cambridge, last part of 6th stanza.)

5.3 'Sociological' issues

The questions addressed in this section are generally amusing rather than profound. They concern quirky issues of the sort one hears about in relaxed settings—cocktail party banter, for the most part. In each case I simply record the things I've heard on both sides of the question; in a few cases, I inject my own views, or explain how one side has gained more currency with time.

5.3.1 *Why did it take so long?*

When thinking of the human side of KAM theory, one of the first questions that comes to mind is the timing: Why, after Poincaré's enormous strides in the late 19th century, did it take more than a half century to take the next big step, to find invariant tori in nearly integrable systems? But just a little reflection is enough to see this question as naïve, and even to turn it on its head: How—under the circumstances—were the tori conceived of and found so quickly?

After all, Poincaré's shadow extended very far, and his belief that the Lindstedt series did not converge (so that results like KAM could not hold) was widely known (see the quotes **Q**[Poi3], **Q**[Poi4], **Q**[Poi5] in Appendix E). In almost every subject he touched, Poincaré had shown himself to be prescient, able to sense the important direction of development and move toward it, or at least indicate the path for later researchers. It was no easy feat to overcome his authority and see that his prescience had failed in this case, especially since his (and Boltzmann's and Birkhoff's) followers had built additional obstacles out of the ergodic hypothesis.

There were also other obstacles. Europe nearly destroyed itself in the two world wars that fit neatly into the span between Poincaré's death in 1912 and Kolmogorov's announcement of KAM in 1954. The same period saw many of the best mathematical minds drawn away from classical questions and into the astonishing progress taking place in relativity theory, quantum mechanics, and nuclear physics. And of course, there were still real mathematical subtleties. Siegel's treatment of small divisors and Kolmogorov's use of a generalized Newton's method look relatively straightforward to us now, but at what points could they first have been conceived?

Looked at in this light, against the grain of Poincaré and his successors, in the midst of calamity and a revolution in physics, it seems amazing indeed that the invariant tori could have been conceived before their existence was rudely forced into mathematicians' consciousness by way of the computer-generated pictures that began to appear in the 1960s. To see them with the mind's eye first is to me the real measure of Kolmogorov's genius.

Nevertheless, it's amusing to speculate what might have been had Poincaré had a different hunch about the convergence of the Lindstedt series. If he really believed they converged, it's hard to think that the proof could have been beyond his reach. But is a late 19th or early 20th century KAM theorem really conceivable? What would it look like?

5.3.2 *Why so few Americans?*

Another curious feature of research in KAM theory is the dearth of Americans[7] working in it. This was no surprise at the outset; after all, the subject grew out of Poincaré's work, was born of Kolmogorov's (and Siegel's) genius, and pioneered by their protégés. All of the initial activity was exclusively European (Moser worked in the U.S. for about two decades and became an American citizen, but was born and raised in Germany). But usually, when a great scientific discovery occurs outside the U.S., American researchers are quick to pile on, sometimes even surpassing the originators. So now more than a half-century after the ICM in Amsterdam, it seems legitimate to wonder why KAM theory never gained much foothold in the U.S.[8]

Some point to the diverging traditions between U.S. and European schools in dynamical systems, with the U.S. being dominated by 'Smale's School' since Stephen Smale's spectacular successes[9] beginning in the late 1950s. In this view, Americans have concerned themselves with the sorts of topological questions that are or were more to Smale's taste, leaving the technical or 'hard analysis' issues of KAM theory to Europeans.

---

[7]With the usual apologies to researchers in other parts of the Americas for using 'American' here as a synonym for 'born, raised, and working in the U.S.'

[8]The most significant exception to the general rule may be Michael Herman (1942–2000), who did outstanding work in the theory of circle maps (see Appendix B, Part B.2), and also obtained the optimal smoothness conditions for twist theorems on the annulus (the type of KAM theorem first proved by Moser; cf. §4.7 (a) above). However, Herman's status as an American was not entirely clear: he was born and raised in the U.S., but decamped to France for graduate studies and remained there for the rest of his career, eventually becoming a French citizen and going by 'Michel' more often than 'Michael.'

[9]To list just some of the highlights of his long career, Smale showed the existence of sphere eversions in three space, proved the Poincaré conjecture in dimensions five and higher (cf. footnote 31, p. 42, above), established the h-cobordism theorem, and produced the example now known as the *Smale horseshoe* in dynamical systems.

One could also wonder why Moser—a consummate analyst and eventually an American[10] himself—did not establish an American school of KAM theory.[11]

Others say that KAM theory appeals primarily to sentimental European physicists and mathematicians, who see in it a link to their glorious past (as described above in §3.2). Hard-nosed American physicists, by contrast, would like to see evidence of KAM's usefulness before investing the time and effort necessary to work on it. This latter view is of course tendentious, and is connected with other somewhat controversial views of the differences in mathematical styles and tastes across various cultures; see Appendix C for a brief discussion of some of these differences in dynamical systems.

In the end there may not be any special reason for the lack of Americans working in KAM theory. Perhaps this book will give an impetus to some of them to do so.

5.3.3 *Did Kolmogorov prove KAM?*

On the one hand, it seems silly to ask whether Kolmogorov proved a KAM theorem. He stated a (very surprising) theorem, outlined how to prove it, and both the statement and proof-outline were later found to be correct in detail. What more could you want?

On the other hand, no one denies that the detailed proofs supplied by Arnold and Moser (of somewhat different results) were themselves major achievements, even if they lack the imaginative leap required to seek the invariant tori or curves in the first place, which is without a doubt the biggest innovation. And there was very likely skepticism and suspense in the years 1954–62 as researchers waited for rigorous proofs.[12]

But what gives this discussion or 'mock mini-controversy' a continuing resonance is again the way that it feeds into discussions of ways of doing mathematics across cultures. Western mathematicians often accuse

---

[10]Moser became a naturalized U.S. citizen in 1959.

[11]Moser's most successful American student was probably Charles Conley (1933–1984), a highly creative and original researcher (some say even more original than his advisor) who again demonstrated the American penchant for a topological flavor of dynamical systems in his work.

[12]One rumor holds that when Kolmogorov announced his theorem in Amsterdam, C.L. Siegel was in attendance and was quite skeptical, at least initially. However, this is gainsaid on p. 79 of [Dav85], where it's claimed that Siegel could not attend the ICM because he had no passport with which to travel at the time. It seems his (relatively new) American passport was revoked in 1954 after a three-year absence from the U.S.

Russians of failing to supply all the details of their proofs; Russians in turn complain that Westerners take Russian results, fill in a few details, then claim them as their own (see Appendix C below, and the interview with Vladimir Arnold in [Zdr87]).

In the case of the KAM theorem, some defenders of Kolmogorov have been vociferous in their assertions that he 'essentially' proved it, knew how to prove it, proved it on a blackboard in Moscow, and so forth.[13] Yet Kolmogorov himself was apparently never very concerned with this issue, and was happy to read Arnold's and Moser's proofs, and to share his name with theirs in the acronym that became the standard way of referring to the result.

5.3.4 *How hard is the proof?*

Strong arguments can be made in both directions concerning the difficulty of proofs in KAM theory; here is a quick look at some of them.

(a) KAM proofs are easy

From the purely mathematical viewpoint, KAM methods—the techniques used in proving KAM theorems—are not especially advanced or difficult. In fact, putting aside the *symplectic geometry* setting that isn't really essential, the proofs are downright elementary. The usual approach can be broken down into two basic parts: (i) solution of the homological equation, which overcomes the small divisor problem at first order in ε, and (ii) iteration of the same procedure to all orders in ε, using a 'rapidly convergent' technique such as the generalized Newton's method in *Banach spaces*.

If one harbors no prejudice against the Cantor-like sets arising in part (i), or against the rapid convergence needed in part (ii), then these elements can also be passed over in elementary fashion. Part (ii) can be further broken down so that the convergence is assured by use of an 'iterative lemma,' which resembles nothing so much as a subroutine in a computer program (similar estimates are checked at each step to ensure that the next

---

[13]At the 1997 'Arnoldfest' in Toronto, Vladimir Arnold wrapped up his first lecture with the following words (recorded on p. 16 of [Ar99]): "Moser criticized that a proof of the theorem in the case of analytic Hamiltonians was never published by Kolmogorov. I think that Kolmogorov was reluctant to write the proof, because he had other things to do in the years still remaining of active work—which is a challenge, when you are 60. According to Moser, the first proof was published by Arnold. My opinion, however, is that Kolmogorov's theorem was proved by Kolmogorov."

step can proceed). Putting everything together involves many estimates, a little calculus, a little complex variables (to take advantage of analyticity, if available), and checking the all-important convergence (again, this is similar to ensuring that a subroutine can be successfully called by a computer program).

In the end, although the conceptual benefits of abstractness may be considerable, KAM proofs require none of the abstract settings or 'big machinery' associated with some important mathematical theorems (such as the Ricci flow used in Perelman's proof of the Poincaré conjecture, or the algebraic geometry of modular forms used in Wiles' proof of Fermat's last theorem, to name just two recent monumental mathematical achievements). With care and in the right setting, no part of the proof needs to go beyond an undergraduate mathematics curriculum (even in the U.S.).

(b) KAM proofs are hard

Here I won't contradict anything written above under 'KAM proofs are easy.' But let me ask a simple question: I wonder how many ordinary mathematicians could sit down with the statement of a KAM theorem and an outline of the proof, and fill in all the details alone in the space of a few hours?[14] I suspect that not very many could do it (unless they'd done it before, with the luxury of references).

Despite the elementary nature of individual steps in the proofs, and despite occasional statements that the proofs are 'trivial,' in fact the global details of KAM proofs can be formidable, even in the simplest cases (as the reader may verify, for example, in [Way96]). This is especially true for otherwise mathematically literate scientists who have never been through— or mastered on their own—an elementary course in rigorous analysis (one that begins with 'delta-epsilon' arguments and progresses to delicate convergence proofs). This last group could include a significant fraction of research physicists, even some theoretical physicists who are easily conversant in advanced mathematical topics in geometry or algebra. Elementary rigorous analysis is certainly not beyond their ability, but many will simply never have spent the time required to master the techniques, and until they do, the 'hard analysis' in KAM proofs could seem all but impenetrable. I think a lot of the reputed difficulty of KAM proofs stems from this

---

[14]In the opening part of his proof of Kolmogorov's theorem [Ar63a], Arnold invites readers to do essentially this, writing "All the basic ideas are set out in §1; it is my hope that the expert reader will be able to construct the proofs from them." A similar invitation is offered in the last section (§5) of the introduction to [Ar63b].

unfortunate fact.

(c) <u>Conclusions</u>

Clearly some people find KAM proofs easy, others find them hard, and a case can be made for either position. In particular, some of the quotes in the introduction which bemoan the difficulty of KAM theory seem entirely justified for those approaching the subject without much background. But it seems to me that any *a posteriori* discussion of the difficulty of KAM proofs misses the main point. As with so many surprising or revolutionary results, the really difficult thing about KAM theory was to see it in the first place, and to see it where Poincaré—among others—did not. And finally, the difficulty of the proofs is certainly no pretext for avoiding an understanding of the main ideas in KAM theory.

5.4   How much celebration is called for?

Near the beginning of this book (cf. §1.2), I pointed out that in mathematics and physics papers of the 1980s and 90s it became so common to see the term 'celebrated KAM theorem' that one might think the adjective 'celebrated' had been permanently attached as part of the theorem's name. In some cases, one could also detect irony, or get the feeling that the author didn't see much cause for celebration. So now that we know something about KAM theory, it seems natural to look more carefully into its celebrity, and the ways it may or may not be deserved (the extent of the 'hype,' to put it simply). By now, the rough outlines of both sides of this question are clear; I will summarize them here, then describe my own reasons for advocating a further celebration of sorts.

5.4.1   *Quick summary of the usual arguments*

Enthusiasts celebrate KAM theory because it answers questions that puzzled the founders and greatest theorists of classical mechanics, which discipline lies at the very heart of the scientific method. In particular, it resolves the paradox described in §3.12.5 by a synthesis of integrability and nonintegrability, as illustrated in the cartoons of §§5.1.1 and 5.1.2. Beyond the mathematics and its context, there's also a strong element of human drama in the way the KAM story unfolded. Poincaré's breakthroughs brought him tantalizingly close to seeing KAM-like results, and it's tempting to think that he would have done so in only slightly different circumstances. As it happened though, he not only failed to do so, but he expressed carefully worded doubts that such results could be true (cf. §3.12.5 and Quotes **Q**[Poi4] and **Q**[Poi5], App. E). This and the ensuing rush to ergodicity added great drama to Kolmogorov's announcement of the persistence of invariant tori at the 1954 ICM in Amsterdam. When all of this is combined with the way KAM theory fits into age-old questions in celestial mechanics and newfangled questions in statistical mechanics, it's easy to see cause for celebration.

On the other side of the question are those who have looked carefully into the physical applications of KAM theory to see if it lives up to the early promises of enthusiasts. It must be admitted that they've been disappointed somewhat. Yes, it's true that, against expectations, KAM theory shows that some $n \geq 3$ body problems have eternally stable solutions; and yes, it shows that generic Hamiltonian systems are not ergodic. These are great conceptual advances. Yet KAM theory does not show that our own physical solar system is stable, nor does it go very far toward resolving the deep-seated questions at the foundations of statistical mechanics (not yet, at least). In fact, if one seeks rigorous KAM results, they often appear to be limited to models of rather simple systems (one might almost say 'toy systems'), and even these are obtained only after strenuous effort. For these reasons (and others detailed above in §5.2), celebrations might understandably be dampened.

In the end, perhaps it's possible to sum up the division between those who advocate celebration and those who don't by saying that the former are those whose tastes run to the conceptual and philosophical; the latter are those more interested in practical applications. (And it may not be wholly coincidental that this is also a standard way of describing the division between 'continental' versus 'Anglo-Saxon' ways of thinking.)

5.4.2 *A plea for KAM theory as a basic part of classical mechanics*

I'd like go a little beyond the usual arguments and argue in favor of yet more celebration, or perhaps something slightly different. I wish that the nearly ubiquitous 'celebrated KAM theorem' could be replaced by 'the widely known (and partially understood) KAM and Nekhoroshev theorems.' My reasons for wishing this are simple. Although there is no lack of specialists in the field (except maybe in the U.S.; cf. §5.3.2), and there are many other physicists and mathematicians familiar with the essentials of KAM theory, below this level there is a big knowledge gap. Classical mechanics is the foundation for much of the physical sciences,[15] and so a great many people need to know something about it at a level appropriate for their work. Graduate students in physics need to know more than just a few examples of integrable systems, yet many textbooks do not go beyond this, even on a descriptive level, as was already mentioned in the introduction.

At minimum, it seems that scientists working in or around applications of classical mechanics would want to have an overview of KAM theory roughly like that laid out in §5.1 above. Those who need quantitative stability results would want to familiarize themselves with specific KAM or Nekhoroshev theorems and typical ways of applying them (as in 'The generic application,' §7.3.1 below), and they would want to be aware that the thresholds of validity in such theorems might be orders of magnitude smaller than the true thresholds for existence of invariant tori (or for long-time stability, in the case of Nekhoroshev's theorem).

Further afield, there are those who think and write about science, as popularizers for a general public, or as historians, or critics, or philosophers of science. Very often, they seem not only to lack a basic understanding of KAM theory and its context, they seem to be unaware of its existence, even when they write about subjects in which it is vitally important.

Two examples will suffice here to make my point. The first is the somewhat odd omission of the KAM story from what was by far the most popular book to appear on chaos theory in the U.S. (*Chaos*, by James Gleick [Glei87]).[16] This omission and lamentations about it were among the original motivations for the present book, though I have by now become more appreciative of Gleick's work.

The second example does not have the same high profile as the first, but to my mind is more serious. It concerns the article 'The Idol of Stability' [Tou99] by the highly distinguished philosopher and historian of science Stephen Toulmin (1922–2009). By brutally oversimplifying, the main thrust of the article may be described as follows. For some time, social scientists,

---

[15]One does not need to romanticize classical mechanics or its history to appreciate its central importance in science.

[16]I am certainly not the first or only one to point to omissions of core mathematical material in [Glei87]; a vigorous discussion of this fact erupted following the book's surprising success (see Part D.3 of Appendix D, below). The closest thing I found to a mention of KAM theory in the main text of [Glei87] is the following comment on p. 182: "A scientist accustomed to classical systems without friction or dissipation would place himself in a lineage descending from Russians like A.N. Kolmogorov and V.I. Arnold."

economists, anthropologists and so on have been criticized for trying to impose on their subjects a framework from the hard sciences that does not really fit. They seem especially drawn to classical mechanics and to Newton's *Principia* as a model of rational, deductive science in its most successful incarnation. Toulmin takes a new approach to this, arguing that "no work has been more deeply misunderstood" than *Principia*. He does this by following the story of the *n* body problem (especially the three body problem) and its companion stability problems through the 18th, 19th and into the early 20th centuries. He describes the historical background, the differing interpretations of Newton and Leibniz, and follows the line of our story above (§§3.7–3.9) closely enough to include Weierstrass, Dirichlet, Hermite, Mittag-Leffler, King Oscar II, and of course Poincaré. Incredibly, however, the story ends there, with the conclusion that, because of the 'chaos' discovered by Poincaré at the heart of the three body problem, the dream of stability (of the *n* body problem or solar system—the two are somewhat conflated) remains only a dream, and the 'idol of stability,' regularity, and rationality to which other scholars looked as a prototype for their own subjects was based on 'the model of a physics that never was.'

Now I do not say that KAM theory—the woefully missing part of the story in [Tou99]—nullifies Toulmin's claims or obviates his argument totally. But it seems nevertheless to form a vital part of the story, and to be so relevant that its omission can only be described as bewildering. The heritage of *Principia* indeed seems deeply misunderstood, even sometimes by those who make it their business to understand.

I do not mean to single out Gleick or Toulmin for chastisement; each has made very significant contributions. But the aforementioned examples are not untypical in the larger literature, and they point to a need for something—if not celebration, then at least a wider dissemination of basic knowledge of the heritage of *Principia*.

Chapter 6

Other Results in Hamiltonian Perturbation Theory (HPT)

The main focus of this book is classical KAM theory. But a perspective view of KAM can only be had by knowing where it sits within Hamiltonian perturbation theory (or HPT). For this reason, I give here a rough overview of HPT.

In the article [Loc99], P. Lochak divides HPT into three basic parts: results pertaining to geometric stability, classical stability, and instability. This has always seemed like an eminently reasonable classification scheme to me, and I will describe it and use it here.

Geometric stability refers to the invariant sets in phase space (such as *invariant tori* or *invariant manifolds* generally) to which orbits are confined for all time, thus providing stability over infinite time intervals. KAM theory belongs to the class of geometric stability results in HPT, but we'll see that KAM tori are not the only such geometric structures in nearly integrable Hamiltonian systems.

Classical stability is the sort of stability arising out of classical Hamiltonian perturbation theory. In other words, when faced with the small divisor problem (cf. §3.11 above), one way to proceed is to 'go non-classical,' ultimately leading to the set $\mathcal{T}$ of KAM tori; this may be interpreted as transformation to integrable form on a Cantor-like set. But up to now we've ignored the other way to proceed. One can instead 'go classical,' requiring transformations to be defined on sets with interior. This leads to stability (nearness of solutions' actions to their initial values) over finite—but very long—time intervals. There's a trade-off with respect to geometric stability: one loses the infinite time intervals, but gains a nice, classical set of initial conditions for which results hold (in fact most versions of Nekhoroshev's theorem—described below—hold for essentially *all* initial conditions). Though perhaps less esthetically satisfying than KAM theory, under the right conditions (long enough time intervals, big enough thresholds, etc.) this is probably the kind of stability result best suited to physical applications.

Instability in HPT is basically everything else. In systems far from integrable, solutions are not subject to either kind of stability mentioned above, and may behave 'chaotically' even over relatively short time intervals. Near integrability, solutions may be subject to classical stability over long time intervals, but display a weak drifting behavior beyond these intervals.

We now turn to a more detailed discussion of these three regimes of HPT.

6.1 Geometric HPT: KAM theory, cantori and Aubry-Mather theory

With its invariant Lagrangian (and other) tori, KAM theory belongs to the class of geometric stability results in HPT, and for that reason I won't revisit it, or dwell long on other kinds of geometric stability here. But everyone interested in HPT should at least be aware of some of the other geometric structures that persist under perturbation in Hamiltonian systems (beginning with the *periodic orbits* of the Poincaré-Birkhoff theorem, and the *stable* and *unstable invariant manifolds* of *hyperbolic periodic orbits*, and of *normally hyperbolic invariant tori*).

Perhaps the most famous structures discovered in the post-KAM era are so-called cantori. A cantorus may be viewed as the remnant of a disintegrated KAM torus that nevertheless retains some of its previous structure as the perturbation is increased beyond the torus's existence threshold. (As the remnant of a previously existing, recently departed torus, a cantorus is also sometimes called a 'ghost torus.') We've already seen that the set $\mathcal{T}$ of invariant tori has a 'radial' Cantor-family structure. By contrast, a single cantorus has an 'azimuthal' or 'circumferential' Cantor-set structure which is illustrated in Fig. 6.1.

Cantori may be seen by numerical computation (see Fig. 6.2 (b), p. 130), but their existence was uncovered mathematically by way of 'Aubry-Mather theory.' This theory asserts that given an area-preserving twist map f of an annulus (or cylinder), for every *rotation number* α in the rotation interval[1]

---

[1] If C', C'' are the inner and outer boundary circles of the annulus that is the domain of f, and if f has rotation numbers α' and α'' on C' and C'' respectively, then the rotation interval of f is the set of all numbers between α' and α''.

Fig. 6.1 Schematic of a cantorus cut by a surface-of-section, showing its azimuthal Cantor-set structure.

of f, there exists an f-invariant 'Mather set' M of rotation number α. This M may be a periodic orbit, or an invariant curve, or—surprisingly— an invariant Cantor set (called an Aubry-Mather Cantor set, or, in the catchier terminology introduced in [MacMP84], a cantorus[2]). The theory further asserts that each cantorus lies inside (i.e., is a subset of) a closed *Lipschitz continuous* curve, and that the ordering of successive iterations on orbits in the cantorus is the same as the ordering of orbits rigidly rotated by α on the circle.

For twist maps, Aubry-Mather theory complements and completes KAM theory in a number of interesting respects. In KAM theory for twist maps, one gets the existence of invariant curves of a certain *smoothness class*; but the required hypotheses are strict: the twist map must be of a certain smoothness class, it must be a perturbation of an integrable twist map, and the rotation number of the invariant curve must satisfy rather stringent number-theoretic (*Diophantine*) conditions. By contrast, the hypotheses of Aubry-Mather theory are much weaker: smoothness of the twist map is not required (it need only be a *homeomorphism*), the map need not be a perturbation of an integrable twist map (though it does need to be 'monotone'; see [Mat82]), and there is no restriction at all on α (apart from lying in the rotation interval of f). There is a sense in which the Aubry-Mather sets are '*weak*' or '*generalized*' solutions of (the map-analog of) the homological equation (3.3), while KAM invariant curves are '*strong solutions*' (cf. [Mos86]). In fact, Aubry-Mather theory clearly shows how KAM invariant curves break up as the perturbation strength of the map

---

[2]R. MacKay and J. Meiss have said that I.C. Percival was the one who devised the term 'cantorus.'

increases beyond the curves' threshold of existence: they disintegrate into Cantor-like sets by losing smoothness.

Many of the basic results of Aubry-Mather theory were set out as theorems and conjectures in J.N. Mather's article [Mat82]. Soon thereafter, S. Aubry and P.V. LeDaeron published a paper [AubrL83] in condensed matter physics where they studied the so-called Frenkel-Kontorova model in a mathematically rigorous way. This model has a number of physical interpretations, but Aubry and LeDaeron were concerned with its use in determining the energy ground states of idealized crystals. This led them to show that the ground states are represented by the periodic and *quasiperiodic* orbits of an area-preserving twist map, and to further show—as Mather did—that these orbits exist for any rotation number. However, their methods of proof were different from Mather's, and the approaches have since been combined into what is now called Aubry-Mather theory.

The proofs rely on 'variational theory' which will not be described in any detail here. However, the reader may already be familiar with an elementary version of variational theory used in classical mechanics. In that case, one writes down the *Lagrangian* of a system, and the *action functional* associated with putative trajectories. Then assuming 'Hamilton's principle of stationary action,' one uses 'calculus of variations' to derive the Euler-Lagrange equations for the action of true trajectories of the system. In most cases, the true trajectories of the system correspond to minimum values of the action. Something similar occurs in Aubry-Mather theory, but rather than minimize an action functional over trajectories, one minimizes a more general 'measure functional' over sets to find sets that are invariant. It seems that Mather got the basic idea of his method from I.C. Percival's numerical method [Perc79] based on variational principles, but related methods may be traced back through G.A. Hedlund, M. Morse, and ultimately to J. Hadamard, all of whom used variational methods to explore the behavior of geodesic flows on surfaces (in all likelihood making Hadamard the first to see Cantor sets in dynamical systems).

A nice review of early Aubry-Mather theory appears in Moser's paper [Mos86]. As is often true in dynamics, the theory is fairly complete in the lowest-dimensional case (here $n = 2$), and has also been partly generalized to higher dimensions, by Mather in [Mat91] (based partly on earlier work on 'minimal measures' [Mat89]), and also notably by C. Golé [Golé92].

More generally, Aubry-Mather theory stimulated a renewed interest in variational methods for Hamiltonian systems. These techniques (often grouped together as 'Mather's methods') have proved important in finding other invariant structures, such as island chains, and especially non-Lagrangian invariant tori (i.e., invariant tori of dimension different from n) in n-degree-of-freedom nearly integrable Hamiltonian systems. Mather's methods have also become essential tools in efforts to understand *Arnold diffusion* (see §6.3.2).

Finally, Aubry-Mather theory is also related to more recent developments in partial differential equations. As mentioned above, an Aubry-Mather set can be viewed as a type of *weak solution* of the homological equation of KAM theory. At about the same time Aubry-Mather theory was developed, investigation into other sorts of weak solutions culminated in the discovery of so-called 'viscosity solutions' to a certain class of Hamilton-Jacobi equations. This field of research and its connections to Aubry-Mather theory are known as *weak KAM theory* (see Part D.7.2 of the reader's guide in Appendix D for further references).

6.2   Classical HPT: Nekhoroshev theory

Many of the invariant structures of geometric HPT give rise to stability on infinite time intervals. The behavior of trajectories that don't have the good fortune to reside on a nice invariant structure may be 'messy.' In the next section, we'll see that chaotic behavior may occur in two cases: (i) when the perturbation is large, and (ii) even when the perturbation is arbitrarily small, provided we wait long enough. But in this section, we'll see that 'classical perturbation theory'—the approach to HPT of the pre-Kolmogorov era—can give stability results for *all* trajectories of most nearly integrable systems, provided the perturbations are small enough, and provided we accept *finite* rather than infinite stability times.

6.2.1   *Nekhoroshev's theorem*

The first comprehensive results in this direction were laid out in the 1970s by Arnold's student N.N. Nekhoroshev [Nek71], [Nek73], [Nek77], [Nek79] and have since evolved into a sprawling collection of theorems and applications loosely called 'Nekhoroshev theory.' This has been variously described as a physicists' KAM theory, as classical HPT pushed to its limits, and as the crowning achievement of classical HPT. Theorems of this sort show *adiabatic invariance* of the *action* variables for analytic nearly integrable systems (3.1) satisfying certain *steepness* or 'convexity' conditions, which replace the nondegeneracy conditions of KAM theory. For the nearly integrable system (3.1), adiabatic invariance means that the actions $I = I(t)$ stay near their initial values $I(0)$ over some (long) time interval. To be more precise, we state a

Prototype Nekhoroshev theorem. *Suppose that the nearly integrable Hamiltonian $H(\theta, I, \varepsilon) = h(I) + \varepsilon f(\theta, I, \varepsilon)$ is analytic and that the unperturbed part $h = h(I)$ is steep (or convex, or quasiconvex) on some nice domain. Then there's a threshold $\varepsilon_0 > 0$, and positive constants R, T, a, and b such that whenever $|\varepsilon| < \varepsilon_0$, for all initial actions $I(0)$ in the domain (and far enough from the boundary) we have*

$$\|I(t) - I(0)\| \leq R\varepsilon^b \qquad for\ times \qquad |t| \leq T\exp(\varepsilon^{-a}). \qquad (6.1)$$

I won't define 'steep' here;[3] I'll just say that it's a generalization of 'convex.' To say that $h = h(I)$ is convex on a domain means that its Hessian matrix $\partial^2 h / \partial I^2$ is (uniformly positive, or uniformly negative) definite on the I-domain; and h is 'quasiconvex' if it is convex on every fixed energy level $h(I) = E$ (over a range of appropriate energies $E_1 < E < E_2$).

Some commentary will help place this theorem in relation to KAM theorems. First, note that the conclusion $\|I(t) - I(0)\| \leq R\varepsilon^b$ is weaker than— but still closely analogous to—the conclusion that trajectories lie on invariant tori, and that it is further weakened by holding for finite times rather than for infinite times. But to compensate for this, there is also an important gain: namely, the conclusion holds for *all initial conditions* in the domain, rather than only for those in the Cantor family $\mathcal{T}$ of KAM theorems. It's this last fact that makes Nekhoroshev's theorem a (hypothetically) better choice in physical applications than the KAM theorem, especially when one realizes that exponential time intervals (of length $T\exp(\varepsilon^{-a})$) may be 'practically as long as infinity' (e.g. as long as the age of the universe[4]), depending on the parameters involved.

---

[3]See the glossary for a definition of *steep*.

[4]Age-of-the-universe times are of course very impressive, but much more modest times are often sufficient in practical applications. For example, in the dynamics of particle accelerators, one is interested in stability times on the order of a dozen hours.

6.2.2 *A brief history of Nekhoroshev theory and its applications*[5]

An obvious question regarding the development of Nekhoroshev theory is, Why didn't it happen earlier, before KAM theory? After all, from the mathematical viewpoint, the basic results are less sophisticated and more classical than KAM, and conceivably could have been proved in the 19th century. This question is hard to disentangle from the question of why KAM itself wasn't developed earlier, and the answers are no doubt as idiosyncratic as any of the reasons why theories develop historically rather than logically.

In any case, even if detailed proofs for Nekhoroshev theory had to wait until nearly 1980, the general proof-strategy and the results themselves were suspected almost as soon as KAM was announced. First Moser [Mos55] and then J.E. Littlewood [Lit59a], [Lit59b] conjectured that generically in analytic nearly integrable Hamiltonian systems, the timescale required to observe an $O(1)$-displacement in the action variables was beyond all orders in inverse powers of the small parameter; Arnold later conjectured a similar result [Ar68]. Nekhoroshev's theorem affirms these conjectures, and shows further that the timescale required for $O(1)$-displacement is exponentially long in an inverse power of the small parameter.

One reason for the delays between conjecture, announcement, and proof of Nekhoroshev's theorem lies in the details of the proof. In its original form, Nekhoroshev's proof was closely related to Arnold's proof of KAM, but was even messier and more complicated, and the threshold of validity $\varepsilon_0 > 0$ was even smaller (so once one looked into the details, it was in no way a theorem that invited physical applications). The proof in its entirety took many years (1971–1979) and many densely filled pages of estimates, and can only be described as a herculean *tour de force*. Apart from the inherent difficulty of being first, the proof is complex mainly because Nekhoroshev insisted on finding conditions on $h = h(I)$ (the *steepness* conditions) weak enough to ensure that the results would hold for generic Hamiltonian systems. This was an important mathematical achievement,[6] especially within the Russian school where the concept of genericity was first developed and used extensively.

Despite the prohibitive complications, by the mid 1980s a group of Italian researchers loosely centered in Milan and Padua (call them the 'Milano

---

[5]Portions of this (and the next) subsection have been adapted and expanded from §2.2 of my article [Duma93].

[6]J. Moser once told me that getting the theorem to hold for generic systems was Nekhoroshev's greatest achievement.

group') noticed that Nekhoroshev's proof could be significantly simplified by restricting attention from the broad class of 'steep' h to the smaller class of 'convex' or 'quasiconvex' h [BenGG85b].[7] The beauty of this observation is that, even if the mathematically interesting property of genericity is lost, what remains includes many examples that interest physicists. (For example, a number of physical systems—including basic models in statistical mechanics—have Hamiltonians with kinetic energy terms corresponding to $h(I) = \frac{1}{2}(I_1^2 + \ldots + I_n^2)$. Such h have spherical energy surfaces, and are the 'most convex.' On the other hand, many problems in celestial mechanics have Hamiltonians with h that are steep but not convex or quasiconvex.)

Within a few years, the Milano group was using their simplified methods to address delicate stability questions in physics that had not previously received rigorous treatment. This 'first wave' of applications included estimates of the size of the region surrounding the Lagrange points of the Sun-Jupiter system which enjoy stability times on the order of the age of the universe [GioDFGS89], [CelG91], investigations into the problem of holonomic constraint in mechanical systems [BenGG87–89], and a vindication and extension of some of the previously forgotten ideas of L. Boltzmann and J.H. Jeans concerning very slow relaxation times for high frequency degrees of freedom in statistical systems [BenGG87] (as related, for example, to the ultraviolet catastrophe in classical physics and the problems concerning equipartition of energy and the behavior of classical polyatomic gases with various internal degrees of freedom; see §7.2 below for more detail). These latter ideas also lead toward an understanding of the Fermi-Pasta-Ulam experiments mentioned above in §3.12.4 and described in more detail below in §7.2.3.

Also in the 1980s, techniques developed by the Milano group were used to prove specialized Nekhoroshev-like theorems for trajectories near resonance (rather than for all trajectories). In Padua and Rome, G. Benettin and G. Gallavotti developed estimates for systems of weakly perturbed harmonic oscillators [BenG86], showing that trajectories remain trapped near resonance for exponentially long times. I used related methods in my own work [Duma93] to show that certain motions of charged particles in crystals (called 'channeling motions') remain stable for times exponentially long in the particles' energy. This 'channeling theory' also depends in an essential way on stability near resonances. Together, these results prefigure the point stressed later by P. Lochak that stability can be stronger near resonances

---

[7]Though not strongly emphasized, this observation was already present in Nekhoroshev's work.

than elsewhere.

As in KAM theory earlier, the first wave of applications of Nekhoroshev theory pointed to the need for optimal results. If the theory was to be used to answer subtle questions in statistical physics (i.e., for systems with many degrees of freedom n), the estimates needed to be sharp, especially in terms of their dependence on n. Lochak emphasized this point and, partly based on his reading of Chirikov's heuristic but influential paper on instability [Chir79], also conjectured that in the quasiconvex case, the optimal exponents were $a = b = O(1/2n)$ [Loc90] (the first exponent a controlling the timescale is the more important one). More specialized estimates also show that, at a resonance of *multiplicity* m, the exponent $a = O(1/2(n - m))$.

Lochak's exponents were soon found to hold (but not yet to be optimal) at about the same time by J. Pöschel [Pös93], and by Lochak and A.I. Neishtadt [Loc92], [LocN92]. These papers were not only important for the better exponents they produced, but perhaps even more so for the new proofs they employed. Pöschel significantly improved the 'geometric' part of Nekhoroshev's original proof, while Lochak introduced a radically simpler proof method that avoided small divisors altogether (see §6.2.3 below).

Lochak's efforts in particular inspired a number of later developments. In France in the mid 1990s, L. Niederman began to use Lochak's methods in a series of papers improving and applying Nekhoroshev theory. The paper [Nie96] examines the stability of the semi-major axes in planetary problems, while [Nie04] combines both Nekhoroshev's and Lochak's methods to improve the stability exponents a and b in the steep case (unfortunately, an error in the proof-sketch of Lemma 4.4 of this latter paper has been discovered, and a corrigendum is in preparation). In [Nie06], [Nie07] Niederman extends Nekhoroshev-like stability results to certain non-steep cases, and also gives a more intrinsic, geometric characterization of steepness.

In the early 2000s, J.-P Marco and D. Sauzin achieved the remarkable and unexpected result of extending Nekhoroshev estimates to *Gevrey-smooth* Hamiltonian systems [MarS03] (it was previously thought that analyticity was needed in order to get exponentially long stability times). They also produced examples of Arnold diffusion (see below §6.3.2) showing that their estimates are optimal.

In the meantime, the Milano group assimilated many of the new methods and trained a number of young researchers who moved to other parts of Italy, and in some cases to Spain or France, so that by now it might be better to call it the 'Mediterranean group,' especially considering other researchers in Barcelona who began work more or less independently. The production of this large group—and other loose affiliates around the world—is vast and varied, and I will only list a few representative topics here.

In 1994 in Barcelona, A. Delshams and P. Gutiérrez showed just how close the relation between KAM and Nekhoroshev theory is by producing analytic lemmas that can be used interchangeably to prove either theorem [DelsG94], [DelsG96]. At about the same time, in a piece of work rare for its origin on U.S. soil [PerrW94], A.D. Perry and S. Wiggins of Caltech showed that KAM tori are 'very sticky' by getting exponentially long estimates of the time needed for trajectories to move from initial distance r to distance $2r$ from an invariant torus (given that $r > 0$ is small enough to start with). Soon after, in Milan, A. Morbidelli and his advisor A. Giorgilli got related results showing that there is a kind of natural interpolation between the KAM and Nekhoroshev theorems: on a family of open domains shrinking down onto the set $\mathcal{T}$ of invariant KAM tori, they obtained longer and longer stability times that were in fact 'superexponential' in the small parameter ε [MorG95a], [MorG95b]. A snapshot of much of the work from this period (and its applications) is nicely summarized in Giorgilli's proceedings [Gio98] of the talk he gave at the 1998 ICM in Berlin.

This quick overview of work in Nekhoroshev theory could be continued and greatly enlarged, but I'll end here by simply listing a few other directions of research and representative articles where many further references may be located. First, Nekhoroshev-like estimates for *symplectic* or volume-preserving maps were spurred by the need for a theory to support the numerical technique of symplectic integration, and by the need for estimates of the stability times of charged particle beams orbiting in accelerator storage rings. These were first supplied by A. Bazzani, S. Marmi, and G. Turchetti [BazMT89] by including *symplectomorphisms* into Hamiltonian flows; more direct methods (with proof methods adapted to maps) were more recently given by M. Guzzo [Guz04], who has also worked on other applications of Nekhoroshev theory, such as celestial mechanics (asteroid stability) and rigid body rotation (tops). Also related to rigid body rotation are the stability properties of Riemann ellipsoids (self-gravitating fluids) investigated by F. Fassò and D. Lewis [FaL01]. Finally, in his PhD thesis at Caltech, A.D. Blaom (*né* A.D. Perry) took up one of Nekhoroshev's original suggestions by extending the latter's theory to perturbations of convex, noncommutatively integrable systems. He also generalized and geometrized the theory to apply to the case of perturbations of systems with nonabelian symmetries, interpreting the resulting exponential estimates in terms of momentum maps and co-adjoint orbits, and illustrating them by applications to Euler-Poinsot rigid body motion [Bla01].

6.2.3 *Remarks on the proofs in Nekhoroshev theory*

The simplest way to describe Nekhoroshev's original proof strategy is to say that it employs something like the transformations used in Arnold's proof of KAM (to bring the original Hamiltonian into a *normal form* like (4.7)), but it stops short of completely eliminating the remainder in the normal form. Since it covers an *open set* of initial conditions, in addition to the 'nonresonant' normal form of KAM theory, other normal forms corresponding to the 'resonant' parts of phase space must also be used. Unlike the nonresonant normal form, resonant normal forms still have some angular dependence (even after excluding the remainder).

So why does the strategy stop short of eliminating the remainders? Because, as successive terms are eliminated and the remainders become smaller, the domains (on which the transformations to these normal forms are defined) also become smaller. (For example, in the limit, the domain for the nonresonant normal form shrinks down to the set $\mathcal{T}$ of KAM tori.) Nekhoroshev's strategy stops the successive transformations to normal form just as the domains threaten to come apart; in other words, it stops at the last place where the domains still cover all of phase space. In this way, one obtains the 'best possible' collection of normal forms covering all of phase space, and these have remainders that are exponentially small in the perturbation parameter ε. Getting these normal forms is called the 'analytic' part of the proof; one then proceeds to the 'geometric part.'

It's in the geometric part of the proof that the steepness (or convexity) hypothesis comes into play. So long as a trajectory remains inside the domain of a normal form, it is constrained to remain exponentially close to a 'plane of fast drift'—the lower dimensional plane in action space that corresponds to the combination of angle variables that remain in the normal form. (For the nonresonant normal form, this plane degenerates to a point.) Nekhoroshev's steepness hypothesis assures that the contact between resonant surfaces and the planes of fast drift is (at least weakly) *transverse*; this gives bounds on the size of the intersections between the resonant surfaces and planes of fast drift. In this way steepness controls the passage of trajectories through resonance; without it, trajectories could quickly suffer large deviations in action as they moved unimpeded along resonances. On the other hand, the complicated passage through resonance phenomena are largely eliminated by strengthening the steepness hypothesis to convexity (convex functions are the 'steepest'; the corresponding energy surfaces contact the planes of fast drift in a single point). In that case it can be shown that trajectories remain trapped, to a certain level of approximation, in a single resonance for exponentially long times.

In P. Lochak's approach to the proof, the whole apparatus of resonances, small divisors, and Diophantine conditions—the usual approach to KAM—is discarded, and is replaced by 'simultaneous approximation' methods. The analytic part of Nekhoroshev's proof, with its hierarchy of normal forms governing all n orders of resonance, is replaced by a collection of only one kind of normal form: those based on single-phase averaging to high order around each of the periodic orbits (with sufficiently short period) of the unperturbed system. Using Dirichlet's basic theorem on simultaneous approximation, one can find enough short-period periodic orbits to patch together normal forms with domains covering all of phase space. The usual global estimates of Nekhoroshev theory then follow, and one can also look for more specialized local estimates near various resonances.

Lochak's method vastly simplified the proof of Nekhoroshev's theorem, and has led to a great deal of further progress, especially by Niederman and Blaom, as noted earlier. Lochak also hoped that simultaneous approximation methods could be used to prove KAM theorems, and was encouraged in this direction by H. Rüssmann. In the 1990s I was employed for a short time on this project, but our efforts at that time were not successful. Quite recently, however, A. Bounemoura and S. Fischler announced and proved results [BounF13] showing that KAM theorems are indeed amenable to simultaneous approximation methods.

6.3   Instability in HPT: Chirikov diffusion, Arnold diffusion and other mechanisms

We know that the discovery of persistent invariant tori served to synthesize the seemingly contradictory coexistence of stable and chaotic motions in nearly integrable Hamiltonian systems. But up to now, we've only discussed the unstable motions as they were understood in the pre-KAM era (basically Poincaré's wild 'trellis,' and subsequent over-enthusiastic conjectures of ergodicity). In this section we'll look very quickly at some of the ways that instability in HPT is understood in the post-KAM era.

Broadly speaking, instability in nearly integrable Hamiltonian systems has now been separated into two parts: the far-from-integrable—or Chirikov—regime, and the near-integrable—or Nekhoroshev—regime. Although Chirikov's and Nekhoroshev's main results go back to the 1970s, this nomenclature using their names is relatively recent, appearing in a series of papers starting around 2000 (see for example [GuzLF02]).

The Chirikov regime comprises systems whose perturbations are bigger than the Nekhoroshev threshold,[8] so that KAM tori are sparse or nonexistent and chaotic behavior of various kinds predominates in phase space. Here the basic mechanism for large-scale instability is 'resonance overlap,' first posited by B.V. Chirikov as early as 1959, then more fully described by him and others at the end of the 1970s and beginning of the 80s. Many researchers call this sort of instability 'Chirikov diffusion.'

In the Nekhoroshev regime, systems are close to integrable, KAM tori fill a large part of phase space, and any large-scale instability must manifest itself over very long times, beyond the exponentially long timescales on which Nekhoroshev theory guarantees stability of the actions. At present, variations of only one mechanism of instability (Arnold's mechanism) are thought to exist for systems arbitrarily close to integrable. Instability in the Nekhoroshev regime often goes loosely (many would say misleadingly) under the name of 'Arnold diffusion.'

Starting here I give an overview of these topics, concentrating on Arnold diffusion. In this latter case, not only do I compress decades of work into a few pages, but I also try to convey something about the controversies that erupted—without naming names.

6.3.1 *The Chirikov regime and Chirikov standard map*

It's tempting to think that for Hamiltonian systems far from integrable, where neither KAM nor Nekhoroshev theory applies, behavior might be uniformly chaotic, perhaps ergodic as conjectured earlier by Birkhoff, Fermi, and followers of Boltzmann (indeed, in certain cases it may be possible to find *ergodic components* of substantial extent). But modern numerical experiments generally reveal a bewildering and complex mixture of order and chaos sometimes called a 'mixed system.'

---

[8]Again, the thresholds ε_0 for both Nekhoroshev and KAM theory are notoriously difficult to compute precisely. They are usually estimated numerically, or computed (i.e., badly underestimated) rigorously; only with special methods applied to particular systems do the numerical and rigorous values agree closely.

The system in which all of this has been most extensively studied is the so-called *Chirikov standard map*.[9] This discrete dynamical system is a map from the cylinder[10] $\mathbb{T} \times \mathbb{R}$ to itself; in cylindrical coordinates (x, y), one version of it reads

$$(x, y) \mapsto \left(x + y, \; y + \frac{k}{2\pi} \sin(2\pi x)\right) \tag{6.2}$$

where x is computed $mod\,1$ and $k \in \mathbb{R}$ is a parameter. Of the many physical systems modeled by Eq. (6.2), the most fundamental is probably the 'kicked rotator,' in which x represents the (scaled) rotation angle of a physical body, y its (scaled) angular momentum, and k the strength of an impulsive 'nonlinear kick' given to the body with each iterate of the map.

Mathematically, the standard map is interesting because it is one of the simplest systems to display full KAM-like behavior. In the unperturbed case ($k = 0$) it is integrable: each circle $C_\omega := \{(x, y) \,|\, y = \omega\}$ is invariant, with points x_0 that are rigidly rotated to $x_n = x_0 + n\omega$ after n iterates. For small nonzero values of k, a version of the KAM theorem (Moser's twist theorem, in essence) applies to ensure that invariant circles with highly nonresonant *rotation number* ω persist (they're deformed, but do not break). Those circles with rational (or insufficiently irrational) rotation numbers break up and organize themselves into the usual complex of *hyperbolic* and *elliptic periodic points*, with *stable* and *unstable invariant manifolds* emanating from the *hyperbolic points* to form 'resonance islands' around the *elliptic points*, and heteroclinic tangles around the hyperbolic points (see Figs. 3.2b above, and 6.2 below).

An essential question concerning the standard map is, How large is k when the 'last invariant circle' is destroyed? Once this threshold or critical value $k_c > 0$ is known, it's natural to ask how 'chaotic' the map is for $|k| > k_c$, and how the chaos is manifested. Can the momentum values grow unboundedly, and if so, how and at what rate?

The story of how these and other questions were first answered for the standard map begins with the pioneering work of B.V. Chirikov, who began using a form of his 'resonance overlap criterion' as early as 1959

---

[9]This system appears to have been first introduced in 1944 (in the context of particle accelerator dynamics) by V.I. Veksler [Veks44] and also later by J.B. Taylor in unpublished work. It was then rediscovered by B.V. Chirikov [Chir79], and also studied by J.M. Greene [Gre79]. Chirikov called it the 'standard map'; others usually call it the Chirikov standard map, but also sometimes the Taylor-Chirikov standard map, or even occasionally the Taylor-Chirikov-Greene standard map.

[10]In fact, it can also be seen as an annular twist map in polar coordinates, but it's more natural to view it on the cylinder (or on $\mathbb{T}^2$, by also taking $y \, mod\,1$), since it's then 'exact-symplectic,' or area-preserving.

[Chir59], and continued to develop and refine it in the following decades (most famously in [Chir79]). Working with perturbations of area-preserving twist maps, Chirikov saw that an invariant circle is destroyed when it is 'squeezed' between the 'chains of resonance islands' formed by the stable and unstable manifolds of resonant orbits inside and outside of the circle. He was able to measure how the 'links' of a chain grow with the perturbation parameter, and in turn estimate the critical size of the parameter for which inner links overlap outer links.

The accuracy of the overlap criterion was sharpened substantially with the 'residue method' introduced by J.M. Greene [Gre79]. Working with the standard map, Greene analyzed orbit dynamics using a combination of continued fractions and numerical evidence; the result was a simple criterion for the break-up of invariant circles that had some affinity with Chirikov's, but was more mathematical and more accurate, giving an estimate for k_c of about 1.

Taking inspiration from both Chirikov and Greene, D.F. Escande and F. Doveil soon developed group renormalization methods to describe the standard map at Greene's critical parameter value [EscD81]. Meanwhile, in 1984 J.N. Mather gave what is probably the first rigorous result, using his own variational methods to show the non-existence of invariant circles in the standard map for $|k| > 4/3$ [Mat84]. At about the same time, the renormalization techniques were extended and improved by Doveil, Escande, Greene, and other authors, so that by 1985, R.S. MacKay and I.C. Percival were able to produce a computer-assisted proof [MacP85] showing rigorously that no primary invariant curves exist in the standard map for $|k| > 63/64$.

Turning now to positive results, in 1988 A. Celletti and L. Chierchia used computer assistance to rigorously construct analytic curves in the standard map for values of $|k|$ up to 0.65 [CelC88]. Not long after, R. de la Llave and D. Rana were able to show rigorously [delaLR90] (again with computer assistance) that at least one primary invariant circle exists in the standard map for $k = 0.91$. As they point out, this is more than 90% of the 'converse KAM' result of MacKay and Percival. Finally, MacKay revisited results from his 1982 thesis under Greene and used renormalization techniques [Mac93] to compute numerically $k_c \approx 0.971635\ldots$; this is now accepted with confidence as the value of k at which the most noble nontrivial invariant circle (with rotation number $\omega = (1 + \sqrt{5})/2$) is 'critical.' This threshold k_c is of course very large compared with the 'submicroscopic' values of ε_0 in early KAM theorems.

Once the question about realistic values of k_c is answered, other questions arise. When $k > k_c$ so that the last primary circle is gone, orbits can potentially migrate across arbitrary vertical distances: given any two fixed y values, we can imagine an orbit starting at one value and eventually reaching the other, since no invariant curve blocks the way. Outside the bounds of 'secondary circles,' the entire cylinder becomes a 'zone of instability.'

But even after the last primary invariant curves are destroyed, the cantori that remain in their place present a partial barrier to unimpeded y-motion, since trajectories must pass through the gaps in a given cantorus. It turns out that the 'flux' of trajectories passing through each gap is regulated by a mechanism called a 'turnstile' [MacMP84], so-named because of the 'lobes' that appear to rotate with successive iterations of the map.

Just looking at the pictures in Fig. 6.2 gives an idea of the formidable problems faced by those who would describe motion in far-from-integrable systems. Roughly speaking, in order to get rigorous mathematical results for these systems, one has to resort to 'coarse-grained' and/or statistical methods; in other words, one subdivides phase space into appropriate components and shows that trajectories or sets of trajectories enter or leave at certain rates, on average (thus defining an average flux in or out of a component); or one shows that trajectories remain inside a given component on average (so that the component is a minimal invariant set, to a certain level of approximation).

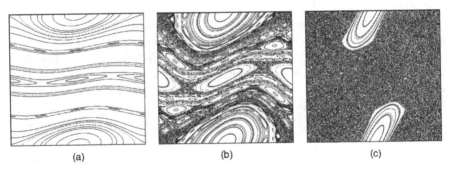

(a)   (b)   (c)

Fig. 6.2   Numerical portraits of the standard map's phase space for the parameter values (a) $k = 0.3$, (b) $k = 0.971635 \approx k_c$, and (c) $k = 3$. Here the x-coordinate runs horizontally, wrapping once around the cylinder, while the y-coordinate runs vertically over one period. In (a) we see an abundance of primary invariant circles (wrapping around the cylinder). At the critical value, in (b), the last primary invariant circle has just become a cantorus. In (c), no trace of primary circles remains; the system is predominantly chaotic, with 'islands of regularity.' These portraits show the behavior typical of mixed systems. [Figures courtesy of B. Vaughan]

The highly irregular geometry of the components makes this task very challenging. Components or their boundaries are often fractal-like structures, such as islands surrounded by chains of smaller islands and so on (much of which was first studied by Birkhoff; cf. §3.12.2). An idea of the difficulties faced in this business may be had just by listing the names given to some of the components or structures. In addition to turnstiles and their lobes, we have 'channels,' 'chimneys,' 'loops,' 'tangles,' and so on (I have also heard 'pinball machines' and 'bumpers' used informally).

I won't venture to describe these things further here; rather, I direct the reader to J. Meiss's clearly written review article [Mei92] (still one of the best introductions), and also to the book by S. Wiggins [Wig92], which is replete with illustrations and examples.

6.3.2   *The Nekhoroshev regime and Arnold diffusion*

The main message of KAM and Nekhoroshev theory is that, under relatively mild conditions, integrable-like stability predominates in a nearly integrable Hamiltonian system $H = h(I) + \varepsilon f(\theta, I, \varepsilon)$ when ε is small (i.e. when the system is close to integrable). Small ε guarantees an abundance of KAM tori (on which motions are perpetually stable), and motions not restricted to KAM tori are controlled by Nekhoroshev estimates of the form (6.1), so they are also stable over very long times. We call the range of small $\varepsilon > 0$ for which this sort of stability holds the Nekhoroshev regime.

Yet even in the Nekhoroshev regime, for systems with more than two degrees of freedom there is the possibility of a slow 'drift' over very long timescales. It turns out that, not only do KAM and Nekhoroshev theory leave open the possibility of instability, but such instability actually occurs. A mechanism by which this happens—variants of which remain the only ones known—was proposed by Arnold in a now-famous paper published in 1964, and so both the mechanism and the resulting instability are loosely termed 'Arnold diffusion.'[11] This long subsection describes Arnold's mechanism and a small part of the large body of research that grew out of it.

(a) Arnold's original 1964 paper

Arnold's original paper [Ar64] is unusual in several respects. First of all, it's very short (4 pages in the original Russian, scarcely more in translation). Second, it employs a slightly special vocabulary in which, for example, the

---

[11] The terminology 'Arnold diffusion' was introduced by B.V. Chirikov in [Chir69].

usual terms 'stable and unstable manifold' are replaced by 'arriving and departing whisker.'[12] Third, there is a slightly humorous footnote pertaining to certain Melnikov integrals[13] (cf. p. 584 of the English translation) which says that "The analogous integrals in [V.K. Melnikov's article [Meln63]] are wrongly calculated." The humor lies in the fact that one of Arnold's own Melnikov integrals is itself incorrect (there is a missing term, but it doesn't affect the results in any essential way).

Finally, apart from these quirky trivialities of presentation, the article has certain mathematical peculiarities that are not immediately apparent, but which have given many headaches to researchers who took up the job of teasing out the details of Arnold's 'simple' model over the last half-century. At the end of his review of the paper in *Mathematical Reviews* (MR0163026 (29 #329)), J. Moser sums it up nicely in an understated way: "The details of the proof must be formidable, although the idea of the proof is clearly outlined." In retrospect, these words have a kind of prophetic ring, as we'll see.

(b) The Arnold instability mechanism

Arnold's model of instability in [Ar64] is the following Hamiltonian system:

$$H(\phi_1, \phi_2, I_1, I_2, t) = \tfrac{1}{2}(I_1^2 + I_2^2) + \varepsilon(\cos\phi_1 - 1)\big[1 + \mu(\sin\phi_2 + \cos t)\big]. \quad (6.3)$$

First note that this system is non-*autonomous* (i.e., has explicit time-dependence), and so is often said to have 'two and a half' degrees of freedom. (It may also be written as a three-degree-of-freedom *autonomous* system, but the important point is that it has more than two degrees of freedom.) Second, note that the system has *two* (small) parameters ε and μ, a point whose ramifications were not fully appreciated at the time of its first appearance.[14]

---

[12] Arnold's use of the word 'whisker' may have something to do with his using the image of a cat's face to illustrate the mixing and recurrence properties of a simple hyperbolic map of the torus, now called 'Arnold's cat map'; cf. p. 6 of [ArA68]. (Some say Arnold got the idea of the cat from the first letters of 'continuous automorphism [of the] torus.')

[13] The Melnikov integral (or function) is the key tool in the so-called Poincaré-Melnikov-Arnold method for detecting the transverse intersection of invariant manifolds. The Melnikov function gives a 'signed distance' between invariant manifolds, so its simple zeros indicate transverse intersection. Poincaré first used a version of the method in [Poi90], Melnikov refined it in [Meln63], and Arnold used it again in [Ar64]. Since then, the method has been further refined and extended to many other contexts; see §4.5 of [GucH83–02] for a concise description and references.

[14] Arnold's use of the two parameters ε and μ reveals his close reading of Poincaré (whom Arnold cites—together with Melnikov—in a terse footnote of [Ar64]). Poincaré

In fact, it was not until three decades after Arnold's paper that L. Chierchia and G. Gallavotti [ChiG94] clearly identified the special class of systems to which Arnold's example belongs. They called such systems 'a priori
unstable,' because *hyperbolicity* is already present before nonintegrability
appears. As we'll see very shortly, this is because the parameters ε and
μ separately control hyperbolicity and nonintegrability. As emphasized by
P. Lochak [Loc99], this separation of the two phenomena greatly simplifies certain technical problems connected with proofs of instability, but in
the end, one wants to go further and eventually understand more typical
systems where the two phenomena occur together.

So even though we already know that Arnold's example is not as representative of typical systems as we might like, how is instability manifested
in it? Probably the best way to get an idea quickly is to look at a diagram. Any such figure must be very schematic, since three dimensions
simply aren't enough to show everything that's going on. Figure 6.3 shows
a diagram[15] of the phase space of system (6.3) in three cases of increasing
complexity.

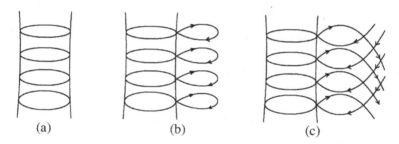

(a)   (b)   (c)

Fig. 6.3   Highly schematic diagram of the invariant manifolds in Arnold's model (6.3)
in three cases: **(a)** $\varepsilon = \mu = 0$, **(b)** $\varepsilon > 0$, $\mu = 0$, and **(c)** $\varepsilon > 0$, $\mu > 0$. The last
case (c) shows the fully developed 'transition chain' of Arnold's mechanism, in which,
using Arnold's language, the 'departing whisker' of one 'whiskered torus' intersects the
'arriving whisker' of another whiskered torus.

The unperturbed system $H = \frac{1}{2}(I_1^2 + I_2^2)$ is very simple, with a phase
space[16] entirely foliated by two-dimensional invariant tori (see Fig. 6.3a).

introduces the same parameters in the *Méthodes nouvelles* [Poi92–99], first in Section 225
(toward the end of Vol. 2) then again in Section 401 (end of Vol. 3) in his study of 'doubly
asymptotic solutions.'

[15]I first saw a picture like this in the article [HolM82]; richly detailed illustrations of
related phenomena may be found in the books by S. Wiggins [Wig92], [Wig90–03].

[16]As an autonomous system in two degrees of freedom, the unperturbed system would be

We then use ε and μ to perturb the system in two stages. In Fig. 6.3b, we see that taking $\varepsilon > 0$ 'switches on' the hyperbolicity of the system; in other words, the tori develop (coincident) stable and unstable invariant manifolds, or homoclinic loop-bands ('arriving and departing whiskers' in Arnold's terminology), indicated in the figure by directed loops attached to the circles. Finally, we see in Fig. 6.3c that taking both $\varepsilon > 0$ and $\mu > 0$ (with[17] $\mu = O(e^{-1/\varepsilon})$) causes the loop-bands to break into separate stable and unstable invariant manifolds which intersect each other transversely (therefore breaking the integrability of the system; cf. §3.11). The truly interesting point, however, is that not only do the stable and unstable manifolds from the same torus intersect each other, they also intersect invariant manifolds attached to nearby tori. This is the crucial feature of Arnold's model which permits him to establish the existence of a 'transition chain': a set of invariant tori spread over a large expanse of phase space and linked by transverse intersections of their respective unstable and stable invariant manifolds. Very near the chain are guiding channels through which unstable orbits of the system may travel large distances along resonances and through the thicket of invariant tori. Note that these unstable orbits do not move in (or 'on') the invariant manifolds (precisely because they are invariant), but instead lie very close to—or are 'guided' by—them. Although it abuses the standard terminology slightly, one often says that the unstable orbits *shadow* orbits in the invariant manifolds (but of course, the unstable orbits ultimately stop shadowing orbits in one stable manifold and begin shadowing those in another as they move from 'link to link' along the transition chain).

To be more explicit about the flow along the transition chain (cf. Fig. 6.4), consider the system (6.3) with $\varepsilon > 0$ and $\mu > 0$ of the right magnitudes so that a transition chain is established. We first imagine an orbit starting very close to an invariant torus of the chain. At first swept along by the flow on the torus, our orbit winds around the torus for a time until it is 'picked up' by the flow of the unstable invariant manifold emanating from the torus. The orbit then follows (or shadows) an orbit in this unstable manifold until it reaches the vicinity of a point of intersection of this manifold with the stable manifold of a second invariant torus. The

---

completely integrable, but to accommodate the time-dependent perturbation, it should reside in an 'extended' phase space, with an additional time dimension.

[17]Arnold sets $\mu = O(e^{-1/\varepsilon})$ here so that he can (easily) use the Poincaré-Melnikov-Arnold method for showing that the homoclinic loop-bands split and intersect transversely when $\mu > 0$. Use of this method is much more problematic in typical cases with only one small parameter.

orbit then switches (or transfers) to the stable manifold, following one of its orbits until it is drawn almost to the surface of the second torus. It then begins to wind around the second torus, and the process just described—the flow along one link of the chain—is repeated. Moving from link to link along the chain, our orbit may travel long distances from its initial location.

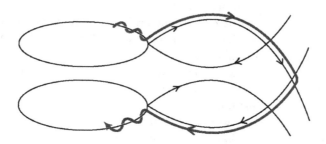

Fig. 6.4   Highly schematic diagram of a transition chain in Arnold's model of instability, with an unstable orbit traversing one 'link' in the chain.

The process of switching or transferring from one invariant structure to the next may seem mysterious, and indeed, the precise way it occurs is highly sensitive to the orbit's initial conditions. Yet once the transition chain's existence is established, we know that such orbits must exist by simple continuity of the flow. That the switching (precisely where and when it occurs) is so sensitive to initial conditions makes it seem like a random process, and soon after Arnold proposed this model, the instability in it was conjectured to be a kind of diffusion. The terminology 'Arnold diffusion' quickly became the standard name for the instability. This is unfortunate, since later efforts to show rigorously that it is diffusion-like have not been very successful.

(c) Significance of the instability mechanism, and Arnold's conjecture

The instability mechanism sketched out by Arnold in 1964 did not immediately cause a stir. Yet with time, researchers began to appreciate its significance. Although it relies on the cleverly constructed and very special system (6.3), many hoped and foresaw that it was a special case of instability to be found commonly in more general systems. Most researchers assumed that (i) a similar sort of instability would be shown to occur generically in nearly integrable Hamiltonian systems, (ii) it would be shown to be diffusive in some rigorous sense, (iii) it would be detected in numerical experiments, (iv) its maximum and average 'speed' would be understood theoretically, and (v) the theoretical and numerically measured speeds would eventually agree within reasonable limits.

Perhaps the most interesting feature of this generalized instability of which Arnold's mechanism was thought to be the harbinger—the feature that distinguished it from other instability processes—was that it should persist in systems arbitrarily close to integrable. Combining this with other insights, Arnold foresaw that, no matter how small the $\varepsilon > 0$, in a generic nearly integrable Hamiltonian system $H = h(I) + \varepsilon f(\theta, I, \varepsilon)$ with more than two degrees of freedom, there would always be initial conditions leading to a significant ($O(1)$ as $\varepsilon \to 0$) drift in the action variables. He also foresaw a still stronger form of instability in which neighborhoods of any two points on a given energy surface would be connected by a drifting orbit. Here is Arnold's precise formulation of what has come to be known as Arnold's conjecture[18] restated for the system above:

Arnold's Conjecture. (cf. [Ar94], Section 1.8, p. 17) For any two points I', I'' on the connected level hypersurface of h in action space, there exist orbits connecting an arbitrarily small neighborhood of the torus $I = I'$ with an arbitrarily small neighborhood of the torus $I = I''$, provided that $\varepsilon \neq 0$ is sufficiently small and f is generic.

Similar statements (but usually less precise) may be found in other parts of Arnold's work (e.g., [ArKN06], [Ar68]); the earliest conjecture-like statement I know of is in §2.1 of the Appendix to [Ar63b], which appeared a year before the 1964 paper.

Of course, the time needed to realize the drift in Arnold's conjecture must stretch rapidly to infinity as ε approaches zero, as required by Nekhoroshev's theorem. In fact, as pointed out by P. Lochak, the relation between this generalized instability and Nekhoroshev's theorem could signify the 'completion' of HPT. That is, if the drift could be shown to be as rapid as permitted in the Nekhoroshev regime (cf. §6.2.2), it would prove that the estimates in Nekhoroshev's theorem were optimal. Or as Lochak put it, the basic elements of HPT would then 'touch one another,' with no further gaps to be filled (at least in terms of the average speed of instability). Since this idea is related to Chirikov's work [Chir79] and was made

---

[18] Arnold was quite fond of conjectures and open problems, and a number of them bear his name in the various branches of mathematics where he worked.

explicit by Lochak [Loc90], we can loosely refer to it as the 'Chirikov-Lochak conjecture.'

Although the Chirikov-Lochak conjecture has (essentially) been vindicated, in other respects, research into Arnold's mechanism and Arnold's conjecture has proved quite daunting, engaging scores of investigators over the last half-century. In describing this very messy business since Arnold first opened the door to it, I'm going proceed as follows. First, to lend some coherence to the story, I'll point out some of the major obstacles encountered along the way. Then I'll briefly list some of the leading researchers and their work, in two stages, or 'waves.' These waves are mostly my own invention, as they were separated by a bit of drama that I watched from the sidelines during the last few years of the 20th century. I'll describe this drama briefly below, because it says something about the nature of research.

(d) <u>Overcoming obstacles to understanding instability</u>

Both in checking the details of Arnold's mechanism and in trying to prove his conjecture, researchers ran up against a host of problems, some of which I'll now describe. This description may aid the reader somewhat in the survey of results that follows.

One of the sketchiest parts of the argument outlined in [Ar64] concerns the use of what Arnold calls 'obstructing sets.' Once the transverse intersection between the departing whisker of one torus and arriving whisker of another torus has been established, the concept of obstructing set is crucial in guaranteeing that trajectories moving very closely alongside the departing whisker will transfer over so as to move closely alongside the arriving whisker of the next torus in the chain. Both in verifying this process in Arnold's mechanism and in extending it to more general settings, researchers were obliged to develop specialized versions of so-called '*lambda lemmas*' and to extend and specialize some of the earlier ideas of C. Conley and R. Easton.

Beyond the verification of Arnold's mechanism in his own model system lie the much more formidable problems associated with extending the mechanism to more general systems, or of finding other mechanisms that might also serve to prove Arnold's conjecture of instability in generic systems.

We've already mentioned Arnold's special use of two small parameters in his model system, how this '*a priori* unstable' set-up allows hyperbolicity and nonintegrability to be dealt with separately, and how it allows the 'ordinary' Poincaré-Melnikov-Arnold method to be used in showing the transverse intersection of departing and arriving whiskers of nearby whiskered tori. In more typical generic systems involving only one small parameter, there are singular perturbation problems arising from the simultaneous appearance of hyperbolicity and nonintegrability. Consequently, the angle of intersection between departing and arriving whiskers becomes exponentially small in the small parameter, requiring much more sensitive versions of the Poincaré-Melnikov-Arnold method for proper detection and measurement.

Another closely related special feature of Arnold's two-parameter model is that the whiskered tori persist as a continuous family for all positive values of ε and μ. In a generic system with only one small parameter, one ordinarily obtains the analogous tori by way of KAM theory. The tori thus obtained are parametrized over a Cantor set, with its characteristic empty spaces or 'gaps' whose sizes grow with the small parameter. The departing and arriving whiskers must somehow bridge these gaps. At first thought, it may seem that this is easily accomplished by making the single parameter sufficiently small, but unfortunately, as mentioned above, the angle of intersection—thus presumably also the 'bridgeable distance'—between whiskers is exponentially small in the single parameter. This obstruction to extending Arnold's mechanism to generic systems is known as the 'large gap problem.'

As if the large gap problem were not enough, researchers also became aware of an even larger sort of gap that appears in the vicinity of low-order double resonances (i.e., resonances of *multiplicity* 2). While the previously mentioned gaps occur as Cantor-set spaces in a single family of tori, at low-order double resonances, a still larger span must be bridged between separate families of tori. This in turn became known as the 'very large gap problem.'

As researchers confronted these and other problems in Arnold's model, they sought to overcome them not only by examining the model more closely, but also by exploring alternative or variant mechanisms of instability. These alternatives may be combined with Arnold's model, studied independently, or both. For example, in addition to the usual n-dimensional Lagrangian tori found by ordinary KAM theory, an n-degree-of-freedom system may also possess secondary tori (of different homotopy), or lower-dimensional (non-Lagrangian) tori, and in certain cases these other tori may be used, together with their invariant manifolds, to bridge the 'large gaps' mentioned above. Roughly speaking, secondary tori naturally occur in the large gaps, and lower-dimensional tori occur in the very large gaps (precisely because of the multiple resonances there).

Finally, though they are not so much a mechanism as a means of proving the existence of instability, variational methods have shown themselves to be one of the most effective alternatives to using Arnold's mechanism explicitly. As the reader may recall from the brief discussion of Aubry-Mather theory above, variational methods seek to establish the existence of something with a given property by showing that it extremizes (or yields a stationary value of) a certain *functional*, such as the action. In Aubry-Mather theory, the 'something' is an Aubry-Mather set (such as a cantorus); in the present case, the 'something' might be a trajectory linking distant neighborhoods, as in Arnold's conjecture.

With this basic overview of some of the problems in the area, we're ready to look at how research has proceeded during the last half-century; in other words I'll give an idea of who has done what, and when.

(e) The first wave of research, and a bit of drama

For more than a decade after Arnold's example, mathematicians were largely quiescent about the subject; the problems surrounding it seemed very thorny. Meanwhile physicists, as usual, saw no reason not to discuss and conjecture, and did so openly in conferences, proceedings, and journals. Mathematicians were finally awakened at the end of the 1970s by the appearance of a long physics paper of uncanny insightfulness; this was Chirikov's article [Chir79], already cited above in §6.3.1 in connection with 'Chirikov diffusion.' As an example of the boldness of this article, I quote the following sentence (from the third page): "The latter [Arnold diffusion] may be called by right a universal instability [...] since it occurs in any system (except the completely integrable ones [...]) and for arbitrarily small perturbation." This sentence certainly expresses then-current views on Arnold diffusion, and has been largely borne out by later developments, but both of its assertions were far from rigorous verification at the time. Yet later articles by Chirikov and V.V. Vecheslavov [ChirV89], [ChirV92] gave more refined and detailed arguments concerning the drift mechanism and especially its speed (expected to be on the order of $e^{-\frac{1}{2n}}$ in systems with n degrees of freedom). This and other efforts by physicists and applied mathematicians finally put mathematicians to work, albeit slowly at first. The story of these efforts could—and should—be the subject of a book. Here I'm going to cite about a dozen papers chronologically in the first wave of research, and describe them each in a sentence or two.

Among the first significant mathematical papers is the one by P. Holmes and J. Marsden [HolM82] demonstrating a straightforward extension of

Arnold's mechanism to n degrees of freedom. Later, in [Dou88], R. Douady uses Arnold diffusion to show delicate stability/instability results for both maps and flows with n degrees of freedom. Building on Chirikov's earlier work, P. Lochak [Loc90] gives a mostly heuristic argument showing that the average speed of Arnold diffusion should be the maximum permitted by Nekhoroshev theory. In a paper that would later turn out to be key to the use of variational methods in this business [Mat93], J.N. Mather uses his eponymous methods to show the existence of orbits connecting distant phase points of twist maps under certain conditions. At about the same time, Z. Xia argues the existence of Arnold diffusion in problems of celestial mechanics [Xia93], [Xia94]. Then, in the prize-winning paper [ChiG94], L. Chierchia and G. Gallavotti provide a detailed study of Arnold's mechanism, seemingly pushing it to its limits within the bounds of *a priori* unstable systems; however, certain results in the paper are later found to have errors [ChiG98]. In [Moe96], R. Moeckel gives another example of Arnold diffusion in celestial mechanics, but now using lower-dimensional, normally hyperbolic invariant transition tori. Jean-Pierre Marco [Mar96] provides the first example of how to treat Arnold's 'obstruction property' rigorously by means of an appropriate *lambda lemma*, and also establishes the best estimates to date of the speed of drift for Arnold diffusion in n-degree-of-freedom systems. In a crucial series of papers [Bes96], [Bes97a], [Bes97b], U. Bessi extends Mather's variational methods and shows how they may be used not only to give the existence of drifting orbits, but also to gauge their speed. Patrick Bernard [Bernar96] analyzes *a priori* unstable systems using Bessi's and Mather's techniques to show that the drift speed may be polynomially fast in such systems having a single parameter controlling nonintegrability (this is analogous to setting $\varepsilon = 1$ in Arnold's model and observing polynomially fast drift in μ). Finally, J. Cresson [Cres97] extends Marco's earlier use of a lambda lemma (showing Arnold's obstruction property rigorously) to cases with tori in the vicinity of resonances of arbitrary multiplicity.

Despite strong efforts from all involved and despite the important groundwork that was laid for future research, the original problem—the elucidation of Arnold's mechanism in 'typical' systems and the proof of Arnold's conjecture—had advanced surprisingly little by 1997 or 98 (but perhaps only [ChiG94] had strong ambitions in this direction). This state of affairs was brought into focus by P. Lochak's 1999 paper [Loc99] (which began circulating as a manuscript in 1996) and especially by the unpublished 'supplement' to it [Loc97], in which the author outlined the outstanding problems and incisively criticized then-current literature. He pointed out that, while most authors' theorems were correct as stated, their interpretations of the results in terms of Arnold diffusion were overreaching. He pointed further to lack of progress on the 'singular perturbation problem' giving rise to exponentially small splitting of separatrices (related to the simultaneous appearance in typical systems of hyperbolicity and nonintegrability); he also highlighted other problems in the way KAM theory and certain 'lambda-like lemmas' were applied.

Researchers' efforts were eventually brought to bear on Arnold's conjecture in a more focused way. Before this happened, however, between roughly 1997 and 2001, I and many others witnessed a dramatic flurry of activity related to other problems surrounding Arnold diffusion, especially the related issue of exponentially small splitting of separatrices. Much of it took place in e-mail exchanges spread over four or five countries and several continents in which criticism of various authors' work was often heated and decidedly not dispassionate. Errors were brought to light, then affirmed or denied; passionate discussions—punctuated by occasional accusations— were had among and between mathematicians. I had seen exchanges of this sort among physicists (including episodes of shouting and fist-shaking at meetings), but this was the first time I saw such drama in mathematics.

The turn of the millennium saw a cooling of passions, and work returned more or less to normal. But what had been a trickle of mathematical papers about Arnold diffusion in the 1980s and 90s has become a flood: there are now well over a hundred articles connected in some way to results in this area (see for example the reference section of [DelsLS06]). The upshot of all this activity is that much progress has been made in the rigorous understanding of Arnold's instability in typical systems, and Arnold's conjecture itself is now believed to be true under certain conditions. Various proofs are already circulating in manuscript form and awaiting validation; some may appear in print before this book does, as we now discuss.

(f) <u>The second wave of research</u>

Armed with results from earlier labors, researchers continue to approach Arnold's conjecture in increasingly sophisticated ways which I survey again here via 'speed review' (with short paragraphs).

In [BoloT99], S. Bolotin and D. Treschev revisit an earlier result of J.N. Mather concerning time-periodic Hamiltonians in which Arnold diffusion may be interpreted physically as giving rise to unbounded energy, but in place of Mather's method they substitute techniques more along the lines of Arnold's mechanism (i.e., KAM theory and the Poincaré-Melnikov-Arnold method).

J. Cresson [Cres00] shows the existence of Arnold diffusion and Smale horseshoe-like behavior in systems with only partially hyperbolic tori. For this purpose he uses a specially designed lambda lemma together with methods from R. Easton's earlier paper [Eas78] on homoclinic phenomena. (Though Easton doesn't explicitly address Arnold diffusion himself, techniques in his papers [Eas78], [Eas81] will be cited repeatedly by future researchers in this area.)

A few years later, the Mather method (variational approach) is used by M. Berti, L. Biasco, and P. Bolle [BertBB03] to get the maximum average speed of drift for a class of *a priori* unstable systems. (A. Bounemoura and E. Pennamen recently provided an updated and improved version of this approach [BounP12].)

Around the same time, another influential paper by J. Mather appears [Mat03] in which Arnold diffusion-type results are stated for a class of *a priori* stable systems (the more difficult, general case). Although limited to systems with three degrees of freedom (and apparently hard to generalize to n degrees), if true as stated, Mather's results appear to come close to proving Arnold's conjecture (for $n = 3$), as they could probably be refined to show that they hold for generic systems in reasonable topologies. Unfortunately, the results have remained without complete proofs. For some time, the author's outstanding reputation made it seem likely that proofs were forthcoming (as already partly evidenced in the preprint [Mat06]); but more recently, some researchers—including Mather himself—have expressed doubts about whether the program can be carried out as originally hoped.

Meanwhile, another direction toward Arnold's conjecture is taken by D. Treschev in [Tre04], where he shows instability for a C^r-generic set of perturbations of two-and-a-half degree-of-freedom *a priori* unstable systems. Treschev avoids Arnold's transition chains (and thus the large gap problem) by using what he calls a separatrix map. Similarly, in their memoir [DelsLS06], A. Delshams, R. de la Llave and T. Seara analyze a fairly general class of two-and-a-half degree-of-freedom systems and show how the very large gap problem may be overcome by using the lower-dimensional tori found in the resonant zones of the gaps. They rely on their own lambda-like lemmas (and a way of managing them called a 'scattering map') to generalize Arnold's obstructing sets. Shortly thereafter, M. Gidea and R. de la Llave [GidL06] simplify and strengthen results of the memoir [DelsLS06] using Easton's correctly aligned windows method [Eas78], [Eas81]. This is taken still further in the papers of Gidea and C. Robinson [GidR07], [GidR09], who extend the gap-crossing trajectories to non-perturbative, non-Hamiltonian systems (i.e, systems not necessarily close to integrable, with stable and unstable manifolds not necessarily of the same dimension).

In their papers [DelsH09] and [DelsH11], Delshams and G. Huguet revisit and refine the techniques used earlier in [DelsLS06]. In [DelsH09], they show that perturbations giving rise to diffusion are dense in the C^2 topology on the space of Hamiltonians (i.e., diffusion is generic in this sense); while in [DelsH11], they simplify their approach and give sufficient, explicitly verifiable conditions for diffusion.

The variational approach pioneered by Mather and Bessi is streamlined and strengthened by V. Kaloshin and M. Levi in their papers [KalL08a], [KalL08b] (the second of which is more pedagogical), and related methods are used by C.-Q. Cheng and J. Yan in [ChengY04], [ChengY08], [ChengY09]. Although the latter authors treat the *a priori* unstable case, their models do possess n degrees of freedom (but instability phenomena are limited to only a few degrees of freedom). Most importantly however, their results hold for generic perturbations.

Still another approach inspired by variational methods is taken by P. Bernard in [Bernar08] and [Bernar11]. Using methods related to the weak KAM theory of J. Mather, R. Mañé, and A. Fathi (cf. §6.1 and Appendix D, Part D.7.2), the first paper establishes the abstract notion of a 'forcing relation,' which implies the existence of connecting orbits between distant points. The second paper uses related methods to show the existence of normally hyperbolic toroidal annuli with length independent of the perturbation strength in *a priori* stable systems with $n+1/2$ degrees of freedom (thus overcoming one of the singular perturbation problems mentioned earlier for such systems).

Different methods are employed by M. Nassiri and E. Pujals in [NassP12], [NassP13], where they show the existence of 'robust transitivity' in a large class of Hamiltonian systems, meaning instability including (but more general than) diffusion which also exists in nearby systems. To achieve this, they study the minimal dynamics of symplectic iterated function systems, and use a new tool they call 'symplectic blending.'

Turning now to the problem of diffusion speed, the most comprehensive treatment to date appears in [LocM05], where P. Lochak and J.-P. Marco give particular examples of instability with (very nearly) the maximal speed permitted by Nekhoroshev theory in the analytic case. They point the way to improvements and generalizations of their methods, which rely on many earlier developments (especially R. Easton's 'windowing' method [Eas78], [Eas81], and updated versions of S. Sternberg's conjugacy theorems [Ster57], [Ster58], [Ster59]). This work follows upon the earlier success [MarS03] of Marco and D. Sauzin in showing that Nekhoroshev estimates are optimal in the *Gevrey-smooth* case. It should also be noted that in a recent paper [Zh11], K. Zhang has demonstrated the effect of resonance upon the drift speed: for an n-degree-of-freedom system with drifting orbit passing near a double resonance, he follows and extends Lochak and Marco's work to obtain the stability exponent $a = 1/2(n-2)$, in accord with Nekhoroshev theory near multiple resonance (cf. §6.2.2 above). The upshot is that the most recent estimates of Nekhoroshev theory—in particular the stability exponents—are now known to be essentially optimal, thus vindicating the Chirikov-Lochak conjecture.

As we come to the half-century mark since Arnold's original paper, it appears that the vastly different and partial results described so far may at last be coalescing into a coherent whole that will affirm Arnold's conjecture in two and a half or three degrees of freedom, and also go beyond it to describe the instability mechanism in further detail. At present, although proofs have not yet been published, they do exist in written form as preprints awaiting validation. One approach, by J.-P. Marco and M. Gidea [Mar13], [MarG13], uses primarily geometric methods, while others due to V. Kaloshin and K. Zhang [KalZ13] and to C.-Q. Cheng [Cheng13] employ variational methods.

Once (perhaps 'if') the Arnold conjecture is affirmed in detail and for sufficiently general systems, research will continue to push further in at least two directions: first, the details of how instability occurs in specific systems will continue to occupy researchers for some time; and second, the possibility of similar phenomena in infinite-dimensional systems (connected, for example, with problems of Anderson localization or delocalization) will no doubt also be examined closely.

Finally, a few words about the state of numerical detection and modeling of Arnold diffusion are in order. Although still difficult, it is now not uncommon to observe instability in systems very close to integrable using specialized numerical methods. A good example is the article [Cord08] by B. Cordani, in which not only instability speeds are measured, but the 'Arnold web' (of resonances, along which orbits 'diffuse') may be visualized (his visualization is reminiscent of J. Laskar's earlier numerical work using very different methods).

(g) <u>Final remarks on the ways of physics and mathematics</u>

The way research in Arnold diffusion has proceeded highlights many of the issues that typically separate mathematicians and physicists when they work in the same area. Mathematicians grit their teeth at the way physicists cavalierly toss out statements requiring mathematical proof, undeterred by the absence of said proof. ("Ah—our mathematical friends will supply that later"; cf. the sentence quoted from [Chir79] above.) And it always seems difficult if not impossible to convince physicists that many instabilities detected numerically are quite likely not Arnold diffusion, but rather some sort of Chirikov diffusion. (This follows almost by definition, since Arnold diffusion can only be seen in the Nekhoroshev regime, where perturbations are small enough to 'squeeze out' larger instabilities. But once perturbations are so small that only Arnold diffusion remains, it is so slow that it's unlikely to be detected numerically, unless the numerical methods are specially designed and especially heroic.) When they hear this last statement, physicists are apt to ask mathematicians why they're concerned with a kind of instability that can barely be detected numerically, or only manifests itself over age-of-the-universe timescales. One answer is that mathematicians are interested precisely because it's very subtle and difficult. To which physicists sometimes retort, "Yes, it must be difficult, since it's taken you a half-century to understand it halfway, and you have still told us very little we didn't know by our own methods after the first decade."

These exchanges can and do continue almost indefinitely, and of course they reflect the differing points of view that animate mathematicians and physicists. The tension between the two seems necessary to move things forward.

Chapter 7

Physical Applications

This chapter sketches out a few applications—and consequences—of KAM theory in physics. I devote the most space here to celestial and statistical mechanics, since those subjects are closely linked with the earlier narrative in Chapter 3. The other material is very cursory, and refers the reader to more detailed references for in-depth treatments.

When discussing applications of KAM theory to physics, it's helpful to remember that in many cases it may be more appropriate to apply Nekhoroshev theory instead. Nekhoroshev theory is closely related to KAM theory (in fact it's sometimes called 'physicists' KAM theory') and is discussed briefly in §6.2 above. I'll therefore also often mention applications of Nekhoroshev theory in this chapter.

7.1 Stability of the solar system (or not?)

After making so much of the role played in KAM theory by celestial mechanics and the stability problem, we now turn directly to the questions "What does KAM theory really say about the stability of the solar system?" and its close companion "What does KAM theory say about the stability of the n body problem?" Though it's tempting to blend or even identify the two questions, especially since the latter is the starting point for any reasonable mathematical model of the former, we'll see that it's best to make a clear distinction.

To be clear from the start: in the most concrete sense, KAM theory says very little about our own physical solar system, with its Sun (solar mass 1), Jupiter (0.001 solar mass), Saturn (0.0003 solar mass), Neptune (0.00005 solar mass), and so on. Although the masses of the planets appear very small, by the standards of rigorous KAM theory—especially the early theory—they are 'astronomical' (!), and KAM theory simply does not apply to the solar system as a whole. But for small subsystems, or for n body problems with smaller planets in more stable configurations, KAM theory has a great deal to say, though the story has unfolded more slowly than is generally supposed.

Because of the long, detailed history of celestial mechanics, the reader must be prepared to negotiate the legacy of several centuries of work by physicists, astronomers, and mathematicians, manifested in a varied and idiosyncratic vocabulary (often involving different names for the same thing), and above all in a vast array of coordinate changes adapted to various purposes. With that said, let's begin.

7.1.1 KAM theory applied to the n body problem

Here is a very abbreviated summary of results on stability of the n body problem using KAM theory. More detail can be found in A. Celletti and L. Chierchia's marvelously concise paper [CelC06], on which this subsection is largely based.

(a) <u>The n body problem as a Hamiltonian system</u>

Although one speaks of the n body problem, when modeling the solar system (or similar planetary system), it's more convenient and traditional to speak of '$1 + n$' bodies, labeling them with indices $i = 0, 1, \ldots, n$. The index $i = 0$ naturally distinguishes the 0th body as the large central one (the sun, or other star), surrounded by n smaller ones (the planets).

Suppose the ith body has mass m_i and is located in space at the point with Cartesian coordinates $u^{(i)} = (u_1^{(i)}, u_2^{(i)}, u_3^{(i)}) \in \mathbb{R}^3$. Its corresponding (conjugate) momentum is then $U^{(i)} = m_i \dot{u}^{(i)} = m_i(\dot{u}_1^{(i)}, \dot{u}_2^{(i)}, \dot{u}_3^{(i)})$. Using these variables (with $|u - v|$ as the ordinary Euclidean distance between u and v), it is not hard to check that the Hamiltonian

$$\mathcal{H} := \sum_{i=0}^{n} \frac{|U^{(i)}|^2}{2m_i} - \sum_{0 \leq i < j \leq n} \frac{m_i m_j}{|u^{(i)} - u^{(j)}|} \qquad (7.1)$$

generates the canonical equations of motion (for $i = 0, 1, \ldots, n$)

$$\dot{u}^{(i)} = \frac{1}{m_i} U^{(i)}, \qquad \dot{U}^{(i)} = \sum_{0 \leq j \leq n, \, j \neq i} m_i m_j \frac{u^{(j)} - u^{(i)}}{|u^{(j)} - u^{(i)}|^3}, \qquad (7.2)$$

on the phase space $\mathcal{M}$ consisting of the 'collisionless' open domain in $\mathbb{R}^{6(n+1)}$ given by $\mathcal{M} := \{u^{(i)}, U^{(i)} \in \mathbb{R}^3 \,|\, u^{(i)} \neq u^{(j)}, \, 0 \leq i \neq j \leq n\}$.

Equations (7.2) are precisely the equations of motion for the $1 + n$ body problem (in units such that the gravitational constant is 1). They can be seen in the form of 2nd-order ODEs in $u^{(i)}$ (as derived directly from *Newton's second law of motion*) by using the first equations $\dot{u}^{(i)} = U^{(i)}/m_i$ to eliminate $\dot{U}^{(i)}$ from the left-hand side of the second equations, replacing it instead by $m_i \ddot{u}^{(i)}$.

(b) The scaled n body problem

Next, we scale the $1+n$ body problem using a perturbation parameter $\varepsilon > 0$ roughly equal to the ratio of the mass of the planets ($i \geq 1$) to the mass of the sun ($i = 0$). To do this, we first change from *barycentric* coordinates (u, U) to 'heliocentric' coordinates (r, R) via $u^{(0)} = r^{(0)}$, $u^{(i)} = r^{(0)} + r^{(i)}$, $U^{(0)} = R^{(0)} - \sum_{i=1}^{n} R^{(i)}$, $U^{(i)} = R^{(i)}$, $(i = 1, \ldots, n)$. We then use $\varepsilon > 0$ to introduce the 'scaled heliocentric' coordinates (x, X) (with rescaled masses $\bar{m}_i$, $i \geq 1$) by $\bar{m}_0 := m_0$, $\bar{m}_i := m_i/\varepsilon$, $x^{(0)} := r^{(0)}$, $x^{(i)} := r^{(i)}$, $X^{(0)} := R^{(0)}$, $X^{(i)} := R^{(i)}/\varepsilon$ $(i = 1, \ldots, n)$. The new 'scaled planetary' Hamiltonian $\mathcal{H}_{\text{plt}}$ is then given by $\mathcal{H}_{\text{plt}}(x, X) := \varepsilon^{-1}\mathcal{H}(x, \varepsilon X)$; or, using the definitions $M_i := \bar{m}_0 + \varepsilon \bar{m}_i$ and $\mu_i := \bar{m}_0 \bar{m}_i / M_i$, the new Hamiltonian is

$$\mathcal{H}_{\text{plt}}(x, X) =$$
$$\sum_{i=1}^{n} \left(\frac{|X^{(i)}|^2}{2\mu_i} - \frac{\mu_i M_i}{|x^{(i)}|} \right) + \varepsilon \sum_{1 \leq i < j \leq n} \left(\frac{X^{(i)} \cdot X^{(j)}}{\bar{m}_0} - \frac{\bar{m}_i \bar{m}_j}{|x^{(i)} - x^{(j)}|} \right)$$
$$= \mathcal{H}_{\text{plt}}^{(0)}(x, X) + \varepsilon \mathcal{H}_{\text{plt}}^{(1)}(x, X). \tag{7.3}$$

The first term $\mathcal{H}_{\text{plt}}^{(0)}$ is precisely the sum of n decoupled two body problems, so this leading term is by itself completely integrable, and we have arrived to a nearly integrable form of the $1+n$ body problem to which we may try to apply KAM theory. This is not to say that the application is straightforward from here—on the contrary, except for the simplest cases, it has proved exceedingly difficult, for reasons connected with the special nature of the problem itself. We begin with the simplest case.

(c) The simplest three body problem: RPC3BP

We know that the two body problem is completely integrable, but that any n body problem beyond it is not. This is true even for the simplest of all three body problems, the 'restricted, planar, circular three body problem,' or RPC3BP for short, which we now describe.

First let's explain the meaning of 'restricted.' Although the RPC3BP is a three body problem, it is only barely so, because the third body is a kind of ghost,[1] or infinitesimal, with no mass. In other words, it is assumed to be so small that it exerts no force whatsoever on the first two bodies; it exists only as a 'sensor' to test the motion of a body of negligible mass under the gravitational influence of the first two bodies with non-negligible mass. (One often imagines an asteroid moving between the Sun and Jupiter, or a high-orbit satellite between the Earth and Moon.) Next, the word 'planar' simply means that all three bodies remain in a fixed plane in space. Finally, 'circular' means that the integrable motion of the first two massive bodies is circular (i.e., they orbit at fixed distances around their common center of gravity, or *barycenter*). We sum up the RPC3BP this way: Given two bodies in circular orbit around their barycenter, what is the motion of a massless test particle restricted to the orbital plane?[2]

To best represent the RPC3BP in a form to which KAM theory applies, we set $n = 2$ and $m_2 = 0$ in (7.3) and transform to 'rotating Delaunay variables' (a special set of action-angle coordinates) $(L, G, l, g) \in \{(L, G) \in \mathbb{R}^2 \mid L > G > 0\} \times \mathbb{T}^2$ in which the Hamiltonian becomes

$$\mathcal{H}_{\mathrm{rcp}}(L, G, l, g; \varepsilon) := -\frac{1}{2L^2} - G + \varepsilon \, \mathcal{H}_1(L, G, l, g; \varepsilon). \tag{7.4}$$

Here the small parameter is $\varepsilon = m_1 m_0^{-2/3}$, which is small when the mass of body 0 is much larger than the mass of body 1. The perturbation $\varepsilon \mathcal{H}_1$ is constructed from $\mathcal{H}_1 := x^{(2)} \cdot x^{(1)} - |x^{(2)} - x^{(1)}|^{-1}$, where $x^{(1)}$ and $x^{(2)}$ represent, respectively, the heliocentric coordinates of bodies 1 and 2 expressed in the rotating Delaunay coordinates (heliocentric meaning centered at the center of mass of body 0). Explicit expressions may be found in [CelC06].

It is this form of the RPC3BP—Eq. (7.4)—to which the KAM theorem applies. In fact, Kolmogorov's original theorem from 1954 requires only minor modification in order to apply here. To be more precise, although the RPC3BP does not satisfy Kolmogorov's nondegeneracy condition, it

---

[1] The word 'ghost' is used here in homage to the Irish philosopher (and Bishop of Cloyne) George Berkeley (1685–1753), who famously called infinitesimals the "ghosts of departed quantities" in his 1734 book *The Analyst*, a critique of the then-new science of calculus. In fact, the questions raised by Berkeley were not fully answered until the foundations of *analysis* were treated rigorously in the 19th century by A. Cauchy, K. Weierstrass and others.

[2] The phrase 'restricted to the orbital plane' may give the mistaken impression of restraining the particle by force, as in the constraints of a mechanical system. But here we mean to simply start the test particle in the orbital plane with no velocity component orthogonal to the plane. By symmetry it will clearly remain in the plane thereafter; no additional 'constraining forces' are required.

does satisfy Arnold's condition (4.9) (already mentioned above in §4.6 (c)), and it is not difficult to reformulate Kolmogorov's theorem so it applies on fixed energy surfaces using Arnold's condition, as Arnold did in [Ar63a] and [Ar63b]. Assuming this reformulation has been carried out, let's see how it applies to the RPC3BP. In describing the result, it might be appropriate to imagine some kind of theatrical drumroll, since the stability of any sort of n body problem beyond the Kepler problem—even this meager advance— had been a Holy Grail and driving force behind celestial mechanics for a long, long time.

Using the actions $I = (L, G)$ with the unperturbed part $h(I) = h(L, G) = (-1/2)L^{-2} - G$ of the Hamiltonian (7.4), we easily compute Arnold's determinant (4.9) as

$$\det \begin{bmatrix} \partial^2 h/\partial I^2 & \partial h/\partial I \\ \partial h/\partial I^T & 0 \end{bmatrix} = \begin{vmatrix} -3L^{-4} & 0 & L^{-3} \\ 0 & 0 & -1 \\ L^{-3} & -1 & 0 \end{vmatrix} = \frac{3}{L^4} \neq 0. \quad (7.5)$$

Under this nondegeneracy condition, we have the following stability theorem:

KAM theorem for the RPC3BP.[3] *Let (L_0, G_0) be such that the above nondegeneracy condition (7.5) holds, and let $E_0 = \{(L, G, l, g) \mid \mathcal{H}_{\text{rcp}}(L, G, l, g; \varepsilon) = h(L_0, G_0)\}$ be the energy surface corresponding to the unperturbed energy $h = h(L_0, G_0)$. Then there is a critical value $\varepsilon_0 > 0$ such that whenever $0 < \varepsilon < \varepsilon_0$, the Hamiltonian (7.4) has a set of analytic KAM tori $\mathcal{T} \subset E_0$ of positive (relative) Liouville measure in E_0.*

(d) Beyond the RPC3BP

We now proceed to the much more difficult terrain of KAM theory for n body problems beyond the RPC3BP, and here again we shift to more impressionistic descriptions.

Returning to the $1 + n$ body problem (7.3), we find that when we introduce action-angle variables[4] for the unperturbed integrable part $\mathcal{H}_{\text{plt}}^{(0)}(x, X)$,

---

[3]It is quite interesting that Arnold's modification of Kolmogorov's theorem can be applied directly to the RPC3BP, but it's not certain who first noticed this. It was probably Arnold, who describes this application very clearly on p. 415 of [Ar78–97]. What is surprising is that Arnold didn't point this out earlier when he first introduced isoenergetic nondegeneracy in 1963 (but perhaps he was hunting bigger game then, including his more general results on the planar three body problem described below).

[4]In fact, it's customary to introduce 'Delaunay action-angle variables' $(L, G, \Theta; l, g, \theta)$

it does not depend on all of the actions. In KAM theory, such an integrable Hamiltonian is said to be *properly degenerate*. A properly degenerate system satisfies neither Kolmogorov's nor Arnold's nondegeneracy condition, and its perturbations may fail to have KAM tori. Getting around this difficulty is not nearly so simple as introducing the isoenergetic nondegeneracy condition (7.5) above. However, in [Ar63b], Arnold proved a KAM theorem for properly degenerate systems[5] with a special structure, which he supposed would include the general $1 + n$ body problem (7.3). This has turned out to be essentially correct, though perhaps not in the way Arnold envisaged originally. A brief outline of the chronology follows.

(e) Brief chronology of KAM theorems for the n body problem

First, in [Ar63b], Arnold proved a KAM theorem for the planar three body problem. This is a generalization of the RPC3BP in which all three bodies remain in one plane, but the largest two bodies are no longer required to be in a perfectly circular orbit around their barycenter, and the third body is allowed to have positive mass. But the second body's orbit around the largest body must still be very close to circular, and there are still very strict upper bounds on the masses of the second and third bodies (see the estimates by M. Hénon discussed above in §5.2.2, p. 103).

Also in [Ar63b], Arnold announced a KAM theorem for the spatial $1+n$ body problem (7.3) (in which the 'exactly coplanar' requirement is replaced by a 'close-to-coplanar' requirement, and with the usual restrictions on the bodies' masses). He sketched the proof of this theorem to a certain level of detail, and the subject remained at this stage until it was discovered that things were not quite right.[6] In fact, the KAM theorem for properly degenerate systems—and its proof-sketch—turn out to be essentially correct; what is wrong is that the theorem does not apply to the $1 + n$ body problem in quite the way Arnold supposed. But all of this took a long time to discover, as the details of reducing the $1 + n$ body problem to suitable form finally became clear. In this light, it's probably better to call Arnold's results on the spatial problem a 'program' (to be developed) rather than a theorem.

---

in which the unperturbed part $\mathcal{H}_{\mathrm{plt}}^{(0)}$ depends only on the L action variables, not on the Gs or Θs; cf. [CelC06].

[5]Arnold calls this theorem 'the fundamental theorem' in Chapter IV of [Ar63b].

[6]The belief that Arnold had sketched the correct approach to the problem, with only details remaining to be filled in, was widespread for a long time: see the remarks at the end of §36 (p. 277) of [SieM71].

The first serious attempt to carry out the details of Arnold's program was successful for the $n = 2$ case (i.e., for the spatial three body problem). This came out of P. Robutel's 1993 thesis under J. Laskar's direction at the Bureau des Longitudes in Paris, which was expanded and published as [LasR95] (giving the computer-assisted expansion of the three-body Hamiltonian in a form to which Arnold's theorem applies), and [Rob95] (with verification and application of Arnold's KAM theorem for $n = 2$).

The precise nature of the snag in applying Arnold's KAM theorem for $n \geq 3$ was discovered by Michel Herman,[7] who took an interest in the problem in the late 1990s. Participants and onlookers who attended Herman's seminars in Paris in those days tell stories about their growing wonder as they watched him unravel the complexities of the subject. It slowly came to light that for $n \geq 3$ the $1+n$ body problem harbored a strange degeneracy—one that had somehow gone undetected in the preceding half-century, and one that was slightly different from the *proper degeneracies* already known. Called the 'mysterious resonance' or 'Herman's resonance,' it clearly voids the application of Arnold's 1963 KAM theorem to the $1+n$ body problem in its full generality (see [AbdA01] for a self-contained discussion of Herman's resonance).

Once it was recognized, surmounting the mysterious resonance was far from easy. Though Herman seemed to have assembled the necessary tools by 1999, his work on the problem ended with his untimely death in November 2000. Thereafter, the task was taken up by Jacques Féjoz, and published in final form in 2004 [Féj04]. Although Féjoz has always maintained that he acted merely as the conveyor of Herman's ideas, many saw the conversion of Herman's seminar notes into complete proofs (with the usual corrections, reformulations, and clarifications) as a herculean feat.

At any rate, the first truly rigorous stability result for the $1 + n$ body problem, which we might call the Herman-Féjoz result, goes roughly as follows:

Stability theorem for the $1+n$ body problem. [Féj04] *Consider $1+n$ point bodies with masses $\bar{m}_0, \varepsilon \bar{m}_1, \ldots, \varepsilon \bar{m}_n$ whose motions are governed by the Hamiltonian system (7.3) above. For all collections $\bar{m}_0, \bar{m}_1, \ldots, \bar{m}_n > 0$ and $a_1 > \ldots > a_n > 0$, there exists a critical value $\varepsilon_0 > 0$ such that whenever $0 < \varepsilon < \varepsilon_0$, in the phase space of (7.3) and in the neighborhood of circular and coplanar Keplerian motions with semi major axes $a_1, \ldots, a_n$, there is a subset of initial conditions with positive Lebesgue measure leading*

---

[7]Also known as 'Michael,' the American version of his first name.

to quasiperiodic motions with $3n - 1$ frequencies (i.e., there is a collection of invariant tori with positive Lebesgue measure).

The proof of this theorem requires a number of specialized results as well as a rather sophisticated overarching mathematical apparatus. The nondegeneracy condition required for the theorem is a slight variation of the one due to A.S. Pyartli [Pya69], and is closely related to Rüssmann's condition (described above in §4.7 (b)). The convergence scheme is an adaptation of R.S. Hamilton's formulation [Ham82] of the Nash-Moser method (also mentioned above in §4.3.2 (g)) and relies to some extent on the more detailed manuscript version of Hamilton's result which was circulated in 1974. Herman's most strikingly original contribution, though (in addition to discovering the 'mysterious resonance'), was to formulate specialized conjugacy theorems (called 'twisted' and 'hypothetical' conjugacy theorems) which amount to something like abstract coordinate changes in which the offending degeneracies disappear, allowing the KAM construction to proceed. A four-page summary of methods used in the proof is available online [Féj07], and full details may be found in [Féj04].

Since the work of Herman and Féjoz, Luigi Chierchia and coworkers have given more direct proofs of stability theorems for the $1 + n$ body problem (here 'more direct' means that the set-up and proof-techniques are closer to what Arnold had in mind). In [ChiPu09], Chierchia and Fabio Pusateri combine a KAM theorem of H. Rüssmann [Rüss01] on an extended phase space with transformation techniques from Féjoz [Féj04] to eliminate the mysterious resonance and establish the existence of a positive measure set of real-analytic invariant tori (the tori in the Herman-Féjoz result are probably also real analytic, but the proof methods show only that they are smooth). Then, in [ChiPi11a], [ChiPi11b], Chierchia and Gabriella Pinzari present what may be the most natural way to apply KAM theory to the multi-body problem. Reaching back to work of André Deprit [Dep83] (who himself reached all the way back to Jacobi's celebrated 'reduction of the nodes' [Jaco42]), they employ a new variation of reduction, using 'regularized planetary symplectic variables' to put the $1 + n$ body Hamiltonian in a form where certain resonances do not appear, and to which a more standard KAM theorem for properly degenerate systems applies. Chierchia and Pinzari are scheduled to present some of these results at a section of the 2014 ICM in Seoul, Korea.

Like the proof of the Herman-Féjoz result, the proofs of Chierchia and coworkers' results are highly technical, and can only be understood and appreciated upon careful reading. Nevertheless, nice overviews appear in the introductions to [ChiPu09] and [ChiPi11a], [ChiPi11b].

7.1.2 *Specialized results for subsystems*

As pointed out several times already, one of the central points of contention between mathematicians and physicists regarding the relevance of KAM theory lies in the absurdly small values of the thresholds of validity (the ε_0 in the theorems above). Mathematicians are often loath to compute these, pointing out that the existence of any $\varepsilon_0 > 0$, no matter how small, is a triumph. On the other hand, some physicists actually want to use these theorems in applications (!), and have been known to compute or estimate the thresholds from the details of KAM proofs. But when they get values like $\varepsilon_0 \approx 10^{-300}$, they don't even bother to enter into any discussions of relevance, as they feel they're already in the realm of knee-slapping jokes.[8]

Mathematicians with a computational bent have responded to this knee-slapping with specialized KAM theorems that apply, not to general n body systems, but to specialized problems involving, say, three bodies, modeling particular subsystems of our solar system. Proofs typically involve a combination of high-order normal forms adapted to the problem (computed via symbolic manipulation on a machine), followed by KAM estimates using 'interval arithmetic,' also done on machines. These machine-assisted proofs yield validity thresholds that are no laughing matter.[9]

A number of researchers have worked on such projects (mostly around the Mediterranean, in Italy, France, and Spain), and they employ methods of both KAM theory and closely related Nekhoroshev theory (see §6.2 above for a description). An early and still relevant result of this type appears in the paper [GioDFGS89] by A. Giorgilli, A. Delshams, E. Fontich, L. Galgani, and C. Simó. Perhaps none have done more in this direction than A. Celletti and L. Chierchia, who also began computer-assisted proofs for KAM theorems in the 1980s. One of their more recent results [CelC07] investigates a truncated RPC3BP model of the subsystem formed by the Sun, Jupiter, and Asteroid 12 Victoria. On a fixed energy level, and for actual astronomical parameter values, they show the existence of invariant tori that permanently trap the motion of the asteroid. Applied KAM theory

---

[8]Recall again that M. Hénon got a value of this order in a back-of-the-envelope application of Arnold's theorem to the solar system; see §5.2.2, p. 103, above.

[9]As noted above in §6.3.1, the theoretical critical value k_c (validity threshold) for the 'last torus' of the standard map has been in good agreement with numerical experiments since around 1990.

has managed to come some distance from its knee-slapping beginnings.

7.1.3   *The physical solar system*

Having some idea of what mathematicians have done with KAM theory and the n body problem, the reader may be interested to know what astronomers think about the stability of the real, physical solar system.

First of all, as mentioned above, KAM theory does not enter the picture on any practical, computational level. It serves as a conceptual guide, and it removes a number of mysteries about 'integrable-like behavior' that would have puzzled astronomers before 1954, but it does not offer practical assistance beyond some of the subsystems described in the previous subsection. For the solar system as a whole, other methods are required entirely.

In their investigations, modern astronomers make use of the best observations and the best computational models available to them, relying on the former to check the latter. Their methods are sophisticated enough to describe not only planetary orbits (considered relatively easy on short timescales), but such subtleties as dissipation due to tides, the effect of resonances among orbital periods, solar oblateness, evolutions in the obliquity ('tilt') of planets, the effects of general relativity, and so on.

For these virtuosos of modeling and computation, reliably predicting planetary orbits is considered a challenge only when it reaches times on the order of tens of millions of years. The modern era of such work may be said to begin with the construction of the 'Digital Orrery,' a small, special-purpose computer built at Caltech in 1984 by G.J. Sussman and coworkers, then transported to MIT, where it was used over a period of seven years to simulate the dynamics of the solar system. One of its most notable discoveries concerned the orbit of (then-planet) Pluto. Working with colleague J. Wisdom at MIT, Sussman used the Orrery to integrate select orbit-elements of the five outer planets over an interval corresponding to 845 million years. They found that the simulated orbit of Pluto was noticeably chaotic; nearby orbits of Pluto diverged exponentially from one another with an e-folding time of only about 20 million years, indicating relatively large *Lyapunov exponents* and a strong exponential sensitivity to initial conditions. Sussman and Wisdom published their results on Pluto [SusW88] in 1988, and many see this as the 'discovery' that the solar system is chaotic.

Not long after Wisdom, Sussman, and others began their work, Jacques

Laskar at the Bureau des Longitudes (BdL) in Paris took a somewhat different approach to simulating the solar system over long times. Lacking access to a dedicated machine like the Digital Orrery, Laskar resurrected the methods of his early predecessors at the BdL by applying Laplace's and U. Le Verrier's perturbation methods to the equations of motion for the solar system. These methods begin by 'averaging' the effect of one orbit upon another (effectively smearing a planet over its orbit until it becomes a ring), but they go much further: Laskar computed perturbation expansions (by machine) with as many as 150,000 terms in some cases. These expansions permitted him to reliably calculate orbit elements using much longer time steps than possible with more direct methods. Using these techniques, Laskar turned his attention to models of the inner planets (Mercury, Venus, Earth, Mars) obtaining the remarkable result that they, like Pluto, had chaotic orbits, with even larger Lyapunov exponents (stronger sensitivity to initial conditions) [Las89], [Las90].

Then, using a new, custom-built successor to the Digital Orrery (the 'Supercomputer Toolkit'), Wisdom and Sussman undertook a more direct integration of the entire solar system in 1992, modeling its evolution very precisely over intervals corresponding to a time of 100 million years. Their results mirrored Laskar's closely, and as they pointed out in their paper [SusW92], similar outcomes from such different methods offered strong evidence of chaos right in the heart of the solar system.

Today, after cross-checks and refinements, the chaotic nature of the solar system is a generally accepted fact among dynamical astronomers. This does not, however, lead automatically to the conclusion that the solar system is unstable; that question, as always, must be interpreted with care.

What can be said in a general way is this. Given phase-orbits of a solar system model starting at nearby initial conditions separated by a distance d_0 in phase space, their distance d at a time t million years later will be roughly $d \approx d_0 10^{t/10}$ (cf. [Las03]). Thus an initial indeterminacy in phase space of order 10^{-10} evolves to order 1 after 100 million years. This is a significant level of chaos given that the solar system is on the order of five billion years old.

More precise statements can be made by focusing on particular planets or groups of planets. Not surprisingly, those with the least chaotic orbits are the giant, or jovian, planets (Jupiter, Saturn, Uranus, Neptune). Though subject to some chaotic excursions, the jovian planets may reasonably expect to maintain their present configurations and orbital parameters over timescales of billions of years. In other words, they have a very good chance of remaining much as they are for the remainder of the solar system's life span, estimated at roughly 4 billion years.

The smaller bodies of the solar system—including the inner, or tellurian, planets—have orbits that are considerably more prone to instability. Earth-dwellers can take some solace in knowing that, as the heaviest of the solar system's small bodies and by virtue of its nice orbital parameters, the Earth may be the solar system's most stable residence after the jovian planets; its orbit possesses only 'average' chaos. In addition, variations in its obliquity (hence also its climate) have likely been moderated by the presence of the Moon, one of the largest satellites in the solar system. Other tellurian planets are not so lucky. Mars has probably experienced very large chaotic variations in obliquity (as much as 60 or 80 degrees) with correspondingly drastic changes in climate. And Venus probably suffered still more radical change: one of the most likely scenarios leading to its present slow retrograde rotation is that it was flipped upside down by chaotic interactions with Mercury. But no planet has a more uncertain future than the latter and innermost. There is a significant probability that over timescales on the order of a billion years—well inside the expected lifetime of the solar system—Mercury could undergo a collision with either Venus or the Sun. In fact, numerical investigations (reported by Laskar and M. Gastineau in [LasG09]) indicate that the probability of such an event is around 1%.[10] In the same note, the authors report the rather sensational finding that in one of their 2501 simulations of future solar system dynamics, angular momentum is transferred from the jovian to the tellurian planets in such a way as to destabilize all the latter, leading to the possibility of collisions between the Earth and Mercury, Mars, or Venus. This destabilization occurs roughly three billion years from now, again within the expected lifetime of the solar system.

So although astronomers at this point give only a probabilistic response to the question of the solar system's stability, their response is that there is a significant chance of instability. What is more certain is that, even when viewed in isolation from outside influence in a perfectly calm phase space of its own, our solar system as a whole lies in a dynamical region outside the stability provided by dense layers of KAM tori. The real, physical solar

---

[10]In in the absence of general relativity, the probability of these collisions rises to about 60% (because of resonance between precessions of the perihelia of Jupiter and Mercury). This has led some to joke that the creator of the universe added general relativity in order to stabilize the solar system; the joke makes reference to Newton's famous remarks about the need for divine intervention to stabilize the solar system (cf. Quote **Q**[New] in Appendix E), so in this way Newton becomes prescient of Einstein's discoveries(!).

system is simply too big, too heavy, with too many moving parts to live in that idyllic realm.

And although mathematicians will never be deprived of the glory of their proofs that certain n body problems admit eternally stable solutions, some of them seem quite reluctant to concede that this is simply not true for the physical solar system. In this respect, to dynamical astronomers, some mathematicians can seem nearly as dogmatic about stability as the Catholic Church of the 16th century.

7.2 Ramifications in statistical mechanics

Though it's not usually accurate to talk about 'applications' of KAM theory to statistical mechanics, the consequences of KAM and Nekhoroshev theory raise basic questions about the foundations of the subject, and have influenced its development to some extent. These questions concern the *ergodic hypothesis* and the related issue of *thermalization*; that is, the process by which a thermodynamical system reaches thermal equilibrium, and by which 'equipartition of energy' takes place (or does not take place), as modeled for example by the Fermi-Pasta-Ulam (FPU) experiment. In this section I sketch some of the main outlines of these consequences and questions, while keeping in mind that the full story of statistical mechanics and its open questions not only fills volumes, but much of it still remains to be written.

Statistical mechanics may be roughly described as the project of deducing the macroscopic properties of matter from the dynamics of its constituent atoms or molecules. It was the first part of physics to make fundamental use of probabilistic methods, and its history is full of drama, not only because of the passionate discussions it inspired about the meaning of probability and *entropy*, or *reversibility*, or 'time's arrow,' but especially since the atomic hypothesis was not universally accepted until the first decade of the 20th century, well after the basic theory of statistical mechanics had been developed. From its beginnings in thermodynamics and kinetic theory, the theory evolved rapidly with the work of three remarkable scientists: James Clerk Maxwell (1831–1879), Josiah Willard Gibbs (1839–1903), and Ludwig Boltzmann (1844–1906). Although each contributed in roughly equal measure to the subject, it is Boltzmann's work that most concerns us here, as he was the originator of the ergodic hypothesis.

7.2.1 *About Boltzmann*

Ludwig Boltzmann was by turns a highly successful, peripatetic and frenetic, and ultimately tragic figure. Born in Vienna during the reign of the Emperor Ferdinand I, younger than Maxwell or Gibbs, he went to the *Gymnasium* in Linz, then back to Vienna, where he finished his thesis on kinetic theory in 1866 under the direction of Josef Stefan. Over the next decade, he moved from Vienna to Graz to Heidelberg to Berlin, back to Vienna, then back to Graz. Despite this 'kinetic' movement, in the same period he (i) re-derived Maxwell's velocity distribution[11] for molecules in a gas in 1866, (ii) formulated an early version of the ergodic hypothesis in 1871, then (iii) re-derived the fundamental equation[12] of kinetic theory and (iv) derived the famous H-theorem,[13] both in 1872. Pausing for more than a decade in Graz, in the years 1872–75 he (v) established his famous relation $S = k \log W$ expressing the entropy S of an isolated system in a given macrostate in terms of its W distinct realizable microstates (the constant of proportionality k is Boltzmann's constant), and in 1884 he used statistical mechanical methods to derive (vi) the relation between the temperature of a black body and the flux density of the energy emanating from the black body at that temperature (this relation[14] had been posited on the basis of experimental data in 1879 by Boltzmann's former teacher J. Stefan). After a stint as president of the University of Graz, Boltzmann moved again in 1890, this time to Munich, where he stayed three years before moving again to Vienna to succeed Stefan in the chair of theoretical physics. In these years, he also worked on and wrote about a concept underlying the H-theorem that he called (vii) the *Stoßzahlansatz*.[15]

---

[11]The distribution of velocities of molecules in a gas (in *equilibrium* at a given temperature) is now called the Maxwell-Boltzmann distribution, one simple formulation of which is as follows. For a gas at uniform absolute temperature T, comprised of molecules each of mass m, the probability density P of molecular speed v is $P(v) = (m/2\pi kT)^{3/2} e^{-mv^2/2kT}$, where k is Boltzmann's constant.

[12]This is the famous transport equation for molecules in a dilute gas commonly referred to as 'the Boltzmann equation' (since Boltzmann later championed it), but in fact first derived by Maxwell in 1866. Maxwell's derivation now looks more 'modern,' but Boltzmann's was more convincing to scientists at the time.

[13]The H-theorem is a mathematical form of the *second law of thermodynamics*, concerning the increasing *entropy* of an isolated, nonequilibrium thermodynamical system.

[14]The relation between flux density j and absolute temperature T is $j = \sigma T^4$, and is now called the Stefan-Boltzmann law (σ is the Stefan-Boltzmann constant).

[15]Boltzmann's word *Stoßzahlansatz* translates literally as 'collision number hypothesis' (or 'collision number assumption'), and is usually now rendered in English as the 'molecular chaos assumption.'

By any measure, the preceding highlights (i)–(vii) of Boltzmann's career show him to have been an extraordinary mathematical physicist, surpassed in the 19th century only by Maxwell. Unfortunately, the last topic mentioned above, the *Stoßzahlansatz*, may already contain elements of the tragedy that was to overtake Boltzmann in his final years. The *Stoßzahlansatz* was a kind of hidden assumption in the hypotheses of the H-theorem, uncovered only after protestations led by J. Loschmidt and E. Zermelo. They pointed out that although Boltzmann claimed to derive the H-theorem by way of classical mechanics, by showing a monotonic increase in entropy, its conclusions violated both *reversibility* and *Poincaré's recurrence theorem*, which must be obeyed by any isolated mechanical system. After much scrutiny, it was discovered that Boltzmann had introduced an element of time-asymmetry by assuming that molecular velocities after collision were uncorrelated with their velocities before collision—this was the *Stoßzahlansatz*.[16]

Although the paradoxes surrounding reversibility and Poincaré recurrence were eventually resolved in this way, the arguments they engendered appear to have unnerved Boltzmann, who spent the rest of his life, it seemed, battling physicists, mathematicians, and philosophers over questions related to probability, time-reversal, entropy, the ergodic hypothesis, and the atomic hypothesis. No one knows for sure how much all of this contributed to the severe bouts of depression that ultimately led him to take his life while on holiday with his family near Trieste in September 1906. But the tragedy is magnified and made more poignant by the fact that in almost all matters of scientific importance, Boltzmann was vindicated by later developments. In the case of the atomic hypothesis, vindication was nearly at hand before his death, as Einstein had published his seminal paper on Brownian motion in 1905 (but J. Perrin's experimental verification and full acceptance of the atomic hypothesis came a few years later).

In many ways, Boltzmann did science in the style extolled by Karl Popper: he formulated scientific hypotheses that 'stuck their necks out,' or exposed themselves to falsification.[17] And falsification sometimes came too

---

[16]Although the word *Stoßzahlansatz* was not used until after Maxwell's death, Maxwell was also well aware of reversibility problems related to what we now call the Boltzmann equation. His investigation of those problems led him to introduce his famous 'demon,' and to discuss what we now call *sensitive dependence on initial conditions*—the main ingredient of *chaos*.

[17]Popper (1902–1994) became famous in philosophy of science in the mid 20th century for pointing out that scientific induction is essentially untenable: no sequence of experimental outcomes compatible with a scientific theory can wholly verify the theory,

hastily. In the next subsection, we'll look briefly at one of Boltzmann's conjectures that has long outlived its author, and has thus far evaded complete falsification by undergoing a number of metamorphoses.

7.2.2 The ergodic hypothesis

There are few parts of mathematical physics that are presently more successful than equilibrium statistical mechanics. As G. Gallavotti puts it, "no fundamental problem [of equilibrium statistical mechanics] seems to have a theoretical description that is in conflict with experimental results" ([Gall99], p. 110). And yet, there are still certain questions in the foundations of the subject surrounding the legacy of Boltzmann's ergodic hypothesis. In this subsection I describe these problems briefly.

(a) <u>Boltzmann's idea</u>

Let us first recall Boltzmann's original discussion from the 1870s and 80s (anachronistically expressed in more modern language). Consider a macroscopic, isolated, many-particle system at equilibrium, or more specifically— to fix ideas—think of the basic example from kinetic theory of a gas of identical elastic particles in a container at a given temperature. The phase space of such a system is enormous; if there are n particles (where n is typically on the order of Avogadro's number $\approx 6 \times 10^{23}$), the phase space will be of dimension $6n$. The $3n$ configuration components range over locations in the container, while the $3n$ momenta have a gigantic range (e.g. the range must contain the cases where one particle has all the momentum while the other $n - 1$ particles are at rest). To say that the system is isolated means that its total energy is conserved, so there is one integral of motion, and we know that the dynamics evolves on the corresponding $6n - 1$-dimensional energy (hyper)surface in the phase space; call it M.

Unfortunately, this is near the end of what we can know with certainty. (If the n body problem with $n = 3$ or slightly bigger seems intractable,

---

yet a single incompatible result is enough to falsify the theory. He then established his 'principle of demarcation' distinguishing truly scientific theories as those that could conceivably be falsified by experiment. This was directly opposed to the program of the *logical positivists*, contemporaries of Popper who sought to give coherence to more traditional notions of verification by scientific induction. It is interesting that the logical positivists were the philosophical descendants of the 19th century *positivists* who hounded Boltzmann; it is also interesting that more than one observer has detected elements of romanticism in Popper's conception of science (which, though quite appealing to some, does not stand up under close analysis much better than scientific induction).

think of solving a kinetic theory problem where n is Avogadro's number.) From here the discussion proceeds probabilistically, leading for example to the Maxwell-Boltzmann velocity distribution (in which the aforementioned states with one particle having all the momentum have vanishingly small probability). Yet in the modern language of dynamical systems, the system must have a deterministic evolution law $\varphi_t : M \to M$ describing its state on the energy surface M at time t, and we can ask questions about this evolution, even if we can't find it explicitly.

Some of the most natural questions concern macroscopic *observables*, i.e., functions of the form $f : M \to \mathbb{R}$ (examples of observables include temperature and pressure). Now if the systems starts in the state $x \in M$ at time $t = 0$, the value of the observable f at that time is $f(x)$; after a time t, the value evolves to $f(\varphi_t(x))$. Since the observable might fluctuate, requiring some finite time to 'observe' properly, it's natural to ask about its average value $\langle f \rangle_T$ over the time interval $[0, T]$, defined as

$$\langle f \rangle_T(x) := \frac{1}{T} \int_0^T f\big(\varphi_t(x)\big) \, dt \, . \tag{7.6}$$

Unfortunately, this is impossible to calculate, since, as just mentioned, we have essentially no hope of knowing φ_t. Boltzmann's insight was to suppose or conjecture that, though we can't know φ_t explicitly, we can imagine that over long times, as particles jostle one another, colliding in myriad possible ways, the system will evolve through more and more states of M, preferring no particular state over another. For very long times T, we can imagine that the evolution will 'explore all (or essentially all[18]) of M,' spending a total time in each part of phase space that is proportional to the volume

---

[18]Most short accounts of the origins of ergodic theory say something like the following. "Boltzmann originally justified the equality in (7.7) by assuming that any trajectory of the system passed through every point of M, and he called this assumption the ergodic hypothesis. Later, when it was pointed out that this was mathematically impossible, others modified the assumption so that every (or 'almost every') trajectory passed 'arbitrarily close' to every point of M (i.e., almost every trajectory was assumed to be dense in M), and they called this modified assumption the quasi-ergodic hypothesis." However, in a footnote on p. 297 of his translation of Boltzmann's book [Bolt96–98], S. Brush points out that although Boltzmann does discuss and use such assumptions in his writings, he (i) does not use the term 'ergodic' (he instead uses 'isodic'), and (ii) does not readily distinguish between trajectories passing through every point and arbitrarily close to every point. Brush adds that part of the confusion arises from misrepresentations of Boltzmann's terminology in Paul and Tatiana Ehrenfest's long article [EhE11] (which otherwise served to greatly clarify early concepts in statistical mechanics). Adding further to the uncertainty is G. Gallavotti's disagreement with Brush over the way events unfolded, and how to interpret them (cf. [Gall95b] and §1.9 of [Gall99]).

of that part. If this conjecture were true, it's not hard to imagine that something like the following equality should hold:

$$\lim_{T \to \infty} \frac{1}{T} \int_0^T f(\varphi_t(x)) \, dt = \int_M f(x) \, dx \Big/ \int_M dx \qquad (7.7)$$

The limit on the left is called the infinite *time average* of f over the evolution starting at x, or just the time average of f; it can be compactly denoted $\langle f \rangle(x) = \lim_{T \to \infty} \langle f \rangle_T(x)$. On the right-hand side of (7.7) we have the 'ensemble average' or 'phase space average' of f—the average of all the values attained by f as x ranges over all states of the energy surface M, without any reference to the time evolution. (And just how to carry out the integration will be clarified below using the 20th-century idea of *invariant measure*.) On a practical level, the average of a real-valued function over a surface is radically simpler to compute than the infinite time average on the left, in which we have almost no knowledge of φ_t; this is the utility of (7.7) that makes it interesting in statistical mechanics.

Now this vaguely formulated conjecture—that "starting at any state x, the trajectory $\varphi_t(x)$ explores all parts of M in such a way that (7.7) holds"—this is the idea that Boltzmann introduced and that others later called the *Ergodensatz* or *Ergodenhypothese* (*ergodic hypothesis* in English). The terminology is fairly loose; it's used to refer either to the behavior of trajectories on M, to an equality like (7.7), or to both. Speaking only of the equality, one often hears it reduced to the slogan "time averages equal space averages." In fact, looking more closely at the history, one already sees equalities of this kind in R. Clausius' treatment of observables, though Clausius did not justify them very convincingly or point to them as a special assumption the way Boltzmann did later.

(b) Mathematical progress: the birth of ergodic theory

It took a long time to develop the first mathematically precise results along the lines of the ergodic hypothesis—six decades, if one counts from Boltzmann's first mention of it in 1871 [Bolt71] to the first rigorous ergodic theorems announced in 1931 (proved by J. von Neumann [Neu32] and G.D. Birkhoff [Bir31]). A lot transpired in that 60-year span (including Poincaré's entire career), but the key mathematical tool which gave proper expression to the ergodic hypothesis was the Lebesgue-Borel theory of measure and integration, introduced just after the turn of the century. The new theory of integration allowed the evolution law φ_t to be abstracted as a measure-preserving flow on a *probability space* (M, Σ, μ). Or, to explain in more detail, if we assume that φ_t is a Hamiltonian flow (not an unreasonable assumption at the molecular level), we know from *Liouville's theorem* that the canonical volume element (*Liouville measure*) on M is preserved by φ_t. We can then use the volume element to define a measure ρ on a natural class Σ of measurable subsets of M, so (M, Σ, ρ) becomes a measure space, with measure ρ invariant with respect to φ_t. Next, assuming the energy surface M has finite measure $\rho(M)$ (again this is reasonable—we expect M to be a *smooth compact hypersurface*), we can normalize ρ so that it becomes the invariant probability measure $\mu = \rho/\rho(M)$ (in other words, $\mu(M) = 1$), and we have the probability space (M, Σ, μ). Finally, we can now give a more precise meaning to the 'space average' appearing on the right-hand side of (7.7): For any μ-Lebesgue-integrable function $f : M \to \mathbb{R}$, we define its space average $\overline{f}$ by $\overline{f} := \int_M f(x) \, d\mu(x)$. We have arrived to the proper mathematical setting for the basic 'Birkhoff pointwise ergodic theorem,' which we now state (for a more concise, three-line statement of the theorem, see *ergodic theorem* in the glossary).

Birkhoff's ergodic theorem. *Consider the probability space (M, Σ, μ) described above, with μ invariant with respect to φ_t. Then given any observable f that is Lebesgue-integrable on M with respect to μ, the time average $\langle f \rangle(x)$ of f along φ_t starting at x (i.e., the left-hand side of (7.7)) exists for almost every x in M (so that $\langle f \rangle$ is Lebesgue-integrable on M). Furthermore, the space average $\overline{\langle f \rangle}$ of $\langle f \rangle$ equals the space average $\overline{f}$ of f.*

Note that this theorem does not have an equality like (7.7) as a conclusion; instead, this follows as a corollary for a certain class of flows φ_t called ergodic, which are defined as follows.

Definition of ergodic flow. *If (M, Σ, μ) is a probability space, the measure-preserving flow $\varphi_t : M \to M$ is ergodic with respect to μ provided every measurable subset A of M which is invariant under φ_t has either $\mu(A) = 0$ or $\mu(A) = 1$.*

This very economical definition says that any measurable subset B of M with $0 < \mu(B) < 1$ must be 'moved around all over the place' by φ_t (if it 'lingers' anywhere, it risks becoming invariant). Any exceptional sets must either have measure zero (and are thus 'negligible'), or full measure 1 (and are thus essentially all of M, having 'nowhere to move'). It is this property of 'moving sets around all over the place' that takes the place of

Boltzmann's trajectories that pass through every point of M. This subtle shift—from moving single points throughout M, to moving measurable sets throughout M (with negligible exceptions)—is key, and allows one to prove the following corollary to Birkhoff's ergodic theorem.

Corollary. *Suppose (M, Σ, μ) is a probability space and the measure-preserving flow $\varphi_t : M \to M$ is ergodic with respect to μ. Then given any Lebesgue integrable function $f : M \to \mathbb{R}$, for almost all $x \in M$, the time average $\langle f \rangle(x)$ exists and equals the space average $\overline{f}$.*

This finally gave precise mathematical expression to Boltzmann's intuition from six decades earlier. However, in many ways the interesting part of the story begins here. Certainly the ergodic theorem was seen as a triumph,[19] and it launched the mathematical discipline of ergodic theory, which continues as a large and vigorous branch of dynamical systems to the present day, with strong links to other parts of mathematics. And we have already seen (cf. §§3.12.4–3.12.5) that in the first half of the 20th century, enthusiasm for ergodic systems built to a point where it was widely believed that generic Hamiltonian systems were ergodic.[20] It's also reasonable to assume that in the 1930s, researchers hoped that the Hamiltonian models of statistical mechanics would soon be shown to be ergodic, and that this would put equilibrium statistical mechanics on a solid foundation. Yet nearly a century later, it has not quite turned out that way.

(c) <u>Modern ergodic theory of Hamiltonian systems—and the nagging question in statistical physics</u>

Although a whole zoo of ergodic flows and maps are known in dynamical systems (many with much stronger statistical properties, such as mixing), very few smooth Hamiltonian systems arising from physics are known to be ergodic on their energy surfaces. And with the arrival of KAM theory in the 1950s and 60s, it was immediately suspected and later proved[21] that

---

[19] According to D. Szász [Szá96], when answering a questionnaire for the American Mathematical Society in 1954, J. von Neumann listed his version of the ergodic theorem among his most important discoveries.

[20] In fact, this belief seems to have persisted well into the second half of the 20th century; see footnote 70, p. 59, above.

[21] To be more precise, in the papers [MarkM70] and [MarkM74] (with very catchy titles), L. Markus and K.R. Meyer prove that, in the C^∞ topology, there is a generic set G of C^∞ Hamiltonian systems such that for any $H \in G$ it is false that H is ergodic on a dense set of its energy manifolds. In their proofs, the authors use the fact that $H \in G$ has an *elliptic fixed point* surrounded by KAM tori forming a set of positive measure; this set blocks ergodicity by being invariant. (See also §7.3.1 below.)

generic smooth Hamiltonian systems are not ergodic.

This seemed to put (the descendant of) the ergodic hypothesis in a situation analogous to the paradoxical existence of integrable behavior in nonintegrable systems described in §3.12.5. In other words, though the Hamiltonian models of statistical mechanics did not appear to be ergodic in the strict mathematical sense, in many cases practitioners of statistical mechanics got along just fine by assuming that systems behaved ergodically, and they often used the slogan "time averages equal space averages" with impunity. This, in simplified form, is the nagging question that still remains in the foundations of statistical mechanics, and which some physicists simply dismiss as uninteresting or irrelevant. However, a number of explanations have been proposed, some of which are nearly convincing (and may yet turn out to be wholly convincing). They're worth examining very briefly, but first, we need to look a little more closely at results on ergodicity of Hamiltonian models of statistical physics.

When considering the consequences of ergodic theory in statistical mechanics, it's important to keep the relevant problems in mind to avoid being drawn too far afield by other mathematical results (which tend to be very interesting in their own right). For concreteness, we recall that in classical statistical mechanics, the Hamiltonian model of a system of n particles, each of mass m, moving in three dimensions, is quite simply

$$H(q, p) = \frac{1}{2m} p^2 + V(q), \tag{7.8}$$

where the conjugate pair $(q, p) \in M \subset \mathbb{R}^{3n} \times \mathbb{R}^{3n}$ has a total of $6n$ components, $p^2 = p \cdot p$, and where the interactions among particles, or between particles and their environment, is modeled by the potential V. Crucially, in statistical mechanics, one wants results for (7.8) that are either independent of n, or valid in the so-called thermodynamic limit, as $n \to \infty$.

One of the first things one learns about the ergodic theory of (7.8) is that it is sensitive to the smoothness of V (i.e., to the *smoothness class* to which V belongs). Ideally one would like to find examples of V of class C^∞ with ergodic flow, but we already know that KAM theory makes this very difficult. On the other hand, the Oxtoby-Ulam result (cf. footnote 68, p. 59, above) suggests that it might be much simpler to find ergodic systems for potentials of low regularity. And although one ordinarily thinks of systems (7.8) with V of at least class C^2, the earliest and still simplest models of kinetic theory—'hardball systems'—may be interpreted as (7.8) with simple but highly irregular V (i.e., $V \equiv 0$ except at interaction boundaries, where V jumps to $V = \infty$). One speaks of these systems as having 'broken flow.'

(d) Ergodicity in hardball systems

Much is known about hardball systems. In 1963, Ya.G. Sinai conjectured [Sin63], and in 1970 proved [Sin70], that the system consisting of two small hard disks moving inside a square (with elastic, specular reflection between themselves, and between them and the sides of the square) is ergodic. This system turns out to be equivalent to a single point particle moving in a square with a single circular scatterer in the middle; this latter form is sometimes called the periodic Lorentz gas[22] (especially when it's filled with many non-interacting point particles, rather than one), and both systems are sometimes called the Sinai billiard. Like J. Hadamard's much earlier result,[23] to which it is closely related, Sinai's results hold for just one particle (in the billiard with obstacle).

Sinai and N.I. Chernov [SinC87] extended this result to higher dimensions (two hard balls in an N-dimensional box, or a point particle bouncing in an N-dimensional box with a spherical scatterer). Since then, much further progress has been achieved (more balls, more and different sorts of scatterers, etc.); yet curiously, the so-called Boltzmann-Sinai conjecture— that any number of (sufficiently small) elastic hard balls moving on the 2- or 3-dimensional torus is ergodic—has resisted rigorous proof up to now, although it appears tantalizingly close and is likely to be proved soon (see [Szá08] for a brief discussion). If so, it will represent a great achievement, both for Sinai and coworkers, and for Boltzmann's legacy.

(e) Ergodicity in 'soft' or smooth systems

Even as he announced his conjectures about hardball systems in 1963, Sinai was aware of the importance of showing ergodicity for systems with more regular potentials. In fact, in the paper containing the hardball conjectures [Sin63], Sinai also gives what is probably the first proof of ergodicity for a

---

[22]The Lorentz gas takes its name from the Dutch mathematical physicist Hendrik Lorentz (1853–1928), who proposed it as a model for a gas of completely ionized electrons moving in (and 'bouncing off' of) an array of fixed atoms. (Interactions between individual electrons in the gas are neglected.)

[23]At the end of the 19th century, before ergodic theory had a name, Hadamard developed the first results [Had98] for abstract flows. His idea may be explained with reference to (7.8) as follows. Take a single particle ($n = 1$) in two dimensions rather than three, set $V \equiv 0$, and replace the ordinary kinetic energy term $(1/2m)p^2$ by a metric tensor for the 'two-holed doughnut' (the simplest smooth compact 2-manifold with constant negative curvature). This models the geodesic flow of a single particle moving on the abstract doughnut surface. Hadamard found that this system had strong statistical properties (including what we now call ergodicity), and studied them with symbolic dynamics techniques that would eventually lead to the general concept of Markov partitions.

non-hardball physical particle system (the potential V is continuous, but has a jump-discontinuity in the first derivative). This was the start of a long series of efforts to get similar results for increasingly realistic model potentials. Here I touch on some of the high points of that series. I should say at the beginning that a 'soft' system has a continuous potential (V is of class C^0 or higher), while a 'smooth' system has a potential with continuous derivatives of all orders[24] (V is of class C^∞).

In 1976, using his own methods, I. Kubo [Kub76] got ergodicity results similar to Sinai's, but for 'compound' soft potentials (i.e., several soft scatterers). Later, A. Knauf [Kna87] used formidable geometric techniques to show that regularized Coulombic potentials (i.e., Coulombic potentials in which the singularities have been 'surgically removed' and replaced with 'smooth caps') could produce ergodic flow. To complicate matters, at about the same time P.R. Baldwin [Bal88] obtained results on a class of soft potentials proving non-ergodicity for certain parameter values, and gave numerical evidence indicating possible ergodicity for other parameter values. A notable advance occurred when V. Donnay and C. Liverani showed [DonL91] that attracting, repelling, and 'mixed' smooth potentials could all give rise to ergodic flows; and in the case of mixed potentials (i.e., possessing both attracting and repelling regions), they were able to give the first example of a 'physical' Hamiltonian (i.e., of the form kinetic + potential energy) with a smooth potential giving rise to ergodic flow. Still later, D. Turaev and V. Rom-Kedar [TuR98] obtained a subtle but interesting negative result, showing that there is a large class of systems with smooth potentials that are not ergodic, yet in a certain sense they are arbitrarily close approximations of hardball systems which are ergodic. Donnay [Don99] used similar methods to show, under certain conditions, the non-ergodicity of a system of two point particles interacting via smooth Lennard-Jones-type potentials[25] of finite range on a two-dimensional torus. In both of the last results, KAM-like techniques play an essential role in blocking ergodicity, and it is the smoothness of the systems that permits the application of KAM theory.

Looking back at these results (including many that are not described above), one is struck by their variety; there are examples of both ergodic

---

[24]Sometimes when authors say that a potential is smooth, they mean that it is of class C^k for some finite $k \geq 1$; in this case it's better to say explicitly that V is C^k-smooth, but abbreviation is common.

[25]Lennard-Jones-type potentials are convenient mathematical approximations to the (empirically determined) interaction potentials between atoms and/or molecules; they are also used to model the interaction of charged particles with atoms or molecules.

and non-ergodic smooth or soft potentials, and examples that display both behaviors over a range of parameter values. This complicated situation was to be expected, as we already knew that ergodic smooth Hamiltonian systems are exceptional. Yet the situation is in some ways worse than it appears. If the goal is to find ergodic examples of smooth potentials that are standard physical models of particle interactions, then the examples above fall short: essentially all of them have undergone some sort of 'surgery' or have 'finite range' (i.e., $V \equiv 0$ outside of convex curves or surfaces which form the boundaries of obstacles modeled by V). A smooth potential with obstacles having finite range is not radically different from a hardball system: although particles are not specularly reflected from obstacles, they 'interact' with obstacles, then move rectilinearly in the space between them.

In fact, for the smooth Hamiltonians one wants to use to model single-particle interactions in physics, KAM theory does seem to prevent the type of strong statistical properties one sees for hardball systems. But if we go back to our original model (7.8), we remember that we are interested in results for large n, or independent of n, or valid in the thermodynamic limit $n \to \infty$.

(f) Systems with large n, and the current status of the nagging question

To finally approach the 'nagging question'—why systems modeled by Hamiltonians of the form (7.8) seem to behave ergodically, yet the models themselves are often non-ergodic—we need to consider both what we know about single particle systems, as well as the case of large n.

We've seen that hardball systems offer encouragement, since they are ergodic for small n, and widely believed to be ergodic for any n. Unfortunately, as soon as such systems are made more realistic by smoothing, the situation becomes considerably more muddled, and the applicability of KAM theory appears to block ergodicity in many cases of interest.

A number of ideas have been offered to explain this situation. One recurring idea is that the strictly Hamiltonian formalism does not correctly model the physics. Variants of this explanation have some venerable proponents. For example, S. Smale has proposed [Sma80] that non-Hamiltonian perturbations of Hamiltonian systems could be the correct framework for statistical physics, and that their use might resolve problems surrounding the ergodic hypothesis. In his book [Macke92], M.C. Mackey suggests that statistical physics might be best modeled by non-invertible semi-dynamical systems (this stems from his work associated not directly with the ergodic hypothesis, but rather with the second law of thermodynamics).

However, one of the most widely held views—and one that stays within the Hamiltonian framework—is that the obstruction of ergodicity by KAM tori becomes negligible with increasing n, and vanishes in the limit $n \to \infty$. To be more precise, let ρ_n be the *relative measure* of the energy surface M that is occupied by KAM tori, where n is the number of particles in the system. A simple, widely held conjecture is that $\rho_n \to 0$ as $n \to \infty$. If true, then with increasing n, it would permit a larger and larger *ergodic component* in the complement of the KAM tori on M, so the flow could conceivably become ergodic in the thermodynamic limit.

A closely related, more nuanced approach uses the idea of 'ϵ-ergodicity.' Here one gives up the requirement of ergodicity on the entire energy surface, demanding instead that the flow be ergodic on most of the energy surface—more precisely, on a component having *relative measure* $1 - \epsilon$ ($0 < \epsilon \ll 1$). Proponents of this view argue that ϵ-ergodic systems display 'thermodynamic-like' behavior that is essentially indistinguishable from the behavior of fully ergodic systems, provided ϵ is small enough. In this view, the paradox surrounding the ergodic hypothesis might disappear even before the thermodynamic limit (i.e., for large but finite values of n). A nice account of this ϵ-ergodicity conjecture appears in [FriW11].

There is support for these latter conjectures; yet there is also evidence against them, or there are yet other mitigating factors. In [Szá96], D. Szász cites different numerical experiments that support and oppose the conjectures, while in [Broe04], H. Broer references the PhD thesis of H. de Jong as lending support by showing exponential decay of ρ_n with n for a model system of weakly coupled pendulums.

But even if $\rho_n \to 0$ as $n \to \infty$, it does not follow that the Hamiltonian flow becomes 'ergodic enough' on the energy surface M or a large portion of it. It appears that, even if the obstruction to ergodicity provided by the presence of KAM tori themselves disappears in the thermodynamic limit, other problems may remain. This however is most clearly seen not in equilibrium statistical mechanics, but in the much trickier business of the approach to equilibrium, where a version of the nagging question arises concerning systems that are expected to 'thermalize' but do not. This is discussed in the next subsection. At this point, it is worth noting that, more than a century after Boltzmann's first statement of the ergodic hypothesis, its legacy remains unsettled. The nagging questions still nag, and even proliferate.

7.2.3   *Equipartition of energy, FPU, and the ultraviolet catastrophe revisited*

In this subsection we approach the problem outlined above from a closely related but slightly different angle, by following the trail of research into the so-called Fermi-Pasta-Ulam (FPU) paradox. What started in the 1950s as an innocent numerical experiment with one of the world's first computers has evolved into a theoretical quest to understand the approach to *thermal equilibrium* in crystals and related structures, beginning with one-dimensional models. The story begins slowly, with a relatively simple paradox, then later becomes more complicated, as each explanation of the paradox is revealed to be not quite the answer it seemed to be initially. Although a full resolution seems tantalizingly close, most researchers agree that it has still not been achieved. It appears however that KAM and Nekhoroshev theory may have a role to play.

(a) Background

Quantum mechanics is said to have begun in the last months of the year 1900 when Max Planck looked into the thorny issues surrounding radiation emitted from a classical black body. Later, in 1911, P. Ehrenfest would use the term 'ultraviolet catastrophe' to describe the problem in classical physics that Planck had resolved in those final months of the 19th century. In fact, the classical expression for the radiative energy emitted by an ideal black body (either as a function of wavelength or of frequency ν), was derived by Lord Rayleigh (J.W. Strutt) and J.H. Jeans, is called the Rayleigh-Jeans law, and is untenable because it blows up as $\nu \to \infty$. It was derived under the assumption of equipartition[26] of energy among the 'continuum of resonators' comprising the black body. Planck found an ad-hoc solution to the catastrophe by assuming that energy was distributed not continuously with respect to ν, but rather discretely, in multiples (by ν) of a fixed quantity he denoted by h (now called Planck's constant). By his own account, he arrived at this solution reluctantly, by borrowing Boltzmann's ideas about entropy, as well as so-called 'coarse-graining' methods used by Gibbs and Boltzmann to study phase space ensembles in statis-

---

[26]The idea of equipartition—that in a system of many degrees of freedom at equilibrium, each degree of freedom should share an equal portion of the system's kinetic energy—goes back to the work of J. Waterston in the mid 19th century, and was further built upon by R. Clausius, Maxwell, Boltzmann, and others. It is a theorem for classical Hamiltonian systems satisfying certain assumptions, but there are also examples of Hamiltonian systems that do not obey equipartition.

tical mechanics.[27] A few years later, in a series of papers [Ei05], [Ei06], [Ei07], A. Einstein asserted that Planck's assumption was more than an appearance-saving device, that it instead reflected fundamental laws about the interaction of energy and matter. By taking quantization of energy to be fundamental, he explained in a unified way the photoelectric effect, the ultraviolet catastrophe, and the discrepancy between measurements of specific heats and their classical models. Thus was the quantum revolution fully begun.

But before quantum mechanics swept aside classical mechanics in these matters, there was a brief interval during which the classical view was defended (or more precisely, its apparent defects explained), most notably by Jeans, but also less explicitly by Boltzmann. In two articles [Je03], [Je05], Jeans argues heuristically and classically that 'full thermalization' might represent an equilibrium state that would not be reached on natural timescales. Writing about the classical problem of specific heats by considering the exchange of energy between the translational and internal modes of motion[28] in molecular dynamics, he argued that energy would move to the higher frequency internal modes at an extremely slow rate; in fact he wrote explicit expressions of the form $T \sim T_0 e^{\tau \omega}$ for the 'relaxation time' (time to reach equilibrium) of a mode resonating at frequency ω (here T_0 is the mean time between collisions of molecules, and τ is the mean duration of collisions). He remarked that realistic relaxation times might reach 'billions of years.' And although Boltzmann was not so explicit, he also writes (in §45 of Part II of [Bolt96–98]) that "over a long period of time rotation will be equilibrated with other molecular motions, so slowly that such energy exchanges escape our observation." (There follows a similar remark concerning the equilibration of vibrational modes; both remarks are distilled from Boltzmann's earlier paper on molecular dynamics [Bolt95].)

(b) The Fermi-Pasta-Ulam (FPU) experiment

It's not surprising that Boltzmann's and Jean's ideas passed with little notice in the heady days of quantum theory that followed. However, they began to seem relevant a half-century later, when E. Fermi, J. Pasta, and S.M. Ulam performed the numerical experiment mentioned above in §3.12.5.

---

[27] For an engaging account of Planck's work, see [Kra00].

[28] In molecular dynamics, the internal modes of motion include rotation (about axes of symmetry) and vibration (along or among the bonds joining the constituent atoms). Since they absorb thermal energy, these internal modes are important in determining the specific heat of a particular substance.

With access to the best electronic computing machine of the time, Fermi and coworkers were keenly interested to see how a classical system would 'thermalize,' or evolve toward equipartition. Working in 1953 with the aid of computer programmer Mary Tsingou on the MANIAC I computer at Los Alamos, they modeled a one-dimensional crystal lattice with weak nonlinear couplings between atoms. More precisely, they modeled first 32, then 64 linear oscillators coupled together by weak quadratic or cubic nonlinearities. They gave the system an initial configuration with kinetic energy concentrated in the lowest frequency mode, and they allowed the system to evolve over a few hundred oscillations. They expected to see the energy transfer slowly to the higher frequency modes, and eventually spread evenly throughout all available modes as dictated by the equipartition principle. And at first, they did see something roughly like that; but when the system was allowed to run for longer times, they were astonished to see almost all the energy (97%) return to the lowest mode. Writing in the now-famous technical report [FerPU55], the authors remarked in an understated way that "...the results of our computations show features which were, from the beginning, surprising to us." The series of numerical computations reported by the authors came to be known as the FPU experiment (leading to the FPU 'paradox'), and it touched off a slow-motion search for explanations that has continued to the present day. For an engaging historical account, see the book [Weis97]; here I'll compress the story into a few paragraphs, with emphasis on the role played by KAM and Nekhoroshev theory.

It was not long before researchers began to see parallels between KAM theory and the FPU experiment, since both showed the surprising persistence of integrable-like behavior in perturbations of integrable systems. But a rigorous connection remained elusive. The first meaningful explanation came a decade later, in 1965, when N. Zabusky and M. Kruskal showed that the leading-order continuum approximation of the FPU system was the Korteweg-de Vries (KdV) equation, for which they numerically demonstrated the existence of soliton[29] solutions [ZabK65]. The authors argued that the recurrent behavior seen in the FPU experiment was a discrete analog of the solitons in the KdV equation. Though significant as an explanation of FPU, Zabusky and Kruskal's work was an even more important breakthrough for the KdV and similar equations, and it soon led to a new theory of integrability for infinite-dimensional systems and PDEs (and

---

[29]Soliton was the word coined by Zabusky and Kruskal to describe the stable, localized nonlinear waves they discovered in solutions of the KdV equation.

eventually also to KAM theory for PDEs). This integrability theory is now seen in retrospect as one of the most important developments of the 20th century in PDEs and applied mathematics.

The next contribution toward explaining FPU was more firmly rooted in KAM theory. Russian researchers F. Izrailev and B.V. Chirikov argued that the FPU problem was a perturbation of an integrable system (of uncoupled linear oscillators), and that the recurrent behavior represented quasiperiodic solutions confined to KAM tori [IzC66]. Although they did not rigorously apply a KAM theorem to the FPU system, they reasoned that this should be possible, in which case the total initial energy E would be the perturbation parameter. If so, then there should be a critical or threshold energy $E_c > 0$ such that KAM tori cease to exist for $E > E_c$, thus permitting thermalization as originally expected. The authors used Chirikov's resonance overlap criterion (see §6.3.1 above) to estimate E_c, and found satisfactory agreement with results from renewed FPU-like experiments. In other words, thermalization indeed occurs in the FPU problem for E roughly larger than E_c, which is to say the paradox disappears in this higher energy range.

A little later, Izrailev and Chirikov also conjectured that for systems similar to FPU, but now with n degrees of freedom, the 'specific critical energy' $E_c(n)/n$ should decrease to 0 with increasing n; that is, the FPU paradox should disappear in the thermodynamic limit. This is closely related to the conjecture mentioned toward the end of the previous subsection concerning the vanishing of the relative measure of KAM tori in the thermodynamic limit, so it's perhaps not surprising that this conjecture also remains in an uncertain state: it appears to be true in certain cases, and not true in others, depending on how the energy is initially distributed, or on the type of potential used to model atomic interactions (one can for example replace the interactions of the original FPU experiment by more realistic Lennard-Jones potentials, as in [BoccSBL70], where numerical experiments suggest that $E_c(n)/n$ remains bounded away from zero as $n \to \infty$).

Continued numerical exploration and analysis of FPU-type problems has also revealed another complicating phenomenon, namely the appearance of 'metastable states,' where systems seem to display two timescales: an intermediate timescale, on which packets of low-frequency modes are thermalized but high-frequency modes remain 'frozen'; and a much longer timescale, on which the high-frequency modes are finally also thermalized and the system settles to a true equilibrium. This is remarkably reminiscent of Boltzmann and Jeans' ideas about the very long times required for thermalization of high-frequency modes in molecular dynamics. Numerical evidence of metastable states was first reported in the early 1980s in a paper [FuMMPPRV82] where they were also first described as such, but it took more than twenty years for other numerical and theoretical studies to catch up and verify their existence in other cases [BercGG04], [BamP06]. The reader will not be surprised to learn that other researchers have reported systems and/or initial conditions that do not display metastability [Ben05b], [PettiCCFC05].

In summary, the behavior of FPU-like systems is far from simple. Rather, depending on initial energy, initial conditions, interaction rules, etc., they appear capable of displaying almost any combination of the behaviors just described. This complex state of affairs was described in detail by the 20 papers appearing in a 2005 'focus issue' (Issue 1, Volume 15) of the journal *Chaos* devoted entirely to the FPU problem. (These papers treat not only the issues discussed here, but also such topics as heat conduction, 'chaotic breathers,' Bose-Einstein condensates, and more.)

More recently, rigorous mathematical support for certain scenarios in FPU-like systems has appeared. In [BamP06], the authors use resonant normal forms in a way analogous to Nekhoroshev theory to show the existence of metastable states. And in [Rin06] and [HenrK08], the existence of KAM tori in FPU-like systems is rigorously demonstrated by using integrable systems based on the Toda lattice in the first paper and on Birkhoff normal forms in the second.

(c) Concluding remarks on HPT, FPU, and the ultraviolet catastrophe

There was a time in the 1980s and 90s when it seemed to some as though HPT—and especially the application of specialized variants of Nekhoroshev theory—might provide a classical (as opposed to quantum) explanation for the ultraviolet catastrophe. In other words, the history of physics in which Planck was led to introduce the quantum hypothesis might need to be revisited once the corrected classical model was shown to no longer need Planck's ad-hoc assumption. No one suggested that quantum mechanics was faulty, but some did go so far as to think (or dream) that the underlying deterministic cause of quantum mechanics long sought by Einstein and others[30] might be reached in this way (remarks along these lines may be found, for example in [Pat87] or [Galg93]). At present it seems that these hopes were naïve or at least premature, but it is not hard to understand their origin when one looks at the remarkable early results. And even if

---

[30]This is the so-called hidden variable theory of quantum mechanics.

these results are not as widely applicable as was hoped by some, it also appears that they cannot be 'explained away' as easily as was hoped by others. In following the trail of the FPU paradox, we see again that the foundations of statistical mechanics are not as firm or as clearly laid out as once supposed.

7.3 Other applications of KAM in physics

Celestial and statistical mechanics are the most 'celebrated' parts of physics where KAM theory has been applied or has consequences, but it has had a quieter impact in other parts as well. We begin by pointing out one of the most common ways that KAM theory arises in applications to physical systems modeled by Hamiltonian dynamics.

7.3.1 *The generic application: elliptic equilibria of Hamiltonian systems*

In a physical system modeled by a Hamiltonian, one often[31] finds *elliptic equilibria*. When the eigenvalues of the linear part of such an equilibrium satisfy nonresonance conditions to a certain order, the system can be reduced to *Birkhoff normal form* in a neighborhood there, and can then be rescaled so that it becomes a nearly integrable system with small parameter ε representing the distance to equilibrium. Application of an appropriate KAM theorem then gives a set of KAM tori with positive *Liouville measure* in any neighborhood of the equilibrium (and this set becomes 'denser' as one gets closer to the equilibrium; the equilibrium is in fact an accumulation point of the set of tori). This shows that the equilibrium is probabilistically stable (or topologically stable if it has only two degrees of freedom). If more than probabilistic stability is needed, one can apply Nekhoroshev theory to achieve it, at least on finite time intervals.

Together, KAM and Nekhoroshev theory show a fairly strong form of stability in the neighborhood of such points: if one starts close enough to a nonresonant elliptic equilibrium, there is a positive probability of (eternal) stability; and in the less lucky event of not starting on a KAM torus, one still remains close to equilibrium for a very long time (before drifting away via Arnold diffusion). The nicest exposition of Nekhoroshev theory for

---

[31] Elliptic equilibria occur not only often, but in fact generically in Hamiltonian systems (with respect to the C^∞ topology; cf. [MarkM74]), whence the title of this subsection.

this case is probably contained in J. Pöschel's paper [Pös99]. Important examples of this type of application are discussed or mentioned below.

7.3.2 *Stability of charged particle motions in electric or magnetic fields*

One of the ways the aforementioned generic application frequently arises is in problems involving charged particles moving in electric or magnetic fields. This is the third type of important physical system that is well-modeled by Hamiltonian dynamics.

Already in Chapter II of Arnold's paper [Ar63b] (before the famous application of KAM theory to the n body problem in Chapter III), there is a discussion of *adiabatic invariance* for Hamiltonian systems undergoing slow periodic variation. Arnold uses properly degenerate KAM theory to prove perpetual adiabatic invariance for such systems, then applies the result to show that slow-moving charged particles may be forever trapped by magnetic field lines of correctly converging shape and intensity, at least in the one-dimensional model considered.

The (partial) reflection and trapping of charged particles by convergent magnetic field lines has long been known (it's the basic mechanism by which particles are trapped in the earth's Van Allen belts), but Arnold's result was the first to suggest mathematically that the trapping could be perpetual under ideal conditions. In the 1960s and 70s, physicists gave serious consideration to various magnetic bottle devices as a means of confining plasmas in fusion reactors. For a host of technical reasons, most of these have been abandoned as unworkable, but the closely related tokamak and stellerator designs retain some hope of eventual success.

By contrast, particle accelerators (at CERN, Fermilab and elsewhere) have been working successfully for decades. Their ever-increasing operating energies attest to accelerator physicists' understanding of how to guide beams of charged particles with powerful magnetic fields and radio frequency (RF) generators. Much of the theory behind this understanding is based on HPT, since the most basic models for accelerator dynamics are perturbations of integrable Hamiltonian systems. KAM theory is therefore directly implicated, but here the situation is analogous to dynamical astronomy: KAM stands as a conceptual pillar on which the theory is based, resulting in an ideal of particle beams that remain perpetually stable for well-chosen initial conditions. Actual quantitative models of accelerators employ sophisticated numerical methods called 'tracking codes' that incorporate the individual effects of hundreds of magnetic and RF elements into a kind of Poincaré map that represents the beam's evolution during one orbit around the accelerator. This map is then iterated in order to study its stability properties, and indeed, a rich dynamical picture emerges with a mix of behaviors, including stochasticity, invariant or near-invariant surfaces, resonant islands, and so on. All of this may then be discussed and interpreted within the framework of HPT, though accelerator physicists certainly do not confine themselves to this framework alone, and even when they do, they often use their own specialized vocabulary. The reader interested in this topic will find it discussed not only in textbooks and monographs (such as [ConteM08], [Mic95], [Wie93], [Wie95], or [Wil00]) but also in the proceedings of conferences such as EPAC and JPAC.[32]

7.3.3 *More exotic applications*

The physical applications of KAM discussed up to now involve finite-dimensional, classical Hamiltonian systems—the avowed subject of this book. Beyond this however, lie problems in quantum mechanics or infinite-dimensional systems to which KAM theory or KAM-like techniques are applied with increasing success. A few words about these problems may help orient the interested reader.

One standard way that KAM theory naturally arises in infinite-dimensional systems is in the context of an operator on a function space having a dense point spectrum. Since the corresponding resolvent operator does not exist for eigenvalues in this dense set, perturbation expansions of the resolvent run into small divisor problems very similar to those in KAM theory, and KAM techniques may often be adapted to overcome them. Here I list some of the areas where this or related strategies have been used, or are beginning to be used, with a few words of description and references.

First, in the papers of G. Gallavotti and coworkers [Gall95a], [GallGM95], and in a later paper by J. Bricmont *et al.* [BricGK99], a connection is established between the group renormalization approach to quantum field theory and the direct methods of KAM proofs pioneered by L.H. Eliasson (cf. §4.8 (d)).

In H. Broer *et al.* [BroePS03], the authors exploit some of the earlier work of J. Fröhlich and T. Spencer [FröS83] (again together with group renormalization) to understand the Anderson localization phenomenon

---

[32]These acronyms stand for European Particle Accelerator Conference and Joint Particle Accelerator Conference.

(also called strong localization, or electron localization) as a kind of 'stability pocket' of Mathieu type. They use KAM techniques to overcome problems arising from the dense point spectrum of the Schrödinger operator.

Interesting physical problems with infinite-dimensional phase spaces also arise in classical PDEs. Most recently, C. Mouhot and C. Villani employ KAM methods in explaining the Landau damping phenomenon in systems modeled by the nonlinear Vlasov-Poisson equation [MouhV11]. (A *weak KAM theory* for the Vlasov-Poisson equation developed by W. Gangbo and A. Tudorascu [GanT10] may also prove useful in this effort.) This is a surprising result (Landau damping had long resisted fully rigorous explanation), and it figured prominently in the citations of Villani's work when he received a *Fields medal* at the 2010 ICM in Hyderabad, India.

Appendix A

Kolmogorov's 1954 paper

The following is the English translation by H. Dahlby of Kolmogorov's paper [Kol54] originally published in Russian in *Doklady Akademii Nauk SSSR* (*Proceedings of the USSR Academy of Sciences*) **98**, 527–530 (1954).

The translation was produced under the auspices of the Los Alamos Scientific Laboratory (now Los Alamos National Laboratory) and first published in *Stochastic Behavior in Classical and Quantum Hamiltonian Systems* (Lecture Notes in Physics Vol. 93, G. Casati and J. Ford, Eds.), Springer-Verlag, Berlin Heidelberg, 1979, pp. 51–56.

The translation is reproduced here with kind permission from Springer Science+Business Media, and with grateful acknowledgment to the Los Alamos National Laboratory in Los Alamos, New Mexico.

Author's note. This translation has been very useful to English readers since its first publication. The reader may wish to know the following minor ways in which it differs from Kolmogorov's original article in Russian: (i) The first eight lines following Kolmogorov's name (from Kolmogorov's university affiliation and address, through the heading 'Theorem and discussion of proof') do not appear in the original. (ii) In the statement of Theorem 2 (last page), 'Lebesgue degree' and 'complete degree' would be more faithfully rendered as 'Lebesgue measure' and 'full measure' respectively. (iii) In the sentence immediately following Theorem 2, 'the set M_θ' should be 'the complement of the set M_θ'.

PRESERVATION OF CONDITIONALLY PERIODIC MOVEMENTS WITH SMALL CHANGE IN THE HAMILTON FUNCTION*

Academician A.N. Kolmogorov

Department of Mathematics

Moscow State University

117234 Moscow, B–234

U.S.S.R.

ABSTRACT

This paper is a translation of Kolmogorov's original article announcing the theorem now known as the KAM theorem.

THEOREM AND DISCUSSION OF PROOF

Let us consider in the $2s$-dimensional phase space of a dynamic system with s degrees of freedom the region G, represented as the product of an s-dimensional torus, T, and a region S, of a Euclidean s-dimensional space. We will designate the points of the torus, T, by the circular coordinates $q_1, \ldots, q_s$ (replacing q_α with $q'_\alpha = q_\alpha + 2\pi$ does not change points q), and the coordinates of the points, p, of S we will designate as $p_1, \ldots, p_s$. We will assume that in region G, in the coordinates $(q_1, \ldots, q_s, p_1, \ldots, p_s)$ the equations of motion have the canonical form

$$\frac{dq_\alpha}{dt} = \frac{\partial}{\partial p_\alpha} H(q,p), \quad \frac{dp_\alpha}{dt} = -\frac{\partial}{\partial q_\alpha} H(q,p). \qquad (1)$$

The Hamiltonian function, H, is further assumed as dependent on the parameter θ and determined for all $(q,p) \in G$, $\theta \in (-c; +c)$, but not time-dependent. Moreover, further considerations require fairly significant restrictions on the smoothness of the function $H(q,p,\theta)$, stronger than infinite differentiability. For simplicity, in the following it is assumed that the function $H(q,p,\theta)$ is analytic over the set of variables (q,p,θ).

Summation over the Greek indices is assumed to be from 1 to s. The usual vector designations $(x,y) = \sum_\alpha x_\alpha y_\alpha$, $|x| = +\sqrt{(x,x)}$ are used. A whole number vector indicates a vector for which all the components are whole numbers. The set of points (q,p) of G with $p = c$ is designated by T_c. In Theorem 1 it is assumed that S contains the point $p = 0$, i.e., $T_0 \subseteq G$.

Theorem 1. *Let*

$$H(q,p,0) = m + \sum_\alpha \lambda_\alpha p_\alpha + \frac{1}{2} \sum_{\alpha\beta} \Phi_{\alpha\beta}(q)\, p_\alpha p_\beta + O(|p|^3), \qquad (2)$$

---

*Los Alamos Scientific Laboratory translation LA–TR–71–67 by Helen Dahlby of Akad. Nauk. S.S.S.R., Doklady **98**, 527 (1954).

where m and λ_α are constants and for a certain choice of constants $c > 0$ and $\eta > 0$ for all whole-number vectors, n, the inequality

$$(n, \lambda) \geq \frac{c}{|n|^\eta} \qquad (3)$$

is satisfied.

Let, moreover, the determinant composed of the average values

$$\varphi_{\alpha\beta}(0) = \frac{1}{(2\pi)^s} \int_0^{2\pi} \cdots \int_0^{2\pi} \Phi_{\alpha\beta}(q)\, dq_1 \ldots dq_s$$

of the functions

$$\Phi_{\alpha\beta}(q) = \frac{\partial^2}{\partial p_\alpha \partial p_\beta} H(q, 0, 0)$$

be different from zero:

$$|\varphi_{\alpha\beta}(0)| \neq 0 . \qquad (4)$$

Then there exist analytic functions $F_\alpha(Q, P, \theta)$ and $G_\alpha(Q, P, \theta)$ which are determined for all sufficiently small θ and for all points (Q, P) of some neighborhood, V, of the set T_0, which bring about a contact transformation

$$q_\alpha = Q_\alpha + \theta F_\alpha(Q, P, \theta), \qquad p_\alpha = P_\alpha + \theta G_\alpha(Q, P, \theta)$$

of V into $V' \subseteq G$, which reduces H to the form

$$H = M(\theta) + \sum_\alpha \lambda_\alpha P_\alpha + O(|P|^2) \qquad (5)$$

($M(\theta)$ does not depend on Q and P).

It is easy to grasp the meaning of Theorem 1 for mechanics. It indicates that an s-parametric family of conditionally periodic motions

$$q_\alpha = \lambda_\alpha t + q_\alpha^{(0)}, \quad p_\alpha = 0,$$

which exists at $\theta = 0$ cannot, under conditions (3) and (4), disappear as a result of a small change in the Hamilton function H: there occurs only a displacement of the s-dimensional torus, T_0, around which the trajectories of these motions run, into the torus $P = 0$, which remains filled by the trajectories of conditionally periodic motions with the same frequencies $\lambda_1, \ldots, \lambda_s$.

The transformation

$$(Q, P) = K_\theta(q, p),$$

the existence of which is confirmed in Theorem 1, can be constructed in the form of the limit of the transformations

$$(Q^{(k)}, P^{(k)}) = K_\theta^{(k)}(q, p),$$

where the transformations

$$(Q^{(1)}, P^{(1)}) = L_\theta^{(1)}(q, p), \quad (Q^{(k+1)}, P^{(k+1)}) = L_\theta^{(k+1)}(Q^{(k)}, P^{(k)})$$

are found by the "generalized Newton method" (see Ref. 1). In this note we confine ourselves to the construction of the transformation $K_\theta^{(1)} = L_\theta^{(1)}$, which itself permits grasping the role of conditions (3) and (4) of Theorem 1. Let us apply the transformation $L_\theta^{(1)}$ to the equations

$$Q_\alpha^{(1)} = q_\alpha + \theta Y_\alpha(q), \tag{6}$$

$$p_\alpha = P_\alpha^{(1)} = \theta \left\{ \sum_\beta P_\beta^{(1)} \frac{\partial Y_\beta}{\partial q_\alpha} + \xi_\alpha + \frac{\partial}{\partial q_\alpha} X(q) \right\}$$

(it is easy to verify that this is a contact transformation) and seek the constants ξ_α and ζ and the functions $X(q)$ and $Y_\beta(q)$, starting from the requirement that

$$H = m + \sum_\alpha \lambda_\alpha p_\alpha + \frac{1}{2} \sum_{\alpha\beta} \Phi_{\alpha\beta}(q) p_\alpha p_\beta +$$

$$+ \theta \left\{ A(q) + \sum_\alpha B_\alpha(q) p_\alpha \right\} + O(|p|^3 + \theta |p|^2 + \theta^2) \tag{7}$$

take the form

$$H = m + \theta\zeta + \sum_\alpha \lambda_\alpha P_\alpha^{(1)} + O(|P^{(1)}|^2 + \theta^2). \tag{8}$$

Substituting (6) into (7), we get

$$H = m + \sum_\alpha \lambda_\alpha P_\alpha^{(1)} + \theta \left\{ A + \sum_\alpha \lambda_\alpha \left(\xi_\alpha + \frac{\partial X}{\partial q_\alpha} \right) \right\} +$$

$$\theta \sum_\alpha P_\alpha^{(1)} \left\{ B_\alpha + \sum_\beta \Phi_{\alpha\beta}(q) \left(\xi_\beta + \frac{\partial X}{\partial q_\beta} \right) + \sum_\beta \lambda_\beta \frac{\partial Y_\beta}{\partial q_\beta} \right\} + O(|P^{(1)}|^2 + \theta^2).$$

Thus, our requirement (8) reduces to the equations

$$A + \sum_\alpha \lambda_\alpha \left(\xi_\alpha + \frac{\partial X}{\partial q_\alpha} \right) = \zeta, \tag{9}$$

$$B_\alpha + \sum_\beta \Phi_{\alpha\beta} \left(\xi_\beta + \frac{\partial X}{\partial q_\beta} \right) + \sum_\beta \lambda_\beta \frac{\partial Y_\alpha}{\partial q_\beta} = 0 \tag{10}$$

being fulfilled.

Let us introduce the functions

$$Z_\alpha(q) = \sum_\beta \Phi_{\alpha\beta}(q) \frac{\partial}{\partial q_\beta} X(q). \tag{11}$$

Expanding the functions $\Phi_{\alpha\beta}$, A, B_α, X, Y_α, Z_α in a Fourier series of the type

$$X(q) = \sum x(n) e^{i(n,q)}$$

and assuming for definiteness that

$$x(0) = 0, \quad y(0) = 0, \qquad (12)$$

we get for the remaining Fourier coefficients $x(n)$, $y_\alpha(n)$, and $z_\alpha(n)$ and constants ξ_α and ζ of the equation which are relevant to the determination

$$a(0) + \sum \lambda_\alpha \xi_\alpha = \zeta, \qquad (13)$$

$$a(n) + (n, \lambda) x(n) = 0 \quad \text{for} \quad n \neq 0, \qquad (14)$$

$$b_\alpha(0) + \sum_\beta \varphi_\alpha(0) \xi_\beta + z_\alpha(0) = 0, \qquad (15)$$

$$b_\alpha(n) + \sum_\beta \varphi_{\alpha\beta}(n) \xi_\beta + z_\alpha(n) + (n, \lambda) y_\alpha(n) = 0 \quad \text{for} \quad n \neq 0. \qquad (16)$$

It is easy to see that the system $(11) - (16)$ is unambiguously solved under conditions (3) and (4). Condition (3) is important in the determination of $x(n)$ from (14), and in the determination of $y_\alpha(n)$ from (16). Condition (4) is important in the determination of ξ_β from (15). Since, as $|n|$ increases, the coefficients of the Fourier series of the analytic functions $\Phi_{\alpha\beta}$, A, and B_α have an order of decrease not less than $\rho^{|n|}$, $\rho < 1$, then from condition (3) there results not only the formal solvability of equations $(13) - (16)$, but also the convergence of the Fourier series for the functions X, Y_α and Z_α and the analyticity of these functions. The construction of further approximations is not associated with new difficulties. Only the use of condition (3) for proving the convergence of the recursions $K_\theta^{(k)}$ to the analytic limit for the recursion K_θ is somewhat more subtle.

The condition of the absence of "small denominators" (3) should be considered, "generally speaking," as fulfilled, since for any $\eta > s - 1$ for all points of an s-dimensional space $\lambda = (\lambda_1, \ldots, \lambda_s)$ except the set of Lebesgue measure zero it is possible to find $c(\lambda)$ for which

$$(n, \lambda) \geq \frac{c(\lambda)}{|n|^\eta},$$

whatever the integers $n_1, n_2, \ldots, n_s$ are.[2] It is also natural to consider condition (4) as, "generally speaking," fulfilled. Since

$$\varphi_{\alpha\beta}(0) = \frac{\partial}{\partial p_\alpha} \lambda_\beta(0),$$

where

$$\lambda_\beta(p) = \frac{1}{(2\pi)^s} \int_0^{2\pi} \cdots \int_0^{2\pi} \frac{dq_\beta}{dt} dq_1 \ldots dq_s$$

is the frequency averaged over the coordinate q_β with fixed momenta $p_1, \ldots, p_s$, condition (3) means that the Jacobian of the average frequencies over the momenta is different from zero.

Let us turn now to a consideration of the special case where $H(q, p, 0)$ depends only on p, i.e., $H(q, p, 0) = W(p)$. In this case, for $\theta = 0$ each torus T_p consists of the complete trajectories of the conditionally periodic movements with frequencies

$$\lambda_\alpha(p) = \frac{\partial W}{\partial p_\alpha}.$$

If the Jacobian

$$J = \left| \frac{\partial \lambda_\alpha}{\partial p_\beta} \right| = \left| \frac{\partial^2 W}{\partial p_\alpha \partial p_\beta} \right| \tag{17}$$

is different from zero, then it is possible to apply Theorem 1 to almost all tori T_p. There arises the natural hypothesis that at small θ, the "displaced tori" obtained in accordance with Theorem 1 fill a larger part of region G. This is also confirmed by Theorem 2, pointed out later. In the formulation of this theorem we will consider the region S to be bounded and will introduce into consideration the set M_θ of those points $(q^{(0)}, p^{(0)}) \in G$ for which the solution

$$q_\alpha(t) = f_\alpha(t; q^{(0)}, p^{(0)}, \theta), \quad p_\alpha(t) = G_\alpha(t; q^{(0)}, p^{(0)}, \theta)$$

of the system of equations (1) with initial conditions

$$q_\alpha(0) = q_\alpha^{(0)}, \quad p_\alpha(0) = p_\alpha^{(0)}$$

leads to trajectories not moving out of the region G with change in t from $-\infty$ to $+\infty$, and conditionally periodic with periods $\lambda_\alpha = \lambda_\alpha(q^{(0)}, p^{(0)}, \theta)$, i.e., it has the form

$$f_\alpha(t) = \varphi_\alpha(e^{i\lambda_1 t}, \ldots, e^{i\lambda_s t}), \quad g_\alpha(t) = \psi_\alpha(e^{i\lambda_1 t}, \ldots, e^{i\lambda_s t}).$$

Theorem 2. *If $H(q, p, 0) = W(p)$ and determinant (17) is not equal to zero in region S, then for $\theta \to 0$ the Lebesgue degree of the set M_θ converges to the complete degree of region S.*

Apparently, in the usual sense of the phrase, "general case" is when the set M_θ at all positive θ is everywhere dense. In such a case the complications arising in the theory of analytical dynamic systems are indicated more specifically in my note.[3]

REFERENCES

1. L.V. Kantorovich. *Uspekhi Matem. Nauk* **3**, 163 (1948).
2. J.F. Koksma, *Diophantische Approximationen*, Chelsea 1936. 157pp.
3. A.N. Kolmogorov, *Doklady Akad. Nauk* **93**, 763 (1953).

Appendix B

Overview of Low-dimensional Small Divisor Problems

In this appendix, I look at two lines of research that run alongside full-fledged KAM theory, acting as a kind of 'technical laboratory' in which some of the purely mathematical issues surrounding small divisors may be isolated, simplified, and resolved. Important cross-fertilization has occurred in both directions between these lines and KAM theory, most significantly when Siegel's first solution of a small divisor problem pointed the way for Kolmogorov and KAM, but also soon thereafter when Arnold used Kolmogorov's insight into Newton's method to understand properties of low-dimensional maps (analytic circle diffeomorphisms).

Perhaps not surprisingly, Poincaré has a hand in the beginning of this story (in fact in the beginning of both lines of research discussed here), and there is also a French *Fields medal* waiting at the end—the end as far as we go, that is.

B.1 The linearization problem

The first line of research concerns the old problem of linearizing a mapping of the plane in the neighborhood of a fixed point—in other words, finding an invertible change of coordinates that transforms the mapping to its linear part at the fixed point. This is the subject that first led to a solution of the small divisor problem.

B.1.1 From Schröder's functional equation to the Siegel center problem

The subject begins in the early 1870s with the publication of E. Schröder's paper [Schr71], in which he sets out the following problem:

Let N be a *neighborhood* of the origin in the complex plane $\mathbb{C}$, and $\varphi : N \to \mathbb{C}$ be an *analytic map* with *fixed point* at the origin ($\varphi(0) = 0$). On some *open* disk D centered at the origin, we can expand φ in a *convergent power series* $\varphi(z) = \mu z + O(z^2)$, where μ is a fixed complex number, and $L(z) = \mu z$ is the *linear part* of φ. The problem is, Under what conditions can the map φ be *linearized*? In other words, when is it possible to find an *invertible* analytic change of variables $g : D \to g(D)$ (also with $g(0) = 0$) such that

$$g(\varphi(z)) = L(g(z)) \tag{B.1}$$

on some open neighborhood of 0?

Equation (B.1) is known as Schröder's functional equation, and when it holds we say that the map g *conjugates* φ to L. A closely related problem was treated by Poincaré in his 1879 thesis [Poi79]. For a suitably smooth *vector field* $f : \mathbb{R}^n \to \mathbb{R}^n$ with fixed point 0, Poincaré found conditions on the eigenvalues of $Df(0)$ under which f is conjugate to $Df(0)$ near 0 (which partly answers the question, When can the nonlinear system of ODEs $dx/dt = f(x)$ be linearized at 0?). These questions arise again and again in dynamical systems. The linear part L of φ is very nice (it can be viewed as the *normal form* for φ), so it's natural to ask when φ can be made to appear in that form by means of a simple variable change.

In more modern dynamical systems terminology the origin is classified as the following sort of fixed point of φ: It is repelling if $|\mu| > 1$, neutral (or elliptic) if $|\mu| = 1$, attracting if $0 < |\mu| < 1$, and super attracting if $|\mu| = 0$. The repelling and attracting cases together constitute the *hyperbolic* case. All cases but (a subcase of) the elliptic case are relatively easy to understand, and were dispatched by the early 20th century: In both the hyperbolic case (resolved by G. Koenigs in 1884), and the super attracting case (L.E. Böttcher, 1904), the conjugacy can be explicitly constructed as a *formal power series*, and its convergence checked in some neighborhood of 0 (modern alternate approaches include iteration schemes that converge toward a fixed point in an appropriate function space).

The remaining elliptic case $|\mu| = 1$ splits into two subcases, depending on whether or not μ is a root of unity ($\mu^m = 1$ for some positive integer m). It turns out that when $\mu^m = 1$, the map φ is linearizable if and only if φ^m is the identity. A version of this was proved by L. Leau [Lea97], then simplified and improved by P. Fatou [Fato19–20].

So by the early 20th century, the only unresolved subcase was $|\mu| = 1$ with μ not a root of unity. This is now known as Siegel's center problem, or just Siegel's problem. We can rewrite it as $\mu = e^{2\pi i \alpha}$ where α is an irrational number (since $\alpha \in \mathbb{Q} \Leftrightarrow \mu$ is a root of 1).

The first results in this direction were negative: In 1917, G. Pfeiffer showed the existence of irrational α for which there are φ that cannot be linearized [Pf17]. In the 1920s, H. Cremer went further by showing that whenever α is an irrational number with *continued fraction* expansion $\{p_k/q_k\}$ such that $\sup_{k \geq 0}(\log q_{k+1}/q_k) = \infty$, there is a map $\varphi(z) = e^{2\pi i \alpha}z + O(z^2)$ that is not linearizable [Cre28]. (The condition on the behavior of the continued fraction denominators has since been known as Cremer's condition.) He went on to refine this by showing that for any integer $d \geq 2$, and irrational α with continued fraction denominators satisfying $\sup_{k \geq 0}(\log q_{k+1}/d^{q_k}) = \infty$ (Cremer's condition of degree d), no polynomial *germ* $P(z) = e^{2\pi i \alpha}z + \cdots + a_d z^d$ (with $a_d \neq 0$) is linearizable.

Finally, in 1942 (during self-imposed exile at the new Institute for Advanced Study in Princeton and in the company of A. Einstein and K. Gödel) the German number theorist C.L. Siegel got the first positive result. He showed that if α is *Diophantine*,[33] then any map $\varphi(z) = e^{2\pi i \alpha}z + O(z^2)$ *is* linearizable [Sie42]. This was apparently the first solution of a small divisor[34] problem, and the first use of Diophantine conditions in this context; it paved the way for Kolmogorov's use of similar (higher dimensional) conditions in his 1954 announcement of the first KAM theorem.

B.1.2   *Refinements and optimal conditions for the Siegel problem*

In a series of papers [Bruno65], [Bruno67], [Bruno71] beginning in the mid 1960s, A. Bruno (also sometimes transliterated from Russian as Bryuno or Brjuno) showed that if irrational α has continued fraction denominators q_k satisfying

$$\sum_{k=0}^{\infty} \frac{\log q_{k+1}}{q_k} < \infty \tag{B.2}$$

---

[33]The Diophantine conditions on α used by Siegel are $\left|\alpha - \frac{m}{n}\right| > \lambda n^{-\mu}$, where $m, n \geq 1$ are arbitrary integers, and λ and μ are positive numbers depending only on α.

[34]Recall that in the classical Hamiltonian setting (cf. Eq. (3.5), §3.11), 'small divisors' appear in the form $k \cdot \omega$, where $k \in \mathbb{Z}^n$ and $\omega \in \mathbb{R}^n$; such expressions vanish when the *frequencies* (components of ω) are appropriately *commensurable*. For circle maps, small divisors typically appear in the form $e^{2\pi i n \alpha} - 1$, where $n \in \mathbb{Z}$ and $\alpha \in \mathbb{R}$; these vanish whenever α is a rational number p/q with n a multiple of q.

then any map $\varphi(z) = e^{2\pi i \alpha} z + O(z^2)$ is linearizable. The condition (B.2) is called *Bruno's condition*, and numbers that satisfy it are Bruno numbers, or belong to a Bruno set. The Bruno set of numbers α is slightly larger than the set of Diophantine numbers used by Siegel, but not by much: their difference is a set of *Lebesgue measure* 0.

There is a slight question concerning priority in this last result, as in 1964 (not so long before his death in 1966) T.M. Cherry conjectured it, and conjectured its optimality (i.e., he conjectured that there is no larger set of αs for which φ is linearizable) [Cher64]. There was also a rumor that he may have proved it—perhaps also its optimality—in an unpublished manuscript.[35]

In any case, the optimality of the Bruno conditions for the Siegel problem was proved definitively by M. Herman's student J.-C. Yoccoz around 1987 (but was only announced as a CRAS[36] note [Yo88], while the proof circulated for several years in a thick manuscript before appearing as [Yo95b]).

B.2   Mappings of the circle

We now turn to the second line of investigation into low-dimensional small divisor problems, those involving mappings of the circle. These problems go back almost as far as the linearization problem for planar maps, and are better known in dynamical systems for reasons to be described. It was Poincaré who initiated the systematic study of circle maps and obtained the first significant results. Circle maps arise naturally when one takes a *surface-of-section transverse* to the flow of an integrable Hamiltonian system and looks at the *Poincaré map* (see Figs. 2.2, 3.1).

Let $\mathbb{T} = \mathbb{R}/\mathbb{Z}$ be the circle of circumference 1 (i.e. the 1-torus), and $\pi : \mathbb{R} \to \mathbb{T}$ defined by $\pi(x) = \{x\}$ be the 'universal covering map'; i.e., the natural projection that wraps the real line around the circle. (Here $\{x\}$ is the fractional part of x; i.e. $\{x\} = x - [x]$, where $[x]$ is the greatest integer $\leq x$.) We want to study invertible, orientation-preserving[37] maps

---

[35] According to J. Milnor (see the historical note on p. 133 of the 3rd edition of [Miln99]), Cherry's unpublished notebooks are now kept somewhere in France.

[36] 'CRAS' is short for *Comptes rendus de l'Académie des Sciences*, and 'notes' published there are not usually considered full-fledged publications. Further indignity attaches to CRAS because of its resemblance to the word *crasse* (close in meaning to the English word 'crass').

[37] Loosely speaking, a circle map is orientation preserving if it maintains the lineal order of any set of discrete points on the circle. More precisely, all of its lifts have positive slope (and orientation-reversing circle maps have lifts with negative slope).

$\phi : \mathbb{T} \to \mathbb{T}$ that are of smoothness class C^k and whose inverses ϕ^{-1} are also C^k ($k \geq 0$). When $k = 0$, ϕ is a homeomorphism; when $k \geq 1$ it is also a C^k diffeomorphism.

The first thing one learns about such maps is that each has a well-defined *rotation number* (introduced by Poincaré), the average amount by which it rotates points of $\mathbb{T}$. To define the rotation number, we use a so-called 'lift' ϕ_L of ϕ; this is a map $\phi_L : \mathbb{R} \to \mathbb{R}$ with the property $\phi \circ \pi = \pi \circ \phi_L$. (A lift is essentially the same as ϕ, but acts repeatedly on $\mathbb{R}$ rather than 'once' on $\mathbb{T}$.) The rotation number $\rho = \rho(\phi)$ of the map ϕ is then defined as (the fractional part of) $\lim_{n \to \infty} (\phi_L^n(x) - x)/n$, where ϕ_L is any lift of ϕ and x is any point of $\mathbb{T}$. (It's not hard to see that this definition makes good sense by showing that it's independent of the lift and the initial point x.)

The simplest class of circle maps are the rigid rotations. Rigid rotation by $\alpha \in [0, 1)$ is the map $R_\alpha : \mathbb{T} \to \mathbb{T}$ given by $R_\alpha(x) = x + \alpha \pmod 1$, which clearly has rotation number α.

Poincaré effectively classified circle homeomorphisms ($k = 0$) by showing—among other things—that

(i) If the rotation number is rational $\rho = p/q$, then ϕ has a periodic orbit of period q.

(ii) If the rotation number ρ is irrational, then ϕ is topologically semi-conjugate to R_ρ. (This means that there's a *continuous surjection* $f : \mathbb{T} \to \mathbb{T}$ with $\phi \circ f = f \circ R_\rho$.)

This last result (ii) was built upon by A. Denjoy [Den32], who looked at the case $k = 1$ in 1932. He showed that if ϕ is a C^1 diffeomorphism of the circle with irrational rotation number ρ, and if the derivative of ϕ is of *bounded variation*, then ϕ is *topologically conjugate* to the rigid rotation R_ρ (so now the map f in (ii) above is a homeomorphism, rather than just a surjection). He also showed that the hypothesis of bounded variation on the derivative is necessary by producing examples of C^1 diffeomorphisms with irrational rotation numbers that are not conjugate to rotation. (These examples are called 'Denjoy counterexamples,' and their existence or non-existence has a continuing history for various smoothness classes k, etc.)

The reader can probably see where this is going. There was a hiatus in this direction after Denjoy, but once Kolmogorov found the approach to KAM theory, V.I. Arnold took up the subject for analytic (i.e., C^ω) diffeomorphisms of the circle. Arnold's 1961 paper [Ar61] shows that if ϕ is a C^ω diffeomorphism of the circle sufficiently close to a rigid rotation and with Diophantine rotation number, then it is analytically linearizable (i.e., analytically conjugate to rotation by its rotation number). Although for most observers this paper does not prove a KAM theorem,[38] it is the first place where the basic techniques of KAM theory as outlined by Kolmogorov (Diophantine conditions on the small divisors and a modified Newton's method) were used in a complete proof. (Nice pedagogical versions of this sort of proof appear in §12 of Arnold's book [Ar83–88], and in the first part of C.E. Wayne's paper [Way96].)

Other researchers followed Arnold's basic result with a number of refinements and extensions; I won't detail these here. But this line of investigation took a new direction with the work of M. Herman, who introduced 'global' methods showing that ϕ need not be close to rotation [Herm76], [Herm79]. These results were a surprising affirmation of a conjecture of Arnold, and brought Herman to the forefront of research in this area, where he remained throughout his career.

As with the Siegel problem, it was Herman's student J.-C. Yoccoz (building on Herman's and A. Douady's work—see [Marm01] by S. Marmi for background detail) who obtained what is considered the crowning result for analytic circle diffeomorphisms [Yo02][39]. Yoccoz divides the real numbers into the union of two disjoint sets H and H' (H comprises the 'Herman numbers') and proves that whenever ϕ is an analytic circle diffeomorphism with rotation number $\rho \in H$, then ϕ is analytically conjugate to the rotation R_ρ; while for any $\alpha \in H'$, there is an analytic circle diffeomorphism with rotation number α which is not analytically conjugate to R_α. Again, the most striking part of this result is that no mention is made of how close ϕ is to rotation—it need not be a perturbation, and the result is 'global' in this sense. For this work (and also for completing the Siegel problem, and for work on the local connectivity of the *Mandelbrot set*), Yoccoz was awarded a *Fields medal* at the 1994 ICM in Zürich. To date, this is the highest form of official recognition given to work involving KAM-like results.[40] However, in this low-dimensional setting, Yoccoz made extensive use of continued fractions, and has lamented the fact [Yo94] that they have no truly adequate generalization to higher dimensions.

---

[38]The result is not a 'true' KAM theorem because circle maps have no invariant curves or tori, and don't carry quasiperiodic motions. But the difference is slight, and many writers (see, e.g., §5.2 of [BroeT10]) do consider circle map conjugacies to be part of KAM theory, calling them 'KAM-like results' rather than 'KAM theorems.'

[39]Earlier versions of this paper had circulated as long manuscripts since the late 1980s.

[40]It could also be argued that the work of C. Mouhot and C. Villani [MouhV11]—for which, in large measure, Villani received a Fields medal in 2010—employs KAM techniques to some extent, but their work is not a KAM-like result overall.

Appendix C

East Meets West — Russians, Europeans, Americans

Sociologists often point to the cosmopolitan nature of science, to the way cultural and linguistic barriers that are disruptive of other endeavors are no real impediment to productive work in science. This seems especially true in mathematics and mathematical physics, where much crucial communication takes place in precise mathematical language, and where practitioners seem focused on the work in front of them, rather than on the culture, language, or manner of their coworkers or rivals. And yet, even if minimized, these things play an undeniable role in science, which is after all a human enterprise. One recurring theme in this book is the difference between mathematicians and physicists and the way this difference affects the development and understanding of dynamical systems and KAM theory in particular. In this appendix, I touch on some of the less tangible ways that geography, culture, and style impact these subjects.

In considering issues such as these, we immediately come upon questions that can't be easily answered, or whose answers are relative to the context or culture in which they're asked. Some of the broadest questions are: What roles do culture, style, and personality play in research? How is the debate in mathematics between the formalists and intuitionists[41] related to culture? Narrowing down slightly, we might wonder why French mathematicians reacted against Poincaré, or why in the early 20th century did leadership in dynamical systems pass from France to the U.S. and then to Russia and the Soviet Union. Getting still more specific, we might wonder what roles culture or style played in the discovery of KAM. We

---

[41]For those who study the development of mathematics, these terms have precise meanings, with a number of sub-varieties. But here, I use them very loosely to designate preferred styles: formalists are those (like the Bourbaki) who favor rigor, structure, and precision, while intuitionists prefer inspired ideas, with whatever minimum apparatus is required to express and develop them.

might even ask (as above in §5.3.2) why KAM theory never took firm root in the U.S., or why research in Nekhoroshev theory later migrated to the Mediterranean (mostly Italy).

I cannot answer these questions, but I can report some of the things that are said about them which may interest the reader, and which may shed some light on the ways that dynamical systems and KAM theory developed.

C.1 Cultural stereotypes in mathematics

First I'll lay out some of the basic stereotypes of mathematicians in different countries or cultures. I'll confine myself to just a few, moving east to west, and I won't be very subtle. First we have the Russians (also formerly the Soviets), who are known for their exalted level of research and their intuitionist or even heroic approach to mathematics. They're also known for their proofs, which Westerners point to as occasionally being incomplete or having big gaps. (Russians sometimes retort that the gaps are there as exercises for Westerners.) Next we have the Germans, known chiefly for having their preeminence[42] in science dismantled by the Nazis[43] (to the benefit mainly of the Americans). This usually overshadows the old stereotype of the punctilious Prussian scholar, which however resurfaces occasionally. By contrast, the French are still strongly associated with the Bourbaki. To some, this means precision and elegance; to others, sterile formalism. The British are often assimilated to the Americans (together— sometimes with Commonwealth countries thrown in—the two are 'Anglo-Saxons'), but the British are known to perform at a higher average level. Finally, the Americans are the big upstarts, having come to mathematical research only in the last century or so. They are known for the low level of mathematics in their schools, for the very high level in their best universities (often thanks to foreign talent), for their faith in computing machines, and above all for their emphasis on money. This emphasis is seen as crass venality when it involves researchers with big grants and fat paychecks, or as opportunity when those same researchers are offering jobs to junior researchers of any nationality.

---

[42]To get some idea of German preeminence in mathematics during the century before the Nazis, recall that this was the time of Gauss, Riemann, Weierstrass, and Hilbert, among others.

[43]On p. 753 of the book [Wat02], P. Watson gives a more colorful description, saying that "Germany, the world leader in many areas of thought until 1933, had its brains ripped out by Hitler in his inquisition, and has not yet recovered."

C.2   Cultural and stylistic tensions

There are also tensions between the various mathematical cultures and styles; many of these are caused or influenced by factors outside mathematics. We've already noted (cf. §1.3) the stark difference between the theory-laden 'continental style' of philosophy (in France, Germany, etc.) and the more pragmatic Anglo-Saxon style (in the U.S. and U.K.). This difference extends into popular culture, and into the ways mathematicians and scientists are trained.[44]

Between German and Russian cultures there have always been strong affinities and tensions.[45] Other cultures (e.g. the Polish) have often been justifiably wary of German and Russian collaboration; and yet when KAM theory was founded by two Germans and two Russians, it happened in the aftermath of some of the largest land-battles in the history of humankind between the armies of Germany and Russia.

But the tension that is most relevant to our subject is that between the intuitionists and formalists in mathematics, represented most strongly by the Russians and the French, respectively. Since before Peter the Great, Russian intellectuals and aristocracy looked to Paris for inspiration, and in the case of Poincaré, it seems they were inspired to greater heights than were his own compatriots. This episode in the history of dynamical systems has yet to be fully understood. It was then and later—in the middle third of the 20th century, when Russian and Soviet predominance in dynamical systems was not fully or properly recognized in the West, and when the Bourbaki was at its height—that the most interesting tensions arose.

I can personally attest to some of the ways this was manifested as late as the 1980s or 90s. From time to time in those days at professional meetings of mathematicians working in dynamical systems, during a presentation given by a 'Western' mathematician X, an 'Eastern' mathematician Y (usually but not always a Russian) would stand in the audience and interrupt the speaker, saying something like "This result was already well known in the Soviet Union since the late 1950s. I saw [mathematician Z] give a seminar on

---

[44]The interaction between Anglo-Saxon and continental views is not exclusively one of opposition: When U.S. universities embraced French literary theorists in the 1970s and 80s, some saw echoes of the way educationists had imported 'new math' principles (ultimately descended from Bourbaki) into U.S. schools in the 1960s; both enthusiasms were also later seen by many as having gone too far.

[45]For a short but interesting account of the historical relationship between Germany and Russia, with an emphasis on cross-cultural and intellectual influences, see Chapter 3 of [Joh91].

it at the Steklov Institute in 1959, but in fact his theorem goes further than what you present here." This was usually answered (sometimes by another Western supporter) with "Can you show me [us] the reference after the talk?" usually followed by a "Yes" which politely stopped the discussion until later. But in at least one case, mathematician Y continued "It was not our custom to publish such trivialities," and in more than one case, mathematician X (or a supporter) would add that he had seen the published work of mathematician Z, but did not consider his proofs to be complete.

C.3   Cultural cross-currents in KAM theory

It's certainly not hard to see how some of the development of KAM theory fits into this. Kolmogorov's outline of the proof of KAM can be viewed as an outstanding example of a 'Russian style' proof, but with gaps big enough even for most Russians to acknowledge. Yet there were those with strong views who persisted in hotly discussing the issue (as described above in §5.3.3); I view most of this as akin to the pleasures of talking about a great sports game. As mentioned earlier, Kolmogorov himself seems never to have claimed that he fully proved his theorem, and was pleased to see Arnold and Moser complete that task.

The tension between intuitionists and formalists is also expressed in other ways by the founders of KAM. Although neither Kolmogorov nor Moser seem to have been drawn into the debate, Siegel was known as a fierce critic[46] of Bourbaki and Arnold was perhaps Bourbaki's most vocal and prominent opponent ever.[47] Arnold was also well known for his strong defense of Russian mathematicians whose results he saw as having been unscrupulously appropriated by Westerners[48] and for his criticism of administrators, money-seekers, and zealous educational reformers.[49] All of these opinions were presented in his own strong but wry and ironic style; his humor alone practically invited agreement.

In the intuitionist-formalist debate, it is easy to side with the intuitionists, if only because they seem more romantic or heroic. Who doesn't feel some exhilaration in Arnold's remarks about Newton and Leibniz[50]

---

[46]See, e.g., Siegel's June 1, 1959 letter to André Weil quoted on pp. 218–19 of [Grau94].
[47]See, e.g., p. 30 of [Zdr87], p. 109 of [Ar90], or p. 438 of [Lui96].
[48]See p. 32 of [Zdr87].
[49]See p. 438 of [Lui96].
[50]See, e.g., p. 47 or p. 94 of [Ar90], in which Arnold contrasts Leibniz' formalism with Newton's geometric insight and intuitionist approach.

or about the relationship between physics and mathematics?[51]   The obvious rejoinders, while clearly necessary for balance, seem uninspiring by comparison.[52]

Yet there is a last episode that stands out as a balancing antidote. We recall (§7.1.1) that Arnold's proof of the application of his own KAM theorem to the n body problem contained not just a gap, but something more like an omission, or error. (Arnold freely acknowledged this and encouraged the error's repair.) The initial repair of this problem occupied two mathematicians (M. Herman and J. Féjoz) trained in something close to the Bourbaki tradition, and practicing its art at a high level of abstraction and rigor. It seems that in its own quiet way, Bourbaki still has its place in mathematics.

---

[51]From the opening paragraph of Arnold's article on teaching [Ar98a]: "Mathematics is a part of physics. Physics is an experimental science, a part of natural science. Mathematics is the part of physics where experiments are cheap."
[52]Rejoinders may be found, for example, in the reviews MR1024727 (91c:01018a) and MR1618209 (99k:00011) on *MathSciNet* or in *Mathematical Reviews*.

Appendix D

Guide to Further Reading

Because the present book has in-text references, this reader's guide is not organized according to chapters in the main text. It has its own organization indicated by the headings and subheadings below, and so gives a second, independent approach to the literature that overlaps with—but significantly differs from—that of the main text.

D.1 General references on KAM

D.1.1 *Original KAM articles, and priority*

There does not seem to be any evidence that the founders of KAM theory were ever involved in disputes among themselves over priority.[53] In fact it's not a simple matter to assign priority in KAM theory, because early progress took place in various directions that are not always easy to compare. Here I present a 'consensus view' of the original articles.

Siegel's 1942 proof of his center theorem [Sie42] is only rarely called a 'KAM-like result,'[54] but is universally acknowledged as the first place where small divisors were overcome, thus setting the stage for KAM later.

Kolmogorov's first published announcement of (what became known

---

[53]However, as mentioned above (§C.3), Arnold was always alert to Westerners who attempted to usurp Russian results. In a footnote on p. 2 of [Ar97], Arnold writes "In the American literature of the sixties, one can find papers with proofs of the 'analytic counterpart of Moser's theorem' (which is of course Kolmogorov's original theorem). J. Moser never supported these attempts to attribute Kolmogorov's theorem to him." See also footnote 13 on p. 108, above, for Arnold's view of who proved Kolmogorov's theorem.

[54]The expression 'KAM-like result' is obviously very elastic, used narrowly by some authors and expansively by others. It has also changed over time as KAM theory has changed.

as) the KAM theorem is [Kol54], and is included here as Appendix A. It includes a succinct 'discussion of the proof,' and references Kantorovich's use of a generalized Newton's method [Kan48]. The proceedings of Kolmogorov's address to the 1954 ICM in Amsterdam were published in [Kol57], and include a statement of KAM with a brief discussion as it relates to ergodic theory and dynamics in general.

The first fully detailed use of what were later called 'KAM techniques'[55] was Arnold's 1961 proof of the linearizabilty of certain analytic circle maps [Ar61]. As mentioned above (footnote 38, p. 192), most observers call the main theorem in this paper a 'KAM-like result,' rather than a KAM theorem.

By consensus, the first detailed proof of what is now called a KAM theorem (for annulus maps, and including an invariant circle in the conclusion) was given by Moser in [Mos62]. In one sense (the low dimension of the system), Moser's 'invariant curve theorem' is weaker than what Kolmogorov announced; in two other senses (the finite smoothness of the system, the setting slightly more general than Hamiltonian), it is stronger.

The first detailed proof of a KAM theorem in arbitrary dimensions was given by Arnold in [Ar63a]. This theorem is close to Kolmogorov's, as it applies to analytic Hamiltonians with n degrees of freedom and shows the persistence of n-dimensional invariant tori carrying quasiperiodic motions. Arnold's proof departs somewhat from Kolmogorov's proof-outline, but in doing so, it defined and shaped standard KAM techniques for a long time afterward. Arnold had already begun to discuss the application of KAM to the n body problem the previous year [Ar62], and made the discussion more rigorous in [Ar63b], which includes the first detailed proof of eternal stability for certain orbits of the three body problem. It also presents Arnold's program for applying KAM to more general n body problems, which was only completed roughly half a century later (cf. §7.1.1 above and §D.6.1 below).

D.1.2 *Accessible proofs of KAM theorems*

A number of experts have taken the time to write clear and accessible proofs of KAM theorems. Among the first proofs, Moser's [Mos62] is the most

---

[55]The expression 'KAM techniques' is as elastic and time-dependent as 'KAM-like result.' In the early days, it meant 'Diophantine conditions combined with a Newton-like method for accelerated convergence'; but by now, the tools used to prove KAM theorems have expanded considerably, as has the setting in which they're used.

readable,[56] and there is also a nice discussion in §§31–33 of the book by
Siegel and Moser [SieM71], which proves the existence of invariant curves
for analytic twist maps of the plane. The early (often overlooked) proof
by R. Barrar [Barr66] of a KAM theorem for analytic systems with two
degrees of freedom is quite accessible. Very readable accounts of standard techniques appear in Arnold's (and colleague's) books [Ar83–88] and
[ArA68]. Also very good are the pedagogical papers of J. Pöschel [Pös01]
and C.E. Wayne [Way96]. More recently, L. Chierchia 'completed' Kolmogorov's original proof-sketch in [Kol54] by filling in the details; the result
stands as a very nice self-contained proof [Chi08]. (The first paper showing
in detail that Kolmogorov's method 'worked' was [BenGGS84].) One of
my favorite proofs of Kolmogorov's theorem is the one by J.H. Hubbard
and Yu. Ilyashenko in [HubI04]. This follows [BenGGS84] by using the
Lie method to define canonical transformations (as in the present book),
but it provides more detail. Another discussion and proof roughly along
the lines suggested by Kolmogorov is the recent one by J. Féjoz [Féj11],
[Féj12]: the first article (from a short course given in Milan) provides a
succinct background to KAM in modern language, along with a modern
statement of Kolmogorov's theorem; the second article uses a fixed point
theorem in a functional analysis setting, and manages to demonstrate Kolmogorov's theorem rigorously and elegantly in under four pages. On the
other hand, proceeding in a way almost antithetical to Kolmogorov (using a
'slowly convergent' scheme suggested by H. Rüssmann [Rüss10]), J. Pöschel
proves a special case of KAM by elementary means in an even shorter
space [Pös11]. Further proofs—and discussions of proof techniques—may
be found in many of the references immediately below in D.1.3 and D.1.4.

D.1.3 *Books on KAM theory (what books?)*

There are no truly comprehensive textbooks on KAM theory, and the
sprawling and protean nature of the subject may ensure that there never
will be. In an ideal world, one could imagine Kolmogorov, Arnold, and
Moser (perhaps just Arnold and Moser) writing the definitive KAM textbook, but it did not turn out that way. However, as discussed below,
Moser and Arnold did write at some length beyond the founding articles
cited above in D.1.1. In fact, several of the articles or chapters cited imme-

---

[56]Note, however, that in the mid 1980s, M.B. Sevryuk discovered a gap or 'technical
inaccuracy' in Moser's proof. This gap is discussed and repaired along the lines suggested
by Sevryuk in Moser's note [Mos01], published posthumously.

diately below could be expanded into books, or simply be reprinted alone as (short) books.

D.1.4   *Review articles, monographs, & book chapters on KAM*

Moser wrote two very nice early monographs [Mos68] and [Mos73]; the first includes applications of KAM theory to stability in the restricted three body problem of celestial mechanics, and to the problem of magnetic containment of charged particles (in choosing these problems, Moser echoes Arnold's choices in [Ar63b]). Moser's second monograph [Mos73] emphasizes celestial mechanics, has interesting historical notes, and contains a proof of the twist theorem for annulus mappings of smoothness class C^5.

With colleagues V.V. Kozlov and A.I. Neishtadt, Arnold wrote the third volume of the *Russian Encyclopaedia of Mathematical Sciences*, called *Mathematical Aspects of Classical and Celestial Mechanics*. The first (Russian) edition appeared in 1985; successive editions have been translated into English and it is now in its third edition [ArKN06]. This large volume gives a wide view of modern mechanics, with careful statements of results, historical notes, unsolved problems, and insightful commentary, but without many complete proofs. The sixth chapter (on perturbation theory for integrable systems) has a long section on KAM theory, one of the better compact mathematical surveys of the subject yet to appear. Arnold's own heuristic discussion in Appendix 8 of his earlier text [Ar78–97] is also quite readable.

One of the first good review articles on KAM is the one (in French) by J.-B. Bost [Bos85]. It is written at a higher level of abstraction than was common at the time in KAM literature (but of course, it's a *Bourbaki seminar*), and it remains a useful reference.

Toward the end of Chapter 5 (Sections 5.9–5.12) of the book [Gall83–07], G. Gallavotti presents a detailed discussion and proof (using 'renormalization techniques') of a KAM theorem (called Proposition 15). There are also a number of interesting remarks on the relevance of KAM theory and Arnold diffusion to celestial mechanics.

Chapter 2 of the book [KapP03] by T. Kappeler and J. Pöschel is a model of elegant concision. It lays out the basics of integrability and classical KAM theory in a short space and without proofs, but with care. (The remainder of the book is devoted to KAM theory for PDEs.)

A very good, easily accessible resource is the online *Scholarpedia* entry on KAM theory [ChiM10] by L. Chierchia and J. Mather. It is nicely illustrated and conveys a lot of insight in a short space.

Recently, modern leaders in KAM theory have published a number of long review articles—some nearly as long as this book. Although the authors of these articles take great care to write for a broad audience, nevertheless they write mostly for mathematicians, and the expositions are at an unmistakably higher mathematical level than the present book.

The first of these articles, and in many ways still the most comprehensive, is that by R. de la Llave [delaL01], which has been updated since its first publication (see de la Llave's website). The style is loose and discursive (I said "One thinks of Cervantes" in my review of it in *Mathematical Reviews*), but it is a massive effort, and could be transformed into a very good book with perhaps one more effort.

More recently, a large (10 volume) *Encyclopedia of Complexity and Systems Science* appeared containing a number of articles that treat—or touch upon—KAM theory. These include H. Broer and H. Hanßmann's survey of HPT [BroeHa09] and A. Celletti's entry on perturbation theory in celestial mechanics [Cel09a]. But the entry directly focused on KAM theory [Chi09b], by L. Chierchia, deserves special attention, as it may be the most direct path by which a reader could move from the present book to the next mathematical level. A leading researcher in KAM theory, Chierchia does not provide a wide-angled view of the subject here, but rather focuses on the main ideas of the classical KAM theorems and their proofs. He presents in turn each of the original theorems by Kolmogorov, Arnold, and Moser, accompanied in each case by convincing mathematical arguments in modern language, yet with a minimum of additional apparatus.

More recently still, Volume 3 of the *Handbook of Dynamical Systems* was published, containing one of the most extensive review articles yet to appear on KAM theory [BroeS10]. This is the work of H. Broer and M. Sevryuk, two pioneers who have led KAM theory into new territory in recent years. With coworkers, they have greatly extended KAM theory for reversible, volume-preserving, and dissipative systems. Their earlier monograph [BroeHuS96] with G.B. Huitema (based on their paper [BroeHu95] and on groundbreaking work with F. Takens [BroeHuT90]) presents many of these ideas in highly mathematical form, while the newer article [BroeS10] condenses this earlier work, renders it in more reader-friendly form, and places it in context by historical notes, commentary on applications, and a review of the Hamiltonian setting.

Broer, Hanßmann, and Sevryuk also edited a special issue of *Discrete and Continuous Dynamical Systems* (Series S, Vol. 2, no. 4, 2010) devoted to KAM theory and its applications and dedicated to the memory of N. Nekhoroshev. Their preface [BroeHaS10] describes the articles in that issue, and serves as a brief overview of problems and results in the field at the close of the first decade of the 21st century.

D.1.5 *Expository, historical, & other sources on KAM*

There is a vast range of expository sources on KAM theory, from the very short but informative (e.g. [Mof90]) to the more detailed and historical (e.g. Chapter 5 of [DiH96], or Section 10.3 of [Barro97]).

A number of authors (mostly mathematical physicists) have taken the trouble to write about KAM theory in the language of physics and with applications in mind. M.V. Berry's article [Berr78] covers a lot of ground in chaos theory and related topics, and is replete with illustrations, thereby giving context to the accompanying section on KAM theory. The article by J.M. Greene [Gre93] recasts KAM theory in the language of renormalization operators, which is familiar to many physicists, and was a source of much progress in understanding the structure of KAM tori and the perturbation strength of their thresholds of existence.

On the other hand, expository works on KAM by mathematicians often focus on historical aspects of the subject. In [Mos99a], J. Moser briefly recalls how he came to formulate his twist theorem after taking on the job of reviewing Kolmogorov's ICM paper [Kol57]. In the proceedings article [Mos99b] (written in 1995), Moser discusses M. Herman's KAM-based counterexamples to the quasi-ergodic hypothesis, as well as certain developments in KAM theory for PDEs. And in one of his last papers, the proceedings of his 1998 ICM lecture in Berlin [Mos98], Moser surveys dynamical systems from the historical viewpoint of the late 20th century, using the stability problem for Hamiltonian systems as a loose connecting thread.

In [Sev03], M. Sevryuk (a former student of Arnold) lays out the various results and refinements of KAM theory available up to 2003. Sevryuk's clearly written paper was an invaluable resource in preparing the present book, and forms the basis (with some other recent material) of §4.5 and §4.6 above.

The article [Broe04] is probably the best ultra-short overview of KAM theory available, somehow combining clear statements of results, historical remarks, and applications and ramifications in physics into fewer than a half-dozen pages.

Finally, perhaps the best non-technical, popular account of KAM theory and its origins is contained in the final chapter of the book [DiH96] by F. Diacu and P. Holmes; this includes material based on interviews and private communications not available elsewhere.

D.2   Mathematical background

D.2.1   *Dynamical systems and ODEs*

There are scores if not hundreds of texts and monographs on dynamical systems, especially since the 1970s through 1990s when the popularity of chaos theory demanded them. It's not surprising then, that some are better than others, or that some are better suited as background for KAM theory. Here I mention just a few.

A good general introduction to dynamical systems for a wide but scientifically educated audience is the textbook by M.W. Hirsch, S. Smale, and R.L. Devaney [HirSD04] (an extensively revised, updated version of an earlier much-loved book by Hirsch and Smale). Another solid introductory textbook is the one by J. Meiss [Mei07c], which includes clear treatments of ODEs, continuous dynamical systems, and Hamiltonian systems, supported by applications to physics, biology, chemistry, and engineering. A somewhat more comprehensive introduction with an emphasis on chaos theory and applications may be found in the book by S. Wiggins [Wig90–03], now in a second, expanded edition. Another good book with a slightly narrower, more mathematical view is the one by D.K. Arrowsmith and C.M. Place [ArrP90]; it begins with the basics, then continues with *hyperbolicity*, *structural stability*, bifurcation theory, and ends with discussions of KAM and Aubry-Mather theory. The book [BroeT10] by H. Broer and F. Takens is somewhat special, as the authors are themselves innovators in KAM theory, and they give it considerably more emphasis than is standard in an introductory dynamical systems text. Their book also stresses chaotic behavior, time series, and structural stability. Finally, one of the best overall references at an advanced level is the text [KatH95] by A. Katok and B. Hasselblatt. It is surprisingly comprehensive, combines careful statements and proofs with valuable narrative, and is expertly edited. It is not, however, for absolute beginners.

For readers interested more specifically in ordinary differential equations (ODEs), one of the best general references is the book [Hale69] by J. Hale, and there are many others, including Arnold's conceptual, geometric treatment [Ar73]. Nice presentations of the classical theorems on existence, uniqueness, and continuation of solutions of ODEs may be found at the beginning of Hale [Hale69], while sharper, more specialized versions appear in the first two chapters of E.A. Coddington and N. Levinson [CodL55], and in the book by P. Hartman [Har82].

A more generalized concept of solutions of ODEs is presented in the paper [DipL89] by R. DiPerna and P.-L. Lions; rather than define solutions for ICs consisting of single points, they define them for certain measurable sets. This gives the existence of solutions for vector fields having only Sobolev regularity, and has been important in the theory of PDEs, especially for the Boltzmann equation.

Finally, the reader should be aware of the online *Encyclopedia of Dynamical Systems*, part of the peer-reviewed online encyclopedia *Scholarpedia*. Although at present it is only partially complete, it is growing steadily, with contributions that are carefully prepared and reviewed by leading researchers.

D.2.2 *Classical mechanics and Hamiltonian dynamics*

For several decades, the standard mathematical textbook on classical mechanics has been Arnold's book [Ar78–97]. This book has been praised for its breadth, its coherence, its attention to the subject's origin in physics, its lack of unnecessary jargon, and especially for its intrinsic, geometric approach. It has also occasionally been criticized for the brevity—in some cases the opacity—of its proofs. In any case, even enthusiasts agree that most of us need a more conventional introduction to mechanics (as provided, e.g., by L.D. Landau and E.M. Lifshitz's book [LanL76]) before approaching Arnold's book for the first time.

Although a great classic never goes out of style, if Landau and Lifshitz seems slightly old fashioned, the reader may appreciate the book [JoséS98] by J.V. José and E.J. Saletan (a reworking of an earlier well-known book by Saletan and A.H. Cromer). This is perhaps the most successful of the books that try to 'bridge the gap' between the way mechanics was taught to physicists until quite recently, and the way it has further evolved under the influence of modern dynamical systems.

Having digested Arnold's or José and Saletan's book, one could look into the formidable treatise by R.H. Abraham and J.E. Marsden [AbM78], an expanded version of the first edition [Ab67] by Abraham. This book presents mechanics in its most mathematical incarnation, as a part of differential topology and symplectic geometry, with a strong emphasis on Hamiltonian systems and the n body problem.

There is also a very worthy successor to Abraham and Marsden's treatise, namely the book [MarsR94] by Marsden and T.S. Ratiu. This large volume somehow manages to be simultaneously more modern and more reader-friendly than [AbM78], with a large number of applications, examples, and exercises that make it suitable for self-study.

The authoritative text by Arnold, V.V. Kozlov and A.I. Neishtadt [ArKN06] is something of a cross between a textbook and an encyclopedia, or perhaps it is a Russian answer to Abraham and Marsden [AbM78]. Like the latter text, this monumental work is appropriate for readers with previous training.

A different sort of detailed treatment is given by G. Gallavotti in the book [Gall83–07]. It is written in the author's own special style, and is rich in mathematical detail and commentary, with rigorous treatment of many topics (oscillations of all sorts, dissipative and conservative systems, constraints, etc.) that are difficult to find elsewhere. It is also teaming with exercises for students or enthusiasts.

Although the title of the book [TreZ10] by D. Treschev and O. Zubelevich makes it seem like an introductory text, in fact it is a monograph on advanced topics in Hamiltonian perturbation theory developed by the authors (especially the first). The authors do present introductory chapters on KAM theory and the Poincaré-Melnikov-Arnold method before dealing with more specialized topics such as stochasticity near hyperbolic fixed points, 'continuous averaging,' and behavior of systems at large values of the perturbation parameter.

A very nice introductory summary of Hamiltonian perturbation theory by M.B. Sevryuk will soon appear in the *Celestial Mechanics* volume of the EOLSS-UNESCO encyclopedia [Sev13]. This includes concise discussions of KAM and Nekhoroshev theory.

There are many, many other sources available. One of my personal favorites, which I came across by chance, is the first chapter of [Tak08] by L.A. Takhtajan. This book is primarily about quantum mechanics, but Chapter 1 has a particularly elegant and succinct presentation of modern classical mechanics, also incorporating large parts of the traditional picture. Another favorite is the opening section of [HubI04], called 'A crash course in Hamiltonian mechanics,' and supplemented by an appendix with many nice proofs.

A good way to get an idea of specialized developments in Hamilto-

nian dynamics (up to 1987) is to look at the review article [MacM87a] by
R.S. MacKay and J.D. Meiss at the beginning of their extensive reprint
selection [MacM87b]; this volume contains a number of the articles referenced in the present book, along with many other classics of independent
interest.

Finally, remarks at the end of D.2.1, above, also apply here: the online
Encyclopedia of Dynamical Systems contains up-to-date articles on Hamiltonian systems by leading researchers, and is growing steadily.

D.2.3   *Ergodic theory*

Since the 1980s, the standard mathematical texts in ergodic theory have
been those by Petersen [Pet83]; Walters [Wal82]; Cornfeld, Fomin and Sinai
[CornFS82]; and Mañé [Mañé87], and together they still stand as a comprehensive introduction. Significant new strides in the field are reflected in
more recent books, but I don't list those here, since the connection between
recent developments and the roots of the subject described in this book are
tenuous at best. But for some ideas of where to look at ergodic theory's
roots and its connections with physics, see Part D.6.2 of this guide below.

D.3   Chaos theory

From the mathematical point of view, chaos theory may be seen as a popularized form of dynamical systems and ergodic theory, or as these subjects
applied to real-world problems, and sometimes 'freed from mathematical
constraints.' Yet very often the scientists at work on these applications
feel that they have discovered important principles of chaos theory on their
own, and that their 'bottom-up' approach gives more insight and context
than the corresponding mathematical formulations. The tension between
these points of view and the success of popularized accounts has produced
a large literature, certain parts of which are very interesting.

D.3.1   *The popular side of chaos*

A great deal of popular chaos literature was published in the last three
decades of the 20th century. It appeared in newspapers, magazines of all
sorts, and in books ranging from paperbacks to coffee-table editions with
large-format glossy illustrations. Here I want to focus on the two books
that had the widest impact in the U.S. (and possibly also world-wide). The

first was [Man75–89] by the Polish-born, Franco-American mathematician B.B. Mandelbrot. It appeared in France (1975), was translated and expanded for the U.S. (1977), then expanded again for a second English edition called *The Fractal Geometry of Nature* (1983), while further editions appeared in France up to 1989. No one claimed that it was organized like a textbook, or gave the clearest explanations of the underlying mathematics, or had the best illustrations (those were supplied slightly later in the books of H.-O. Peitgen *et al.*). Yet it undoubtedly expressed something new which enchanted a large readership. Though many readers were puzzled, it was clear to all that Mandelbrot's *fractals* used mathematics in a new way to model features of nature that most scientists had simply ignored before. The book added measurably to the growing awareness of *dynamics* as an effective tool for understanding random, iterative, and nonlinear phenomena. It also engendered a much smaller but not insignificant negative reaction among some mathematicians who found that it lacked mathematical depth.

The second popular book was *Chaos: making a new science* [Glei87] by science writer J. Gleick. Published in 1987, it became one of the best-selling texts on the physical sciences ever to appear in the U.S. (it also sold very well in other countries). Gleick recounts the story of how nonlinear dynamics and computer modeling have reshaped the way physical scientists carry out their work, and even reshaped the way they think about the world. The book has many merits, including short portraits of 'eccentric' scientists (often based on exclusive interviews), and a style brilliantly suited to the subject. It deserves to be read by anyone seriously interested in the physical sciences. Yet it too elicited strong reactions from some mathematicians who felt that the mathematical part of the story had been de-emphasized or even brushed aside, while at certain points, claims of the revolutionary impact of chaos theory had been exaggerated.

D.3.2 *A chaos debate*

In Issues 1, 3, and 4 of Volume 11 (1989) of *The Mathematical Intelligencer*, the reader will find a very interesting debate encompassing (but certainly not limited to) the two books just mentioned. The debate consists of a dozen pieces, ranging in length from one paragraph to five pages. It features, on the 'popular side,' author J. Gleick and mathematician/author B.B. Mandelbrot (supported by letters from mathematicians K. Devlin and R.G. Douglas), and on the 'traditional side,' mathematicians J. Franks,

M. Hirsch, and S.G. Krantz. The discussion revolves around the impact and value of the work of Gleick and Mandelbrot (in Gleick's case, the focus is on the book *Chaos*; in Mandelbrot's case, it extends to both his and others' professional and popular work). The issues raised and discussed are interesting enough—and germane enough to dynamical systems—to survey briefly here.

Franks opens the discussion with his review [Fr89a] of Gleick's *Chaos*. He praises the style and scope of the book, but feels that it greatly over-estimates the achievement of chaos theorists. He sees a parallel between the claims made by proponents of chaos and earlier claims by proponents of catastrophe theory, which were later seen to have been overstated. He thinks that what Gleick calls the 'chaos revolution' would be better termed the 'computer revolution,' with observations of chaos as one of its many aspects. He feels that the book caters to proponents of fractal geometry and denigrates the Bourbaki. More than anything else, Franks points to the book's de-emphasis of the role of mathematics in the rise of nonlinear dynamics and its under-appreciation of rigor and the theorem-proof methodology of mathematical research. In the same issue, Gleick replies [Glei89a] that the computer revolution is one of the central themes of his book, that in it he gives sympathetic and detailed portraits of a number of mathematicians, and that he does emphasize the important and special role of rigor in mathematics. At the same time, he stands by his claim that many mathematicians are disdainful of Mandelbrot's work on fractals. Franks then counter-replies [Fr89a] that although the book does discuss the widespread use of computers in science, its overall emphasis on a 'chaos revolution' is misplaced. We are instead seeing a computer revolution. And although the book is certainly sympathetic to individual mathematicians, it is much less so to the discipline itself and to its methods.

Two issues later (Issue 3, Vol. 11, 1989), letters to the editor from mathematicians Devlin and Douglas are largely supportive of Gleick, pointing out that he has done much more good than harm for mathematics and its attendant sciences. Douglas' letter ends thus: "If [Gleick's book] doesn't tell the story you want, the way you want, then write your own book!" (which served as one of the inspirations for the present book).

These letters are then followed by four more pieces in the same issue. M. Hirsch writes a spirited defense [Hir89] of Franks' position, pointing to instances where mathematical proof was essential to the understanding or advance of ideas in nonlinear dynamics (e.g. Smale's horseshoe) and lamenting the 'missed opportunities' to demonstrate these instances in Gleick's book. Despite these shortcomings, Hirsch praises Gleick's efforts and style, then ends his piece with the sentence "But I wish he had given less publicity to the nonexistent science of chaos and more to rigorous mathematics." In his reply [Glei89b], Gleick seizes upon this sentence, saying that only by averting their eyes from the broader landscape could mathematicians believe that the science whose rise he documents does not exist. He urges mathematicians to open their eyes to their surroundings, and to admit that meaningful and original insights can arise outside the theorem-proof framework.

In the next piece [Man89a], B.B. Mandelbrot presents a different perspective, coming from his background in France, which he left long ago to escape, in his own words, the "stifling influence of the Bourbaki." But Mandelbrot is careful to point out that he was not escaping the notorious rigor of the Bourbaki so much as their "raw political power." He stresses that the Bourbaki did not hesitate to use their power and that for them "the fields to encourage were few in number, and the fields to discourage or suppress were many." He goes further, pointing to the surprising range of Bourbaki's ambitions, extending for example to the way school children were educated. He points out the irony of dynamicists who lionize Poincaré while defending the Bourbaki, who after all proudly appointed themselves to the task of "cleaning up the 'mess' left by Poincaré." In Mandelbrot's view, messiness indeed has a place in science, even in mathematics, and there are times when it is useful to write down and explore ideas and concepts before they are fully formed, or before they properly fit into a top-down, fully self-referential structure such as preferred by the Bourbaki.

In the final piece of Issue 3 [Fr89b], Franks responds a last time to the comments elicited by his original review of Gleick's *Chaos* in Issue 1. He again praises Gleick's book, but reiterates his stance that the core of mathematics is finding and proving theorems, and wishes that such a 'talented writer' had found the means to convey some sense of what mathematics really is to the public, especially since so few mathematicians are able to do so.

Finally, in Issue 4, the reader will find a slightly different series of opinion pieces by S.G. Krantz and B.B. Mandelbrot. Krantz originally wrote a review of two books on fractal geometry by H.-O. Peitgen and coauthors, but because of editorial wrangling, the review was not published as intended, and wound up instead as the first article here [Kran89]. The piece can only be described as an attack on the intellectual integrity of the field of fractal geometry. As the effective leader/founder of the field, Mandelbrot feels obliged to respond afterward [Man89b]. The series ends with a one-paragraph rejoinder by Krantz.

D.3.3 *The aftermath of popular chaos theory*

Taking 1987 (the year Gleick's *Chaos* appeared) as a convenient marker for the height of chaos popularity, now roughly a quarter-century later it is possible to look back and take some stock of what happened. Not surprisingly, from this longer vantage point, there appears to be merit in the views of both chaos enthusiasts and those who saw too much hype in the enthusiasm. But—perhaps sadly—it's probably also true that the more sober, less enthusiastic side has prevailed overall. The insight that sensitive dependence on initial conditions is common in nonlinear systems is indeed very important, but it has hardly been a revolution in science on a scale comparable to relativity or quantum theory. If anything, it has placed strong limits on the reliability of mathematical models and their computer implementations. Researchers in fields as diverse as climatology, finance, biology, cryptography and geology ignore these limits at their peril.

There are few if any 'we-told-you-so' articles celebrating the demise of chaos enthusiasm. Instead, the more sober viewpoint was usually expressed by working mathematicians and scientists during the height of chaos popularity, and has simply emerged in hindsight as the more reasonable assessment. Perhaps the best books along these lines are those by I. Ekeland [Ek88], I. Stewart [Stew89], and D. Ruelle [Rue91]. A nice article from this time period is the one by P. Holmes [Hol90], who uses the Poincaré/King Oscar story as the starting point for discussing modern dynamical systems and its applications. Both at the beginning and end of the article, Holmes points to the pitfalls inherent in using chaos theory too metaphorically.

Articles or chapters in French by leading scientists and historians appear in the collection [DaCC92]; several of these (especially the chapter by S. Diner [Din92]) decry the misrepresentation of nonlinear science in popularized accounts. A decade later, one of the editors of this collection (A. Dahan Dalmedico) collaborated with D. Aubin in a long article [AubiD02] (in English) which studies, among many other things, some of the sociological and cultural reasons for the rise of popular chaos theory.

D.4   History

D.4.1   *The special nature of history of math & physics*

In addition to the usual difficulties associated with historical works, the history of mathematics and physics presents special problems. First, to do a proper job, one needs to be an expert in both historical methodology and in the mathematics or physics under discussion; unfortunately such individuals are very rare. Second, there is the more subtle problem of anachronism: it is always tempting and in many cases virtually necessary to reformulate historical results in a modern language and present them in a modern context. This is an inevitably distorting process, and although it occurs to some degree in all histories, it is most pronounced in the history of science, because, as S. Bochner says, no other subject is so strongly cumulative and directional (cf. Chapter 2 of Bochner's book [Boch66], entitled 'How History of Science Differs from Other History').

But these special problems don't wholly explain why much of the history of mathematics (and physics too, though perhaps to a lesser degree) is so sloppy. Mathematicians take great pleasure in telling stories and anecdotes about their subject and its creators, but they don't always invest the time and energy required to check original sources and put the stories together coherently, in the context of their time of origin. This doesn't mean that there aren't many reliable and accurate historical resources in mathematics and physics, it simply means that one must be aware of pitfalls.

With this as preamble, I can discuss historical sources in a more light-hearted way. While I make every effort to trace serious and reliable sources, like other mathematicians I describe above, my main reason for reading history of science is pleasure.

D.4.2   *Early history of mathematics and astronomy*

To begin at the beginning, to get a real sense of the importance in human history of the solar system stability problem, one really ought to look into the origins of astronomy and astrology in Mesopotamia, see how these traditions helped shape (and were shaped by) mathematics, then continue to the classical Hellenistic era, through the dark ages, and on to the early Scientific Revolution of Copernicus, Brahe, Kepler, Galileo, and finally Newton. I first recommend the most memorable place where I read these things, namely A. Koestler's book [Koe59]. This text comes with the controversies and idiosyncrasies peculiar to Koestler, but it cannot be said to be boring. More standard, sober treatments of this period may be found in the book [TouG61] by S. Toulmin and J. Goodfield, and in many other sources (e.g., [Drey53]). A brilliantly compact, erudite discussion focusing on the role of mathematics in physics and mechanics during the Hellenistic period and the Scientific Revolution appears in Chapters 4 and 5 of Bochner [Boch66], while the friendly narrative in the first four chapters of Peterson [Pete93] covers the early development of astronomy. Newton and Leibniz are treated together in A.R. Hall's book [Hall80], which places their disputes in the context of their time. Finally, Arnold gives his own particular (and controversial) view of events around Newton's time in [Ar90].

D.4.3 *Between Newton and Poincaré*

Only two centuries separate Newton's first edition of *Principia* and Poincaré's prize-winning memoir [Poi90]. Although Arnold famously calls this interval a "desert filled with calculations,"[57] others see it as the most momentous period in the history of classical physics, and of the mathematics that goes with it. For that reason, it's extremely difficult to describe it as a simple story based on a few major characters.

I don't know of—and I doubt there exists—an authoritative, comprehensive overview of this period describing just what one would like to know about the development of mechanics and mathematics. The developments are many, and disparate, and one probably has to piece together the story from many sources. René Dugas' book [Dug55] is detailed and carefully researched (though technical and somewhat narrow). It is possible to pick and choose parts of large survey works such as those by C.B. Boyer [Boy68], or M. Kline [Kli72] (more than 1200 pages long), or perhaps even the last two volumes of the monumental—and untranslated—work by M. Cantor [Can80–08], though this takes one directly into the world of 'bearded gaslight-Victorians' evoked by Bochner in his brief biographical sketch of Cantor on p. 316 of [Boch66].

D.4.4 *Weierstrass and Poincaré's time*

The stories linking Weierstrass and Poincaré (and King Oscar, Mittag-Leffler, Kowalevski, Phragmén, etc.) make for good drama and are told in a number of places, for example in I. Peterson [Pete93], F. Diacu and

---

[57]See p. 30 of [Zdr87], or p. 80 of [Ar90].

P. Holmes [DiH96], and J. Barrow-Green [Barro97]. These texts are listed in order of increasing detail, the last one being a study of Poincaré's work on the three body problem, with emphasis on the mathematical side and with extensive references. The second book [DiH96] is the most reader-friendly; in fact the authors go so far as to 'dramatize' parts of their narrative with the protagonists' imagined conversations or interior thoughts (all such dramatizations are clearly indicated in the end notes).

To get an idea of the standard histories available around the time of Poincaré's death, see the bibliography at the end of Chapter 1 of F.R. Moulton's book [Moul14]. The various 'historical sketches' in Moulton are also very good by themselves, giving a perspective from the beginning of the 20th century.

The words of several protagonists may be found in the many letters collected and reprinted in various sources. For Poincaré's letters to Mittag-Leffler, see [Nab99]; for Weierstrass's letters, see [Mit12], [Koč73], and [Nab99].

D.4.5 *The Painlevé conjecture & the n body problem*

In this text, emphasis is placed on the special role of the n body problem, since Poincaré first showed it to be classically nonintegrable (cf. §3.11), yet it was later shown to be KAM-stable for certain parameter values (cf. §7.1.1). But since Poincaré's time, a number of very interesting results have appeared for the n body problem outside the framework of KAM theory; several of these are described in the nice series of articles [Di93], [Di96], [Di00] by F. Diacu in the *Mathematical Intelligencer*.

The first article [Di93] concerns the famous conjecture formulated by P. Painlevé at the end of the 19th century (that solutions with noncollisional singularities exist in the n body problem for $n \geq 4$) which was finally affirmed in a number of beautiful examples toward the end of the 20th century.

The second article [Di96] describes the series solutions of the n body problem that were obtained by K. Sundman (for $n = 3$) and Q. Wang (for arbitrary $n \geq 3$) toward the beginning and end of the 20th century, respectively, and explains why they don't contradict Poincaré's nonintegrability result.

The last article [Di00] tells the story of the origins of algebraic topology in Poincaré's work on the three body problem, and how developments in that field roughly a century later led to descriptions of certain 'integral manifolds' of the three body problem, thus successfully answering some of the questions that originally motivated Poincaré.

D.4.6 *The Soviet & Russian schools of dynamical systems*

As alluded to several times in this book, in many respects the Soviet and Russian schools in dynamical systems pulled ahead of their counterparts in the West during the 20th century, and this was undoubtedly key to the formulation of KAM theory. Information on the how and why of this development seems scarce; the best sources I found were Arnold's article [Ar97], the chapter entitled 'Schools Amid Turbulence' in [Ya02], the article by S. Diner [Din92] (in French), Section 2.1 of the article by P. Holmes [Hol05], and the collection of articles [ZdrD93–07] (especially the last article in the 2nd edition, by V.M. Tikhomirov). As Diner points out in [Din92], the absence in the West of serious discussion of the Soviet school of dynamics is an 'indecency.'

D.4.7 *History of dynamical systems in general*

The most readable account of the history of dynamical systems is Gleick's book [Glei87], but that statement is subject to a number of qualifications from the mathematical point of view (see §D.3, above). Shorter, more mathematically oriented accounts appear in P. Holmes' articles [Hol05] and [Hol07], and in the article [HasK02a] by B. Hasselblatt and A. Katok focusing on the contributions of J. Moser. The rather long article [AubiD02] by D. Aubin and A. Dahan Dalmedico traces not only developments in dynamics, but cultural and sociological influences on the way its history is written. Recollections from the days when results came hard and fast are set down by some of the key researchers in the collection of articles [AbU00] edited by R.H. Abraham and Y. Ueda. In Part Three of the book [DaCC92] (in French), edited by Dahan Dalmedico, J.-L. Chabert, and K. Chemla, one finds historical discussions of Hadamard's and Poincaré's ideas, and of determinism in physics and mathematics.

D.5 Biography

The world wide web now makes it possible to find basic—and often reliable—information about almost any important scientist. This is extremely useful, but unfortunately the overall quality of multi-author, anonymous articles in web-based encyclopedias such as *Wikipedia*[58] varies wildly. Traditional single-author biographies may contain as many errors as multi-author web-based articles, but they often succeed in bringing personalities and epochs to life in ways that the latter do not.

D.5.1 *General biographical sources*

The most interesting general biographical sources I know of are these: the book by E.T. Bell [Bel37] (often criticized for inaccuracies if not outright fabrication—especially in the case of Galois—but still great fun); the biographical sketches—small idiosyncratic jewels of writing—in the back of S. Bochner's book [Boch66]; A. Koestler's minor masterpiece [Koe59] (criticized for being melodramatic in places, and for its denigration of Galileo; lauded for its treatment of Kepler, which is excerpted as the shorter book [Koe60]); the inspiring historical notes in G.F. Simmon's textbook [Sim72]; and the extraordinarily well written material in B.H. Yandell's book [Ya02], which is organized around Hilbert's problems and their solvers. The new book [HershJ10] by R. Hersh and V. John-Steiner has an appendix with short vignettes on mathematicians; the names listed there often differ (in a refreshing way) from those in other books. Finally, the reader interested in the (mostly French) mathematicians who brought Bourbaki to prominence in the mid 20th century will find nice biographical sketches of them in the books [Mas02] and [Acz06], both of which are briefly discussed below in D.8.

D.5.2 *The principals*

In order of birth date, the principal players in this story are I. Newton (1642[OS]–1727), L. Boltzmann (1844–1906), J.H. Poincaré (1854–1912), C.L. Siegel (1896–1981), A.N. Kolmogorov (1903–1987), J.K. Moser (1928–1999), and V.I. Arnold (1937–2010).

For obvious reasons, Newton is the subject of more biographies than others on this list. The standard biography of Newton is R.S. Westfall's [Wes80]; if that tome is too much, the reader may want to look at the abbreviated version [Wes93]. The short treatment by J. Gleick [Glei03] is worthwhile, and is poetic at times. I also admit to having read, and enjoyed,

---

[58]Another nice website for mathematical biography—with identifiable authors—is the MacTutor History of Mathematics archive at http://www-history.mcs.st-and.ac.uk hosted by the University of St Andrews, Scotland.

one of the books concentrating more on Newton's alchemist-sorcerer side [White99].

Boltzmann makes for almost as interesting a subject as Newton, and two very fine books in English take advantage of this. The lighter of the two [Lin01], by D. Lindley, manages to weave the story of Boltzmann's life together with an understandable account of the upheavals in 19th century physics—not an easy feat in a short book. The second, much weightier entry [Cer98] is by C. Cercignani, one of the 20th century's leading authorities on the Boltzmann equation. A short but interesting account of Boltzmann as the center of conflict over the atomic hypothesis may be found in [Por98]. Finally, the curious reader should also be alert to the many biographies of Boltzmann in German.

Poincaré's life is treated in the books by P. Appell [Ap25] and A. Bellivier [Belli56], and in the elegy by G. Darboux [Dar14], none of which, unfortunately, has been translated from French to English as far as I know. Quite recently, two new scientific biographies in English have appeared. The first and most concise [Ver12], by mathematician F. Verhulst, focuses on Poincaré's work in dynamical systems. The second, more expansive book [Gray13] is by the historian of mathematics J. Gray, and includes rich detail and contextual material.

I don't know of any biographies of Siegel, but from the interesting material on him in [Dav85] and [Ya02], it seems he deserves one. Much of the material in Yandell [Ya02] derives from second- or even third-hand recollections, or from the book [Brau80] by Siegel's student and later friend and companion Hel Braun. Siegel was apparently quite a character: curmudgeonly, opinionated, with a special sense of humor.

I also don't know of a traditional biography of Kolmogorov, though there is enough material to make one in the collection of articles and reminiscences [KolS00]. There is also the *Scholarpedia* article [Vit07] on him maintained by P.M.B. Vitanyi, and a very readable sketch of his life in [Ya02].

The last two principals are perhaps too close to us in time to yet have their stories fully told. But there is a lot of detail about Moser's life in the memorial pieces [Chi09a], [Gio00], [HasK02a], [Lax02], and [MatMNR00] by his former students, peers, coworkers and admirers. The material in Yandell [Ya02] is also illuminating, and interesting anecdotes appear in R. Hersh's forthcoming biography [Hersh13] of Peter Lax (including the story of Moser standing guard over the Courant Institute's CDC 6600 computer as protesters seized it in May 1970).

Arnold is perhaps a special case, as his precociousness, wide range of interests, special sense of humor, and above all his outspokenness made him a subject of interest and controversy for many decades both in and outside the mathematical community. Readers will get a sense of his views in the interviews by S. Zdravkovska [Zdr87] and S.H. Lui [Lui96], as well as by his expository writings [Ar90], [ArV89], [Ar97], [Ar98a], [Ar02], or [Ar06]. A good biographical sketch is again found in [Ya02].

D.6 Applications of KAM (and Nekhoroshev) theory

D.6.1 *Applications to celestial mechanics; stability*

Celestial mechanics and the n body problem make up a very large, highly evolved area of research; for modern background material, see the forthcoming articles by A. Celletti [Cel13], L. Chierchia [Chi13], or J. Féjoz [Féj13], the book chapter [Al02] by A. Albouy, the online article [Chenc07] by A. Chenciner, the book [MeyHO09] by K.R. Meyer, G.R. Hall, and D. Offin, or the broad survey in the first part of the book [ArKN06] by Arnold, Kozlov and Neishtadt. Another (partly historical) path for the newcomer to approach celestial mechanics is by way of the remarkably comprehensive review article [Gut98] by M. Gutzwiller. Gutzwiller uses the moon-earth-sun system to trace the development of celestial mechanics generally and the three body problem in particular. The discussion begins in ancient Mesopotamia, and proceeds carefully through stages, arriving to the most modern treatments of lunar motion available at the close of the 20th century.

For proofs of KAM theorems that apply to the n body problem in the fullest generality to date, see J. Féjoz' article [Féj04] (or the shorter summary [Féj07]); see also the articles by L. Chierchia and coauthors [ChiPu09], [ChiPi11a], and [ChiPi11b], which begin with very readable introductions.

The leading researchers in applications of KAM and Nekhoroshev theory to concrete problems of celestial mechanics and astronomy are A. Celletti and L. Chierchia. Celletti is perhaps the more specialized in this direction; an introduction to her work may be found in the book [Cel09a], which contains many references. Good surveys may also be found in her article [Cel09b] and in the article [CelC06] and memoir [CelC07], both with Chierchia.

A nice, self-contained discussion of Nekhoroshev theory applied to stability of the Lagrange points L_4 and L_5 in the (spatial) circular restricted three body problem is given in G. Benettin's article [Ben05a] (which includes other applications as well).

For discussions of the stability of the physical solar system, see the book by I. Peterson (especially the later chapters), the newspaper article of Moser [Mos75], and any number of articles by J. Laskar [Las89], [Las90], [Las92], [Las94], [Las13], who is a leading authority on the subject.

D.6.2 *Applications to statistical mechanics, ergodic theory*

As mentioned earlier, it usually isn't quite right to speak of applications of KAM and Nekhoroshev theory to statistical physics, since at present there are few statistical mechanical systems to which the theories apply directly. It's perhaps better to say that despite the great success of statistical physics, important questions and uncertainties remain in its foundations. In particular, the status of the ergodic hypothesis—often taken as a cornerstone of statistical physics—remains unclear, and there appear to be systems in which KAM or Nekhoroshev theory invalidate the ergodic hypothesis. One says "there appear to be systems" here, because it's difficult to conceive of model systems that are both physically relevant and amenable to KAM or Nekhoroshev theory. Currently, the most interesting models in this category are Fermi-Pasta-Ulam (FPU) systems and their relatives; but even here the applicability of KAM or Nekhoroshev theory is sometimes uncertain and controversial.

To wade into the literature of this subject is a daunting task. Concerning statistical mechanics itself, there are not only mathematical and physical versions of the literature with their differing approaches and nomenclatures, but, because (as in quantum mechanics) the relationship between the mathematics and physics is not always clear, a third type of 'interpretive' literature arises—the philosophy of statistical mechanics. This is often intertwined with the subject's history, because in many cases the problems have not advanced very far from their earliest formulation and discussion around Boltzmann's time.

Some of the leading authors in the history and philosophy of statistical physics are S.G. Brush, L. Sklar, and J. Uffink. Brush's detailed history [Brus83] is still one of the best introductions, and his later anthology [Brus03] of classic papers is accompanied by commentary and further references. Sklar's book [Skl93] and web-article [Skl09] give an authoritative overview of issues in the philosophy of statistical physics, with special attention to the use of probabilistic methods and to problems associated with time asymmetry and 'time's arrow.' Uffink's compendium [Uf07] (nearly book-length) is historical, mathematical, and very broad: it begins by explaining why any treatment of statistical physics must be partly historical, then proceeds to do precisely that, succinctly covering Maxwell's, Boltzmann's and Gibbs' approaches to the subject, then continuing on to the modern theory. Uffink's web-article [Uf08] focuses on Boltzmann's work, showing the protean nature of Boltzmann's genius as his views evolved over time. Finally, for an up-to-date survey on the foundations of statistical mechanics, with a comprehensive reference list, R.P. Frigg's article [Fri08] is a valuable source.

Moving from general treatments toward the more specific issues surrounding the ergodic hypothesis, the reader may want to approach this thorny subject by way of two short, reader-friendly articles by D.K. Campbell [Cam87] and A. Patrascioiu [Pat87], both in the special issue of *Los Alamos Science* (No. 15, 1987) in memory of Stanislaw Ulam. A more technical and focused discussion of the role of ergodic theory in statistical mechanics is found in J. Earman and M. Rédei [EarR96], which cites KAM theory as a likely obstruction to ergodicity in physically relevant systems. Much more recently, R.P. Frigg and C. Werndl [FriW11] argue against the relevance of KAM theory, saying that physically relevant systems should be 'epsilon-ergodic' (and display thermodynamic-like behavior) in spite of KAM.

In this book, emphasis is placed on FPU-like systems because, among systems in statistical physics, it is for these that some progress in the rigorous application of KAM or Nekhoroshev theory has been achieved. The literature on the FPU problem is very large, comprising hundreds of articles. However, reasonable surveys of the subject and guides to various parts of the literature are also available. There are two books, a popular one [Weis97] by T.W. Weissert, and the more recent collection of articles [Gall08] edited by G. Gallavotti.

Some of the best guides to the FPU literature are found in Issue 1, Volume 15 (2005) of the journal *Chaos*; this is a special 'focus issue' devoted entirely to the topic of FPU. The introductory article [CamRZ05] summarizes the twenty articles in the issue, and the next five articles [Zab05], [Zas05], [BermI05], [CarGG05], [PettiCCFC05] give a very broad and historical overview of FPU and its complexities, applications, and ramifications. Finally, for up-to-date developments, the reader may consult the current versions of the two online *Scholarpedia* articles [DauR08] and [Rin09] maintained on FPU (the second is the more mathematically oriented).

D.6.3 *Other applications*

Although broader applications of KAM and Nekhoroshev theory are given only cursory treatment in this book, they are substantive fields in their own right. As a starting place, the reader may look at the end of Appendix 8 in Arnold [Ar78–97]; Appendix A of A.J. Lichtenberg and M.A. Lieberman [LicL92]; or Sections 6.3.9, 6.4.1 and 6.4.2 of Arnold, Kozlov and Neishtadt [ArKN06].

The study of charged particle motions (in particle accelerators, fusion reactors, free-electron lasers, crystals, zeolites, and many other places) is a vast subject and a big business. The subject of KAM theory is often in the background as a guiding principle, but is not often used directly for practical purposes. To get an idea of how researchers model these motions, the reader could look, for example, at the article [For06] by E. Forest (written in a very lively style) which reviews the use of symplectic integrators in mathematical models of particle accelerators. These special numerical methods are among the most refined currently in use, and show abundant families of KAM tori when their output data are converted to pictorial form.

More comprehensive introductions to particle accelerators may be found in the texts [Wils01], and [EdS93]; intermediate and advanced treatments are contained in the books [For98], [Tz04], and [Mic95].

For more esoteric emerging applications of KAM (to nonlinear PDEs, quantum mechanics, etc.), the reader may consult some of the proceedings of the ICM lectures by leading researchers, for example those by J. Bourgain [Bour95], S.B. Kuksin [Kuk98], and L.H. Eliasson [Eli98], or go directly to state-of-the-art results by the latter two authors, as presented in their paper [EliK10].

D.7 Mathematical topics related to classical KAM theory

Both in its origins and later developments, classical KAM theory overlaps a number of mathematical areas and has growing significance in many of them. For a survey of some of these areas, see R. de la Llave [delaL01] (especially the first introductory section). Here I'll restrict attention to just the four (already large) areas below.

D.7.1 *Low-dimensional small divisor problems*

Unlike many of the topics associated with KAM theory, the mathematics of low-dimensional small divisor problems is especially elegant and self-contained, with connections to other parts of pure mathematics such as holomorphic dynamics and number theory. Nice introductions to the field are found in the monographs by R. Pérez-Marco [Pére92], J. Milnor [Miln99], S. Marmi [Marm01], and in the workshop proceedings by L.H. Eliasson, S.B. Kuksin, S. Marmi, and J.-C. Yoccoz [EliKMY02]. Pérez-Marco takes the reader (in French) on a tour through J.-C. Yoccoz's work on the Siegel center problem; Milnor introduces the basics of holomorphic dynamics and points the way to 'Fatou-Julia-Sullivan theory' (Chapters 4 and 11 overlap and greatly extend Appendix B of the present book); and Marmi gives a detailed introduction to some of the modern techniques of small divisor theory. Yoccoz's article [Yo02] in [EliKMY02] provides an overview of his work on analytic linearization of circle diffeomorphisms, while Marmi and Yoccoz's article [MarmY02] in the same volume describes open problems related to small divisors as of 1998. The comprehensive book by A. Katok and B. Hasselblatt [KatH95] also treats homeomorphisms and diffeomorphisms of the circle in Chapters 10 through 12 (small divisors appear in §12.3), while the Poincaré-Siegel theorem is discussed in §2.8.

Finally, for a comprehensive survey of low-dimensional dynamics *without* an emphasis on small divisors, the reader may consult the book [MeloS93] by W. de Melo and S. van Strien.

D.7.2 *Aubry-Mather & weak KAM theory, KAM for PDE*

The present book is concerned with 'classical' KAM theory: the persistence of invariant tori (or curves) carrying quasiperiodic motions in sufficiently smooth, finite-dimensional Hamiltonian systems or maps of the annulus. Ideas from KAM theory have also taken other paths only touched upon here, yet the significance of these still-evolving fields may eventually rival or surpass that of classical KAM theory.

The founding articles of what's now called Aubry-Mather theory are by J. Mather [Mat82], and S. Aubry and P.V. LeDaeron [AubrL83]. A nice survey of the early theory appears in Moser's article [Mos86], while Chapter 13 of the book [KatH95] by Katok and Hasselblatt gives a slightly more modern presentation.

At about the time Aubry-Mather theory appeared, and following his book [Lio82] on generalized solutions of Hamilton-Jacobi equations, P.-L Lions co-authored two articles [CranL83], [CranEL84] (with M.G. Crandall, then with Crandall and L.C. Evans) on so-called viscosity solutions of Hamilton-Jacobi equations. Viscosity solutions were a new type of *weak solution* for certain classes of PDEs, and since solving Hamilton-Jacobi equations is one means of integrating Hamiltonian systems, the viscosity solutions have come to be seen as a form of *weak KAM theory*.

Later, A. Fathi [Fath97a], [Fath97b] showed that weak KAM theory has deep relations with Aubry-Mather theory. These relations have since been exploited by Evans [Ev04], [Ev08], [Ev09], S.B. Kuksin [Kuk00], [Kuk04], and Fathi [Fath10], so that the theory by now is fairly mature.

Meanwhile, in the same way that the theory of solitons and hierarchies of integrals for wave equations may be seen as an analog of integrable (finite-dimensional) Hamiltonian systems, so there are also analogs of classical KAM theory for PDEs and for infinite-dimensional systems of ODEs. But unlike KAM theory in finite dimensions, in this business one deals with both finite- and infinite-dimensional invariant tori. This theory emerged in the late 1980s and early 1990s with the articles of S.B. Kuksin [Kuk88a], [Kuk88b], W. Craig and C.E. Wayne [CraW93], [CraW94] and the book by Kuksin [Kuk93]. Later, a very explicit KAM-like theorem for PDEs was developed by J. Pöschel [Pös96], while J. Bourgain extended the concept to more general PDEs [Bour99a], [Bour99b]. Further results are available in the monograph by Craig [Cra00], the book by T. Kappeler and J. Pöschel [KapP03], the books by Kuksin already cited [Kuk00], [Kuk04], as well as the recent article by L.H. Eliasson and Kuksin [EliK10].

D.7.3 *Nekhoroshev theory*

Because §6.2 above serves as a fairly broad reader's guide to Nekhoroshev theory, I give here a very short survey of the highlights of that theory. The founding articles from the 1970s by N.N. Nekhoroshev are [Nek71], [Nek73], [Nek77], [Nek79]; these established the theory and showed that it described a generic property of Hamiltonian systems under the rather weak hypothesis of *steepness*. Italian authors [BenGG85b] emphasized that replacing the steepness hypothesis by the stronger assumption of convexity (or quasi-convexity) greatly simplified the proofs in the theory, yet still yielded results that were useful in physical applications.

Later, J. Pöschel [Pös93] reworked the geometric part of the proofs (and also sharpened certain analytic parts) to obtain the optimal stability radius estimates posited earlier by P. Lochak, based on his reading of B.V. Chirikov. At about the same time, Lochak presented a wholly new approach to Nekhoroshev theory based on single-phase averaging [Loc92]; this method also produced the optimal stability radius estimates [LocN92].

Since then, the main innovator in the mathematical side of Nekhoroshev theory is L. Niederman [Nie04], [Nie06], [Nie07], while applications (to celestial mechanics, and to the FPU problem) have been led mostly by Italian authors; see §6.2.2 above, and Parts D.6.1 and D.6.2 of this guide.

D.7.4 *Arnold diffusion*

As in the case of Nekhoroshev theory, a large number of references for Arnold diffusion are already provided above (§6.3.2; though unfortunately many references are also omitted). Here I give the shortest possible reader's guide.

Although it is somewhat cryptic, the original article [Ar64] by Arnold is still worth reading for its sheer audacity and cleverness. And although it is by now somewhat out of date, Lochak's 'compendium' [Loc99] is also valuable, because his criticisms serve not only as a reader's guide to literature through the mid 1990s, but also continue to be relevant even now.

The most comprehensive reference list of (nearly recent) literature I know of is the bibliography of [DelsLS06], while proofs of Arnold's conjecture (cf. §6.3.2 c) in various special cases appear in the preprints [Cheng13], and [Mar13], [MarG13], and [KalZ13], and are likely to be published around the same time as this book.

D.8 Culture, philosophy, Bourbaki, etc.

It is not easy to find good treatments of the effects of culture or philosophy on mathematical research, or vice-versa. An exception to this is the recent book [HershJ10] by R. Hersh and V. John-Steiner, which examines the human side of doing mathematics. Discussions of the way research is done in France are perhaps also an exception because of the attention focused on Bourbaki. Two nice books in this direction are available, one by M. Mashaal [Mas02], the other by A. Aczel [Acz06].

Mashaal's book [Mas02] is very animated, with photographs, sidebars, and colored diagrams highlighting biographical sketches, short history lessons, and mathematical discussions. We meet the founders of Bourbaki, read excerpts from their famous texts, and follow their rise and influence in both French and world mathematics. Most interesting are perhaps the third and ninth chapters. Chapter 3 explains the conditions in France that gave rise to the group (including the generation of young French mathematicians nearly extinguished fighting in World War I), while Chapter 9 outlines the internal tensions and external criticisms that arose as the Bourbaki gained power and influence.

The book [Acz06] by Aczel is a single narrative linking together biographies of Bourbaki's founders and the group's surprising structuralist influence on other subjects such as anthropology and psychology. A strongly recurring theme is the lament for Bourbaki's break with their most brilliant member, Alexander Grothendieck. Grothendieck's estrangement first from Bourbaki, then from mathematics, then finally from society must be counted as one of the great cultural and scientific losses of the 20th century.

Further, shorter discussions of Bourbaki and its influence may be found in articles by V.I. Arnold [Ar02], A. Borel [Bor98], B.B. Mandelbrot [Man89a], and I. Stewart [Stew95], as well as in the books by Yandell [Ya02] and by Hersh and John-Steiner [HershJ10].

It is again regrettable that similar attention has not (to my knowledge) been paid to the cultural aspects of mathematical research in other places and the cross-currents between them. This is most true of research in Russia and the former Soviet Union, where great narratives of some of the 20th century's best mathematics await telling to Western readers. Glimpses of this may be had in the sources cited above in D.4.6.

An in-depth look into the culture and foundations of mathematics might occupy a lifetime. On shorter timescales, one can nevertheless get rough ideas about various approaches to the subject (such as logicism, intuitionism, formalism), and see how these approaches form part of the culture of mathematics and shape the way it is done. As always, basic facts and more may be gleaned from encyclopedic parts of the web, but the reader should also consider some of the introductions prepared by named individuals. See for example the article [LindsP09] by S. Lindström and E. Palmgren, the volume in which it is published [LindsPSS09], or the award-winning book [DavisH81] by P.J. Davis and R. Hersh.

Finally, one of the most interesting and fun-to-read expository texts I've come across is the collection [Rota97] by G.-C Rota. Rota was a first-rate mathematician and also a philosopher. The book just mentioned is partly memoir (with sharply drawn—and controversial—portraits of eminent mathematicians), partly philosophical rumination (reflections on mathematical certainty and proof in the light of Husserl and Heidegger's phenomenology), and partly opinion. Except for certain technical passages, the style is conversational, lucid, and engaging.

Appendix E

Selected Quotations

Quotes are listed alphabetically by author, then chronologically.

J.-L Lagrange:

- **Q**[Lag] (c. 1800) From p. xx of [Delam67]:

Newton était le plus grand génie qui ait jamais vécu et aussi le plus chanceux car nous ne pouvons trouver le système du monde plus d'une fois.

✠ Translation [HSD]: See footnote 6, p. 32, above.

P.-S. de Laplace:

- **Q**[Lap] (1814) From the introduction to [Lap14]:

Une intelligence qui, pour un instant donné, connaîtrait toutes les forces dont la nature est animée, et la situation respective des êtres qui la composent, si d'ailleurs elle était assez vaste pour soumettre ces données à l'Analyse, embrasserait dans la même formule les mouvements des plus grands corps de l'univers et ceux du plus léger atome : rien ne serait incertain pour elle et l'avenir, comme le passé serait présent à ses yeux.

✠ Translation [HSD]: See footnote 16, p. 36, above.

I. Newton:

- **Q**[New] (1706, 1717) From 'Query 31' of *Opticks* (2nd English Edition, 1717) [First appearing in 1706 Latin edition as 'Query 23']:

Now, by the help of these principles, all material things seem to have been composed of the hard and solid particles above mentioned, variously associated in the first creation by the counsel of an intelligent agent. For it became Him who created them to set them in order. And if He did so, it's unphilosophical to seek for any other origin of the world, or to pretend that it might arise out of a chaos by the mere laws of Nature; though, being once formed, it may continue by those laws for many ages. For while comets move in very eccentric orbs in all manner of positions, blind fate could never make all the planets move one and the same way in orbs concentric, some inconsiderable irregularities excepted which may have risen from the mutual actions of comets and planets upon one another, and which will be apt to increase, till this system wants a reformation. Such a wonderful uniformity in the planetary system must be allowed the effect of choice.

J.H. Poincaré:

• Q[Poi1] (1892) From [Poi92–99], Introduction to Volume I, pp. 3–4:

La plupart de ces développements ne sont pas convergents au sens que les géomètres donnent à ce mot. Sans doute, cela importe peu pour le moment, puisque l'on est assuré que le calcul des premiers termes donne une approximation très satisfaisante; mais il n'en est pas moins vrai que ces séries ne sont pas susceptibles de donner une approximation indéfinie. Il viendra donc aussi un moment où elles deviendront insuffisantes. D'ailleurs, certaines conséquences théoriques que l'on pourrait être tenté de tirer de la forme de ces séries ne sont pas légitimes à cause de leur divergence. C'est ainsi qu'elles ne peuvent servir à résoudre la question de la stabilité du système solaire.

✠ Translation [HSD]:

Most of these series are not convergent in the sense that geometers understand the word. No doubt it matters little for the moment, since we can be sure that the calculation of the first terms give a very satisfactory approximation; but it is no less true that these series are unable to provide an arbitrarily long approximation. There also comes a point therefore where they are unsatisfactory. Moreover, certain theoretical consequences that one may be tempted to infer from the form of these series are not legitimate because of their divergence. It is because of this that they cannot be used to resolve the question of the stability of the solar system.

- **Q**[Poi2] (1892) From [Poi92–99], Volume I, Chapter I, p. 32:

Problème général de la Dynamique.

§13. Nous sommes donc conduit à nous proposer le problème suivant :
Étudier les équations canoniques

$$(1) \qquad \frac{dx_i}{dt} = \frac{dF}{dy_i}, \qquad \frac{dy_i}{dt} = -\frac{dF}{dx_i},$$

en supposant que la fonction F peut se développer suivant les puissances
d'un paramètre très petit μ de la manière suivante :

$$F = F_0 + \mu F_1 + \mu^2 F_2 + \dots,$$

en supposant de plus que F_0 ne dépend que des x et est indépendant des y ;
et que F_1, F_2, ... sont des fonctions périodiques de période 2π par rapport
aux y.

✠ Translation [HSD]:

Fundamental Problem of Dynamics.

§13. We are thus led to propose the following problem:
Study the canonical equations

$$(1) \qquad \frac{dx_i}{dt} = \frac{dF}{dy_i}, \qquad \frac{dy_i}{dt} = -\frac{dF}{dx_i},$$

assuming that the function F may be expanded in powers of a very small
parameter μ in the following way:

$$F = F_0 + \mu F_1 + \mu^2 F_2 + \dots,$$

assuming moreover that F_0 depends only on the x variables and is independent of the y variables; and that F_1, F_2, ... are periodic functions with
period 2π in the y variables.

- **Q**[Poi3] (1893) From [Poi92–99], Volume II, Chapter IX, pp. 15–16:

M. Lindstedt ne démontrait pas la convergence des développements qu'il
avait ainsi formés, et, en effet, ils sont divergents ; mais nous avons vu dans
le Chapitre précédent comment ils peuvent néanmoins être intéressants et
utiles.

✠ Translation [HSD]:

Mr. Lindstedt did not show the convergence of the series that he formed
in this way, and, in fact, they are divergent. But we saw in the preceding
chapter how they may nevertheless be interesting and useful.

- **Q**[Poi4] (1893) From [Poi92–99], Volume II, Chapter XIII, pp. 102–103:

§149. Il nous reste à traiter la deuxième question; on peut encore, en effet, se demander si ces séries ne pourraient pas converger pour les petites valeurs de μ, quand on attribue aux x_i^0 certaines valeurs convenablement choisies.

Ici nous devons distinguer deux cas.

[. . .]

Si les séries (2) convergeaient, à cette valeur commensurable de $\frac{n_1}{n_2}$ correspondrait une double infinité de solutions périodiques des équations (1).

Or nous avons vu au n° 42 que cela ne peut avoir lieu que dans des cas très particuliers.

Il semble donc permis de conclure que les séries (2) ne convergent pas.

Toutefois le raisonnement qui précède ne suffit pas pour établir ce point avec une rigueur complète.

✠ Translation [HSD]:

§149. It remains to treat the second question, and we may again, in fact, ask whether these series might not converge for small values of μ, when one gives certain conveniently chosen values to the x_i^0.

Here we must distinguish two cases.

[. . .]

If the series (2) converged, to the commensurate value $\frac{n_1}{n_2}$ there would correspond a double infinity of periodic solutions of equations (1).

Now we have seen in n° 42 that this can only occur in very particular cases.

It therefore seems permissible to conclude that the series (2) do not converge.

Nevertheless the preceding argument is not enough to establish this point with complete rigor.

- **Q**[Poi5] (1893) From [Poi92–99], Volume II, Chapter XIII, pp. 104–105:

Ne peut-il pas arriver que les séries (2) convergent quand on donne aux x_i^0 certaines valeurs convenablement choisies?

Supposons, pour simplifier, qu'il y ait deux degrés de liberté ; les séries ne pourraient-elles pas, par exemple, converger quand x_1^0 et x_2^0 ont été choisis de telle sorte que le rapport $\frac{n_1}{n_2}$ soit incommensurable, et que son carré soit au contraire commensurable (ou quand le rapport $\frac{n_1}{n_2}$ est assujetti à une autre condition analogue à celle que je viens d'énoncer un peu au hasard)?

Les raisonnements de ce Chapitre ne me permettent pas d'affirmer que ce fait ne se présentera pas. Tout ce qu'il m'est permis de dire, c'est qu'il est fort invraisemblable.

✠ Translation [HSD]:

Can it not happen that the series (2) converge when one gives the x_i^0 certain conveniently chosen values?

To simplify, suppose there were two degrees of freedom; might not the series, for example, converge when x_1^0 and x_2^0 were chosen so that the ratio $\frac{n_1}{n_2}$ is incommensurable, and when its square to the contrary is commensurable (or when the ratio $\frac{n_1}{n_2}$ is subject to another condition analogous to the one I just stated somewhat at random)?

The arguments of this chapter do not allow me to affirm that this will not happen. All I'm permitted to say is, it is highly unlikely.

• **Q**[Poi6] (1899) From [Poi92–99], Volume III, p. 389:

Que l'on cherche à se représenter la figure formée par ces deux courbes et leurs intersections en nombre infini dont chacune correspond à une solution doublement asymptotique, ces intersections forment une sorte de treillis, de tissu, de réseau à maille infiniment serrées; chacune des deux courbes ne doit jamais se recouper elle-même, mais elle doit se replier sur elle même d'une manière très complexe pour venir recouper une infinité de fois toutes les mailles du réseau.

On sera frappé de la complexité de cette figure, que je ne cherche même pas à tracer. Rien n'est plus propre à nous donner une idée de la complication du problème des trois corps et en général de tous les problèmes de la Dynamique où il n'y a pas d'intégrale uniforme et où les séries de Bohlin sont divergentes.

✠ Translation [HSD]: See p. 50.

• **Q**[Poi7] (1908) From [Poi08], p. 68:

Une cause très petite, qui nous échappe, détermine un effet considérable que nous ne pouvons pas ne pas voir, et alors nous disons que cet effet est dû au hasard. [...] Mais, lors même que les lois naturelles n'auraient plus de secret pour nous, nous ne pourrons connaître la situation initiale qu'*approximativement*. Si cela nous permet de prévoir la situation ultérieure *avec la même approximation*, c'est tout ce qu'il nous faut, nous dirons que le phénomène a été prévu, qu'il est régi par des lois; mais il n'en est pas toujours ainsi, il peut arriver que de petites différences dans les conditions initiales en engendrent de très grandes dans les phénomènes finaux [...].

✠ Translation [HSD]:

A very small cause, which escapes us, determines a considerable effect which we cannot fail to see, and so we say that this effect is due to chance. [...] But even when natural laws are not secret from us, we can only know the initial situation *approximately*. If this allows us to predict the ulterior situation *with the same approximation*, then we are satisfied, we say that the phenomenon has been predicted, that it is governed by laws. But it is not always so; it may happen that small differences in initial conditions give rise to very large differences in the final phenomena [...].

K. Weierstrass:

• **Q[Weie]** Excerpt from p. 56 of [Mit12] (a letter from Weierstrass to G. Mittag-Leffler dated February 2, 1889):

Nun hätte ich noch ein Desiderium, worüber ich selbst an P. schreiben werde.

P. behauptet, dass aus der Nichexistenz mehrerer eindeutigen (analytischen) Integrale bei einem dynamischen Probleme nothwendig die Unmöglichkeit folge, das Problem durch Reihen von der Form

$$\sum C_{\nu\nu'\ldots} \genfrac{}{}{0pt}{}{\cos}{\sin} (\nu a t + \nu' a' t + \ldots)$$

zu lösen. Diese Behauptung, die von fundamentaler Bedeutung ist, wird ohne Beweis ausgesprochen.

✠ Translation [HSD]:

Now I have another request that I'll write to [Poincaré] about myself.

[Poincaré] claims that from the nonexistence of further distinct (analytic) integrals in a dynamical problem necessarily follows the impossibility of solving the problem by series expansions of the form

$$\sum C_{\nu\nu'\ldots} \genfrac{}{}{0pt}{}{\cos}{\sin} (\nu a t + \nu' a' t + \ldots).$$

This claim, which is of fundamental significance, is stated without proof.

Appendix F

Glossary

Acronyms and Abbreviations

BdL Bureau des Longitudes (Bureau of Longitudes)

CCW, CW Counter clockwise, clockwise

CERN Organisation européenne pour la recherche nucléaire (European Organization for Nuclear Research ['Organisation' was originally 'Centre'])

DE Differential Equation

DS Dynamical System

ENS École Normale Supérieure (Superior Normal School)

ETH Eidgenössische Technische Hochschule (Federal Polytechnic Institute)

Fermilab Fermi National Accelerator Laboratory

FPU Fermi Pasta Ulam

HPT Hamiltonian perturbation theory

IC initial condition

ICM International Congress of Mathematicians

KAM Kolmogorov Arnold Moser

KdV Korteweg-de Vries

LMAJ Liouville Mineur Arnold Jost

NHIM normally hyperbolic invariant manifold

ODE ordinary differential equation

PDE partial differential equation

RPC3BP restricted, planar, circular three body problem

SDOIC sensitive dependence on initial conditions

WW World War

Symbols

$\mathbb{C}$ set of complex numbers

H Hamiltonian function

I action (vector)

J action (vector)

k integer (vector)

M phase space (metric space, manifold)

$\mathbb{N}^*$ set of positive integers $\{1, 2, 3, \ldots\}$

$\mathbb{N}_0$ set of nonnegative integers $\{0, 1, 2, \ldots\}$

$\mathbb{R}$ set of real numbers $(-\infty, \infty)$

$\overline{\mathbb{R}}$ set of extended real numbers $[-\infty, \infty]$

$\mathbb{R}^n$ Euclidean n-space (n-fold Cartesian product of $\mathbb{R}$ with itself)

$\mathbb{T}$ one-torus

$\mathbb{T}^n$ n-torus

$\mathbb{Z}$ set of integers $\{\ldots, -2, -1, 0, 1, 2, \ldots\}$

$\mathbb{Z}^n$ set of integer n-vectors (n-fold Cartesian product of $\mathbb{Z}$ with itself)

γ Diophantine parameter

ε perturbation parameter

θ angle (vector)

τ Diophantine parameter

ϕ angle (vector)

χ Lie generating function

ω frequency (vector)

$\{\, , \,\}$ Poisson bracket

$\| \; \|_A$ uniform norm (over set A)

$\ll$ much less than

$\in$ belongs to, is a member of, in

GLOSSARY ENTRIES

This glossary is intended to be a starting point for anyone reading about KAM theory, both in this book and in the wider literature. A simple definition is given whenever it is available, sometimes with a remark or two that may help place the entry in context. At one end of the spectrum are a number of simple terms (e.g. *class*, *estimate*, *strong*, etc.) listed for the benefit of readers who may be unfamiliar with the way mathematicians use them. At the other end of the spectrum are many technical terms (especially those pertaining to *symplectic geometry* and *differential topology*, e.g. *symplectic manifold*, *tangent bundle*, etc.) for which the entries are 'impressionistic'; the interested reader is urged to consult more advanced texts in these cases.

More generally, the reader may be interested to see that despite its (largely deserved) reputation for precision, mathematical language often shows its human side in a significant number of loosely defined, ambiguous or overlapping terms that various authors use in different ways. These are pointed out whenever feasible below.

Note that when a glossary entry (or some slight variation) is used in the text of an entry, its first occurrence is in *slanted text*. Quotation marks (' ') are often used in the usual way, but they also often indicate that the term between them has a special or technical meaning not given in this glossary.

Absolute convergence. The *series* $\sum_k a_k$ (with real or complex terms a_k) is *absolutely convergent* if the series $\sum_k |a_k|$ *converges*. (Compare the *weaker* notion of *conditional convergence*.)

Action. In physics, *action* appears in many guises (e.g., it is the 'action' in *action-angle variables*), but in most cases it is a *functional* on a given set of *trajectories* of a mechanical *dynamical system*; in other words, to each trajectory of the set, one associates a real number called the action. Most commonly, the action of a trajectory is computed as the integral of the *Lagrangian* along the trajectory over a given time interval (so that the physical units of action are energy×time). There are many other (interrelated) methods for computing action. For example, in an *n-degree-of-freedom Hamiltonian system* with *canonical coordinates* (q, p), one method of setting up action-angle variables (for periodic orbits) is to compute the 'action integral' $\oint p_k \, dq_k$ over a closed orbit for each *degree of freedom* $k = 1, \ldots, n$.

It is not possible in a short space to sketch out the role played by action in the development of mechanics, but at minimum, everyone should be aware of how 'least action principles' preceded Hamiltonian mechanics and modern variational methods, and of why Planck's constant has units of action (cf. *adiabatic invariance*).

Action-angle variables. These are special *symplectic coordinates* $(\theta, I) = (\theta_1, \ldots, \theta_n; I_1, \ldots, I_n)$ that may be viewed as polar coordinates for n-dimensional concentric *tori* in the $2n$-dimensional *phase space* of a *Hamiltonian system*. By the LMAJ theorem (cf. §2.3), the (bounded) motions of a *completely integrable* Hamiltonian system take place on concentric *Lagrangian tori*, and it is always possible to find *action-angle variables* in which the *Hamiltonian* is independent of the angles $\theta_1, \ldots, \theta_n$. Such coordinates are especially useful for viewing *perturbations* of the integrable Hamiltonian for which they were originally defined.

Adiabatic invariance. The adjective *adiabatic* appears to have entered scientific vernacular in the mid 19th century from Greek $\dot{\alpha}\delta\iota\dot{\alpha}\beta\alpha\tau o\varsigma$ ('impassable') to signify thermodynamic processes that take place without loss or gain of heat. The term *adiabatic invariance* is now used broadly in thermodynamics, *classical mechanics*, and quantum mechanics to describe the constancy (or near-constancy) of an *action* variable (or other *observable*) as the underlying *system* slowly changes in time. Adiabatic invariance typically occurs when the system changes over times much longer than the 'natural' timescale for the *dynamics* of the system. In thermodynamics, the natural timescale might be the time to reach *thermal equilibrium*; in classical mechanics, it might be the period of oscillations; in quantum mechanics, it might be the inverse of the frequency difference between energy eigenstates.

Adiabatic invariance figured prominently in discussions at the 1911 Solvay Congress in Brussels, during which H. Lorentz and A. Einstein groped their way toward the founding principles of 'old quantum mechanics' by considering a pendulum whose length varies slowly compared with its frequency (the two were trying to understand the existence of Planck's constant $\hbar$). For the classical pendulum with energy E and frequency ω, as the length slowly varies, one finds that the ratio E/ω is an adiabatic invariant; this is mysteriously analogous to the quantization formula $E/\omega = \hbar(n+1/2)$ for the harmonic oscillator. This observation was extended by A. Sommerfeld and C. Wilson to a general quantization principle in which the adiabatically invariant actions of a classical system correspond to the quantum states of the analogous quantum system. (For more details, see [LocM88] and references therein.)

To give a more mathematically precise definition, we consider the narrower context of a *flow* defined by an ODE $\dot{x} = f(x, \lambda)$ in which the *vector field* f depends on phase variables x and parameters λ, and the dot represents the t-derivative (t is the time). We introduce a slow time-dependence by setting $\lambda = \lambda(\varepsilon t)$, where $\varepsilon > 0$ is a small parameter. Then the real-valued function $A = A(x, \varepsilon t)$ is an adiabatic invariant of the flow provided $\lim_{\varepsilon \to 0} \sup_{0 \le t \le 1/\varepsilon} |A(x(t), \varepsilon t) - A(x(0), 0)| = 0$ for *almost all initial conditions* $x(0)$. (Here $x(t)$ is the flow of the ODE with dependence on the initial condition suppressed.) In other words, for almost all initial conditions, the departure of A from its initial value over the t-interval $[0, 1/\varepsilon]$ goes to zero with ε (while the t-interval stretches to infinity).

Both KAM and Nekhoroshev theory may be used to derive specialized adiabatic invariance results in a number of applications.

Almost all, almost every, almost everywhere, etc. In a *measure space*, a property is said to hold for *almost all* points or *almost everywhere* if the set of points where it fails to hold has *measure* zero. If the measure is not understood in advance, it can be specified as, say, μ by saying that the property holds 'μ-almost everywhere' (sometimes abbreviated μ-a.e.). In KAM theory, when we say that almost all *tori* of a nondegenerate integrable system are *nonresonant*, we mean that the (*relative*) *Lebesgue measure* of the set of *resonant* tori is zero.

Analysis. The large, sprawling branch of mathematics that includes or overlaps calculus, differential equations (both ODE and PDE), *measure* and integration theory, complex variables, and all things pertaining to *limits* and *convergence*.

Analytic, analyticity. Given an *open set* $D \subset \mathbb{C}$, to say that the *function* $f : D \to \mathbb{C}$ is analytic on D (in symbols, $f \in C^\omega(D)$) means that for each point $z \in D$, there is a positive 'radius of convergence' r such that, at each point x in the *open ball* of radius r centered at z, the *Taylor series* of f based at z converges and equals $f(x)$ (in symbols, $f(x) = \sum_{n=0}^{\infty} f^{(n)}(z)(x - z)^n / n!$ for all $x \in B_r(z)$). Precisely the same definition applies when $D \subset \mathbb{R}$ and $f : D \to \mathbb{R}$, in which case f is sometimes called *real analytic* for emphasis. *Analyticity* is also similarly defined for real- or complex-valued functions of severable variables (i.e., with *domain* $D \subset \mathbb{R}^n$ or $D \subset \mathbb{C}^n$). Analyticity is the *strongest regularity* property a function may possess (considerably stronger than mere *smoothness* which it implies), but it also entails certain 'rigidity' properties: e.g., if an analytic function is constant on an open subset of its (*connected*) *domain*, then it must be constant on the entire domain.

Arnold diffusion. In 1964, V.I. Arnold published a short paper [Ar64] in which he outlined a mechanism for long-time instability in *nearly integrable Hamiltonian systems* with more than two *degrees of freedom*. This paper started a new branch of HPT that investigates the nature of such instability, now loosely called *Arnold diffusion*. Perhaps its most intriguing feature is its persistence in *systems* arbitrarily close to *integrable* (i.e., no matter how small the *perturbation*), though Nekhoroshev's theorem shows that its displacement rate must be extremely slow for small perturbations. Despite much progress, there are still unanswered questions concerning precisely how and when it occurs, its rate, and conditions under which it may be *generic*.

Asymptotic expansion (or **series**). An *asymptotic expansion* of a function $f : D \subset \mathbb{R} \to \mathbb{R}$ (or $f : D \subset \mathbb{C} \to \mathbb{C}$) is a (usually *divergent*) series of functions which, when truncated, gives an increasingly better approximation to f as the argument tends to a particular (often infinite) value. We can use *little o* notation to give a more precise definition as follows. Given a limiting value $a \in [-\infty, \infty]$, the (*formal*) series $\sum_{k=0}^{\infty} g_k(x)$ is asymptotic to $f(x)$ as $x \to a$ if for each m, $g_{m+1} = o(g_m)$ and $f - \sum_{k=0}^{m} g_k = o(g_m)$ as $x \to a$. One of the best-known examples is Stirling's series for the factorial (here $n \in \mathbb{N}^*$ takes the place of $x \in \mathbb{R}$): $n! \sim \sqrt{2\pi n}\,(n/e)^n \left[1 + (1/12)n^{-1} + (1/288)n^{-2} + \cdots\right]$, $n \to \infty$. The symbol '$\sim$' is read 'is asymptotic to' (as $n \to \infty$, in this case), and the series on the right is in fact divergent, as are most interesting asymptotic expansions. However, for large fixed n, truncating the series after a few terms gives a good approximation to $n!$, with the error reaching a minimum at the inclusion of a certain number of terms $k = k(n)$. As n increases, the (relative) error decreases rapidly to zero, while $k(n)$ grows slowly. Poincaré was the first to thoroughly investigate and recognize the importance of asymptotic series. His suggestion that the Lindstedt series expansions of solutions of the n body problem were (*almost everywhere*) divergent asymptotic series is at odds with KAM theory, which shows that the Lindstedt series are in fact *convergent* for a set of ICs with positive *Lebesgue measure*.

Atlas. See *topological manifold*.

Autonomous. Not explicitly time dependent (commonly used in reference to *vector fields, ODEs, Hamiltonian systems*, etc.). An *autonomous* first-order ODE has the form $\dot{x} = f(x)$, as opposed to the 'non-autonomous' form $\dot{x} = f(t, x)$.

Baire category theory (after René-Louis Baire). This collection of results begins by classifying sets as small or large in a *complete metric space* (or more generally in a *topological space*). The small sets are those of *first category* (also called *meager sets*), which may be expressed as a *countable* union of *nowhere dense* sets. The large sets, said to be of *second category*, are those that are not of first category, while the *comeager* (or *generic* or *residual*) sets are those that are complements of meager sets. Baire category theory is often used as a means of characterizing generic properties of *dynamical systems* (see *generic, genericity*). R.-L. Baire introduced these notions at the end of the 19th century [Bai99], just three years before H. Lebesgue introduced the integration and *measure theory* that bears his name. Baire theory has proved quite useful in *analysis* and *topology*, where the best-known result, the so-called 'Baire category theorem,' says (in its simplest version) that every nonempty *complete metric space* is a *Baire space*. Perhaps surprisingly, the notions of small and large as gauged by measure theory and Baire theory are independent: there are not only generic sets of *full measure* and meager sets of measure zero, but also generic sets of measure zero, and meager sets of full measure. (Note that Baire category theory is unrelated to the *category theory* of S. Eilenberg and S. Mac Lane.)

Baire space. A *metric space* or *topological space* is called a *Baire space* if the union of any *countable collection* of *closed sets* with empty *interior* has empty interior.

Banach space (after Stefan Banach). A *complete normed vector space*. In other words, a *normed vector space* that is complete with respect to the metric $d(v, w) = \|v - w\|$ induced by the *norm* $\| \ \|$. The most familiar *Banach spaces* are the ordinary *Euclidean spaces* (which are also *Hilbert spaces*), but Banach spaces are most interesting as *function spaces* of infinite dimension, such as L^p spaces.

Barycenter. In celestial mechanics, this is the name for the center of mass of an isolated system of bodies (e.g. planets). Given a *system* of n bodies in which the ith body has mass m_i and center of mass located at (Cartesian) coordinates x_i, the *barycenter* has coordinates $\sum_{i=1}^{n} x_i m_i / \sum_{i=1}^{n} m_i$, and this location remains fixed (in an inertial reference frame) as the system evolves according to Newton's laws.

Base. [1] In *topology*, if $(X, \mathcal{T})$ is a *topological space* with topology $\mathcal{T}$, a *base* for $\mathcal{T}$ is a *subcollection* $\mathcal{B} \subset \mathcal{T}$ such that for every $O \in \mathcal{T}$ and $x \in O$ there is a $B \in \mathcal{B}$ with $x \in B \subset O$. Conversely, $\mathcal{T}$ can be 'generated' from $\mathcal{B}$ as the collection of all unions of members of $\mathcal{B}$. (Example: In $\mathbb{R}^n$ with the usual topology, the set of all *open balls* with centers at rational coordinates and with rational radii forms a base; in fact it is a *countable* base.)
[2] One occasionally sees 'base' used as an abbreviation of *base space*.

Base point. See *base space*.

Base set. The universal set M in a *sigma algebra* (or in a *measurable space*, or *measure space*).

Base space. This term has a number of meanings in mathematics. In *dynamical systems*, one often sees *base space* in reference to the object (usually a *smooth manifold*) on which a *vector field* is defined (i.e., to which the vectors of the vector field are 'attached'). It may also refer to the manifold M to which the *(co)tangent spaces* are attached in a *(co)tangent bundle*. A point in a base space is a *base point*. The terminology comes from the language of 'vector bundles' and 'fiber bundles,' more abstract objects of which *tangent* and *cotangent bundles* are special cases.

Big O. Given *functions* $f : \mathbb{R} \to V$ and $g : \mathbb{R} \to V$, where V is a *normed vector space*, we say $f(x) = O(g(x))$ as $x \to a$ (read '$f(x)$ is *big O* of $g(x)$ as x approaches a') provided there are positive constants M, δ such that for all $|x - a| < \delta$, $\|f(x)\| \leq M\|g(x)\|$. (In words, norm f is *bounded* by a constant times norm g in a *neighborhood* of a.) Similarly, to say that $f(x) = O(g(x))$ as $x \to \infty$ means that there are positive constants M, A such that for all $x \geq A$, $\|f(x)\| \leq M\|g(x)\|$. (Note that the limiting value (a or $\pm\infty$) is often omitted when understood in context.) Here are two nontrivial (related) examples: (i) $\varepsilon\sin(1/\varepsilon) = O(\varepsilon)$ as $\varepsilon \to 0$. (ii) $x\sin(1/x) = O(1)$ as $x \to \infty$.

Bijection, bijective. A *bijection* is a *function* (or *map*, or *mapping*) that is both *injective* and *surjective* (i.e., *one-to-one* and *onto*). The adjective is *bijective*.

Bilinear form. If V is a *vector space*, a (real) *bilinear form* on V is a function $\omega : V \times V \to \mathbb{R}$ such that for each $v \in V$, the functions $\omega(v, \cdot) : V \to \mathbb{R}$ and $\omega(\cdot, v) : V \to \mathbb{R}$ are *linear*. (One says that "ω is linear in each variable separately.")

Binary relation. Strictly speaking, this term refers to a '2-place *relation*' (i.e., a subset of a *Cartesian product* $A \times B$). But one also often sees the expression 'a *binary relation* on the set A,' meaning a relation between two copies of A (i.e., a subset of the Cartesian product $A \times A$).

Birkhoff normal form (after George D. Birkhoff; also often called *Birkhoff-Gustavson normal form*). The n degree of freedom Hamiltonian $H = H(q_1 \ldots, q_n; p_1, \ldots, p_n)$ is in *Birkhoff normal form* (at the origin, through degree k) if it has the form $H = H_k(I) + R$, where H_k is a polynomial of degree k in the variables $I_j = \frac{1}{2}(q_j^2 + p_j^2)$ $(j = 1, \ldots, n)$, and where R represents higher order terms in I. See Appendix 7 of [Ar78–97] for a nice discussion of Birkhoff normal forms and their applications.

Birkhoff's ergodic theorem. See *ergodic theorem*.

Bode's law. See *Titius-Bode law*.

Bound. In mathematics (especially *analysis*), a *bound* is usually an inequality showing that a set or *function* is *bounded*. It may also be the specific number, symbol(s), or function(s) on the side of such an inequality opposite the set or function in question. One also speaks of 'growth bounds' to indicate inequalities that restrict the 'rate' at which a (possibly unbounded) function grows (e.g., the inequality $|f(t)| \le e^{kt}$ establishes an 'exponential growth bound' on f).

Bounded from above [or **below**]. A set of real numbers is *bounded from above* [*below*] if it has an *upper* [*lower*] *bound* (i.e., if there is a number greater [less] than or equal to all numbers in the set).

Bounded function. A *function* whose *range* is a *bounded set* (but see also the special case of *bounded linear map*).

Bounded linear map. If V, W are *normed vector spaces* and $L : V \to W$ is a *linear map*, to say that L is *bounded* means that the image $L(\overline{B}_1(0))$ of the *closed* (unit) *ball* is a *bounded set* in W.

Bounded set. In a *metric space*, a *bounded set* is a set that is contained inside an *open ball* of finite radius.

Bounded variation. Given a *closed* interval $[a, b] \subset \mathbb{R}$ and a *function* $f : [a, b] \to \mathbb{C}$, we say f is of *bounded variation* on $[a, b]$ (or $f \in BV([a, b])$) if the *total variation* $V_a^b(f)$ is finite. The total variation is in turn defined as $V_a^b(f) = \sup_{S \in P} \sum_{k=1}^{N_S} |f(x_k) - f(x_{k-1})|$, where P is the *collection of* all partitions or subdivisions $S = \{x_0, \ldots, x_{N_S}\}$ of $[a, b]$ by finitely many points with $a = x_0 < x_1 < \cdots < x_{N_S} = b$.

Bourbaki (used either in singular or plural). Nicolas *Bourbaki* was the pseudonym adopted by a self-styled group of nine *anciens élèves de l'ENS* (former pupils of the *École Normale Supérieure* in Paris, France) who gathered in Paris in the mid 1930s with the aim of quickly producing a modern *analysis* program (and 'treatise') for training a new generation of French mathematicians. Although the members' anti-Poincaré bias has sometimes been exaggerated, there is some truth in their wanting to break with Poincaré's intuitionist approach. The original group's membership rapidly changed, and its aims grew more ambitious. Eventually (following a pause during WWII) the group produced a kind of multi-volume encyclopedia of mathematics in a decidedly abstract, *structuralist* style that had great influence in France and elsewhere. Despite the austere style of its texts, the group is also known for its occasional practical jokes or pranks. Inevitably, there emerged a lively—and to some extent still ongoing—controversy over whether Bourbaki has a beneficial or detrimental effect on mathematical research. Those arguing its benefits see a necessary clarification and streamlining without which some fields of research could not proceed; those arguing the opposite see rigid desiccation that drains the intuitive element from research. In [Stew95], I. Stewart describes how the influence of Bourbaki and the controversy surrounding it waned substantially toward the end of the 20th century, as mathematicians finally accepted its necessary lessons (especially regarding *rigor*) and learned to move beyond it.

Bruno condition (sometimes transliterated as Brjuno or Bryuno). Named after Aleksandr Bruno following his use of it in [Bruno65], [Bruno67], [Bruno71], this is a 'number-theoretic' condition (to be satisfied, e.g., by the *rotation numbers* of circle *maps* or the rotation numbers of the *linear parts* of planar maps). The irrational number α satisfies a *Bruno condition* (or is a 'Bruno number') if the *convergents* $\{p_k/q_k\}_{k=0}^{\infty}$ of its *continued fraction* expansion have denominators q_k such that $\sum_{k=0}^{\infty} (\log q_{k+1})/q_k < \infty$. Bruno's condition is closely related to—but slightly *weaker* than—*Diophantine conditions*, and it serves the same purpose: it ensures that the α satisfying it are 'far from *resonance*.' Bruno's condition is known to be the optimal condition for linearizability in the Siegel center problem (see §A1.1). There are also 'multi-dimensional Bruno conditions' for *frequency vectors*, and it is conjectured that they are the optimal number-theoretic conditions for the existence of KAM *tori* in certain *systems*.

Butterfly effect. This popular term for *sensitive dependence on initial conditions* (SDOIC) has its origins in a 1972 lecture by E. Lorenz entitled 'Does the flap of a butterfly's wings in Brazil set off a tornado in Texas?' (This followed the increasing success of Lorenz's articles on *chaos* in models of weather systems, beginning with [Lor63].) The butterfly imagery captured the imagination of a broad public: it is mentioned in a number of Hollywood films, including *Jurassic Park* and of course *The Butterfly Effect* and *The Butterfly Effect 2*. There is also hand-wringing and consternation that SDOIC and *chaos theory* generally are misunderstood or misrepresented; see e.g. [Diz08].

Canonical. This adjective is often used in mathematics to indicate a natural or distinguished case. In the traditional picture of *Hamiltonian dynamics*, one constructs *canonical coordinates* (or *canonical variables*) $(q, p) = (q_1, \ldots, q_n, p_1, \ldots, p_n)$ by first finding *generalized coordinates* (or 'configuration variables') $q = (q_1, \ldots, q_n)$ for the *configuration space* Q of a given mechanical system. One then constructs the *Lagrangian* $L = L(q, \dot{q}, t)$, which in turn generates the *generalized momenta* $p = (p_1, \ldots, p_n)$ via $p_k = \partial L / \partial \dot{q}_k$. In the modern mathematical picture of Hamiltonian dynamics, one starts with a *symplectic manifold* (no mechanical system required), and obtains canonical coordinates as a consequence of *Darboux's theorem* (such coordinates are often called *Darboux coordinates*).

Traditionally, one also uses *canonical transformation* to mean a change of variables $(q, p) \mapsto (Q, P)$ from one set of canonical coordinates to another in a Hamiltonian system (so that *Hamilton's equations* remain invariant under the transformation). In the modern formulation of Hamiltonian dynamics, canonical is often replaced by *symplectic* ('*symplectic coordinates*,' '*symplectic transformation*,' etc.), which has a more precise meaning.

Canonical coordinates. See *canonical*.

Canonical equations of motion. See *Hamilton's equations*.

Canonical transformation. See *canonical*.

Cantor set (after Georg Cantor). A subset C of a *topological space* that is perfect and *nowhere dense*. (Here 'perfect' means that every point of C is a *limit point* of C, or equivalently that C is *closed* and has no *isolated points*.) The best-known Cantor set (many say *the* Cantor set) is the so-called ternary set (or middle thirds set) obtained from the unit interval by removing successively smaller open middle third intervals. The resulting set has many interesting properties shared by most Cantor sets: it has *uncountably* many points, but has no *interior* points (i.e., contains no *open* neighborhoods). It has *Lebesgue measure* zero, and *topological dimension* zero, but *Hausdorff dimension* $\ln 2/\ln 3 \approx .62$; since it is also self-similar, it is often called the first (or simplest) *fractal* set. In KAM theory, one encounters 'big' Cantor-like sets (with positive Lebesgue measure), first as the *frequency vectors* of the *tori* that persist under *perturbation*, then also as the sets of persistent tori themselves.

Cartesian product (after René Descartes; also sometimes called *direct product*). Given arbitrary sets A and B, their *Cartesian product* $A \times B$ is the collection of all ordered pairs (a, b), where $a \in A$ and $b \in B$. More generally, given n sets $A_1, A_2, \ldots, A_n$, the n-fold Cartesian product $A_1 \times A_2 \times \cdots \times A_n$ is the collection of all ordered n-tuples $(a_1, a_2, \ldots, a_n)$ with $a_k \in A_k$. Everyone is familiar with the Cartesian product $\mathbb{R} \times \mathbb{R} =: \mathbb{R}^2$ of $\mathbb{R}$ with itself; this is the 'Cartesian plane' of analytic geometry in which the 'Cartesian coordinates' (x, y) of arbitrary points are so useful in representing *relations* between x and y.

Category theory. Introduced during WWII by S. Eilenberg and S. Mac Lane, this highly *structuralist* subject looks at mathematics 'from above' by investigating similar relations among objects in different parts of mathematics. These investigations produced the new terms 'category,' 'object,' 'morphism,' and 'functor' (among others), so we may now say that in the category of set theory, the objects are sets and the morphisms are *functions*; while in the category of *differential topology*, the objects are *smooth manifolds* and the morphisms are *diffeomorphisms*. We may also ask whether there is a functor (a kind of structure-preserving *map*) between these categories. Some have called category theory 'abstract nonsense,' but it has undeniable appeal to structuralists, and has produced significant results, especially in algebraic *topology* and algebraic geometry.

Cauchy sequence (after Augustin-Louis Cauchy). A *sequence* $\{x_j\}$ in a *metric space* (M, d) is said to be *Cauchy* if for every $\varepsilon > 0$ there exists an $N \in \mathbb{N}_0$ such that $d(x_j, x_k) < \varepsilon$ whenever $j, k \geq N$. (In words, terms of the sequence become arbitrarily close to each other as the indices increase.)

Celestial mechanics. In mathematics and mathematical physics, this is the part of *classical mechanics* that deals with the motion of objects (usually idealized 'point masses') interacting by way of Newton's law of gravitation. (More applied versions treat non-point masses and include the effects of forces other than gravity, such as rocket propulsion, various sorts of dissipation, etc.) Celestial mechanics has a special paradigmatic stature in science, as its basic principles have not changed since they were set out in Newton's *Principia*, yet few branches of science can aspire to the near-perfection with which those principles operate. The development of KAM theory grew out of unanswered questions in celestial mechanics that were present almost from the outset.

Center. Another name for an *elliptic fixed point* (the term *center* is more commonly used for *flows* or *Hamiltonian systems* than for *maps*).

Ceres. The Roman goddess of agriculture and the (abbreviated form of the) name given by G. Piazzi to the object he discovered orbiting the Sun on January 1, 1801, very close to the orbit of the 'missing 5th planet' predicted by the *Titius-Bode law*. In fact, astronomers considered *Ceres* to be a planet for about half a century, after which it was reclassified as the largest asteroid in the asteroid belt. In 2006, the International Astronomers Union (IAU) redefined the meaning of 'planet'; the initial draft of this redefinition would have made Ceres again a planet, but in the end, a requirement was added that a planet "should clear its orbit of planetesimals," a feat of housekeeping that Ceres has failed to carry out. Like Pluto, it is now classified as a 'dwarf planet,' the smallest of the five presently recognized by the IAU.

Chaos, chaotic. There are no universally accepted definitions for these terms in mathematics (but there is some broad agreement). *Chaos* appears to have been first coined in the mathematical sense by T.Y. Li and J.A. Yorke in their 1975 paper [LiY75], after which it circulated rapidly, entering widespread popular use and culminating in the best-selling book [Glei87] of that title. Most mathematicians agree that a *chaotic dynamical system* is one with the following ingredients: (i) *sensitive dependence on initial conditions* (SDOIC), (ii) *transitivity* (i.e., a *dense orbit*), and (iii) a dense set of *periodic points*. Some authors omit ingredient (iii); others retain it and even add a *statistical property*, such as *mixing*. See [MartDS98] for a discussion of various definitions. In a broader sense, 'chaos' may be seen as the modern name for types of behavior already well known to *ergodic theorists* since the early 20th century, but rediscovered by scientists working with computer models of *deterministic systems* beginning in the 1960s. For ergodic theorists, there is a broad hierarchy of increasingly chaotic systems (see *statistical properties* below).

Chart. See *topological manifold*.

Chirikov standard map. (after Boris V. Chirikov; other names are also sometimes used: see footnote 9, p. 128 above for a brief discussion.) This *discrete dynamical system* is a *map* from the cylinder $\mathbb{T} \times \mathbb{R}$ to itself; in cylindrical coordinates (x, y), one form of it reads $(x, y) \mapsto (x + y, y + (k/2\pi)\sin(2\pi x))$, where x is computed mod 1 and $k \geq 0$ is a parameter. The standard map is one of the simplest systems to display KAM-like behavior, and has been studied in great detail.

Class. [1] (a) In everyday mathematical usage, this is another word for 'set,' especially a set of objects easily defined by a property that all its members share. (b) However, in most versions of modern set theory, a *class* is something more general than a set, and the distinction between classes and sets is important in resolving certain paradoxes of naïve set theory.
[2] Often used as an abbreviation for *smoothness class*.

Classical mechanics. The large branch of *dynamical systems* that treats *deterministic* mathematical models of physical systems (i.e., models that don't incorporate quantum assumptions, though statistical methods may be used in analyzing large systems). Though largely abandoned as an area of research by physicists with the advent of quantum mechanics, the subject has continued to advance in the mathematical realm (and KAM theory may be viewed as one result of that advance).

Classical solution (of a *differential equation*). See *strong solution*.

Closed ball. In a *metric space* (M, d), the *closed ball* $\overline{B}_r(x)$ (with center x and radius $r \geq 0$) is defined as $\overline{B}_r(x) := \{z \in M \mid d(x, z) \leq r\}$.

Closed manifold. A *manifold* that is *compact* and has no boundary (e.g., a *torus*, or a *sphere*).

Closed set. In a *topological space* X, a set $F \subset X$ is *closed* if its complement $X \setminus F$ is an *open set*.

Closure. If X is a *topological space* and $A \subset X$, the *closure* of A in X (denoted $\overline{A}$, or sometimes $\text{cl}(A)$) is the *smallest closed set* in X that contains A. It may also be defined as the union of A and all *limit points* of A.

Codimension. If M is a subset of an ambient set G, the *codimension* of M is $\text{codim}\, M := \dim G - \dim M$ (whenever the *dimensions* on the right are defined).

Codomain. The collection of possible dependent variables of a *function* (i.e., the set of possible 'outputs'); or the collection of possible second entries in the set of ordered pairs comprising a *relation*. When the notation $f : A \to B$ is used to display all parts of a function, the *codomain* is B. Note that the codomain can be any superset of the image $f(A)$. Note also that, for most mathematicians, if $B \neq C$, $f : A \to B$ and $g : A \to C$ are considered different functions even if $f(x) = g(x)$ for every $x \in A$ (i.e., even if f and g only differ by having different codomains). However, most logicians and set theorists do not distinguish between such functions.

Co-isotropic. Given a *symplectic vector space* (V, ω), a *vector subspace* U of V is *co-isotropic* if $U^{\perp\omega} \subset U$, where $U^{\perp\omega}$ is the *symplectic complement* of U. (And U is co-isotropic if and only if $U^{\perp\omega}$ is *isotropic*.) A *submanifold* N of a *symplectic manifold* (M, ω) is co-isotropic if for each $p \in N$, the *tangent space* $T_p N$ is a co-isotropic subspace of $T_p M$.

Collection. Another name for 'set,' especially one whose elements are themselves sets (or other, non-point-like objects).

Comeager. A synonym of *residual*. See *Baire category theory*.

Commensurability, commensurable. Another terminology to describe a *resonance condition*, but used when speaking of the components $\omega_1, \ldots, \omega_n$ of a *frequency vector* ω rather than of ω itself. One says that the numbers (or *frequencies*) $\omega_1, \ldots, \omega_n$ satisfy a *commensurability* relation (or are *commensurable*, or *rationally dependent*) if there are integers $k_1, \ldots, k_n$, not all zero, such that $k_1\omega_1 + \cdots + k_m\omega_n = 0$.

Compact. A *topological (sub)space* X is *compact* if every *open cover* of X has a finite *subcover*.

Complement. If A and U are sets, the *complement* of A in U, written $U \setminus A$, is the set of all elements of U that are not elements of A (i.e., everything in U outside of A). When the set U is understood (as e.g. the 'universal set'), one simply says 'the complement of A' and uses a simpler notation without U, such as $\setminus A$, or $\sim A$, A^c, etc.

Complete integrability, completely integrable. A *completely integrable Hamiltonian system* is a Hamiltonian system of n *degrees of freedom* with n *independent constants of motion in involution*. (See §2.3 for more detail.) For Hamiltonian systems, complete integrability is sometimes called *Liouville integrability*, or simply *integrability*.

Complete metric space. A *metric space* (M, d) with the crucial property that every *Cauchy sequence* $\{v_k\}$ in M *converges* to an element v of M (i.e., sequences that try to converge do in fact succeed; M has no 'holes').

Complete measure space. A *measure space* (M, Σ, μ) is *complete* if all subsets of *null sets* are *measurable* (i.e., $\mu(A) = 0$ and $B \subset A \implies B \in \Sigma$).

Completeness (of $\mathbb{R}$). The *completeness* property of the real numbers is the fact that every nonempty subset of $\mathbb{R}$ that is *bounded from above* has a *supremum* in $\mathbb{R}$. (And from the algebraic viewpoint, $\mathbb{R}$ is the only 'ordered' field with this property.) This property is equivalent to $\mathbb{R}$ being a *complete metric space*, using the usual *distance* $d(x, y) = |x - y|$.

Conditional convergence. The *series* $\sum_k a_k$ (with real or complex terms a_k) is *conditionally convergent* if it is *convergent* but not *absolutely convergent*. Every conditionally convergent real series has the rather odd property that given any *extended real number* α there exists a 'rearrangement' of the series that converges to α. In KAM theory, the *Lindstedt series* expansions of *quasiperiodic orbits* residing on KAM *tori* are conditionally convergent.

Conditionally periodic flow on $\mathbb{T}^n$ (also called *Kronecker flow* or *linear flow on* $\mathbb{T}^n$). For integer dimension $n \geq 1$, let $\theta = (\theta_1, \dots, \theta_n)$ be standard coordinates on the n-torus $\mathbb{T}^n$ (i.e., $0 \leq \theta_k < 1$, $k = 1, \dots, n$), and recall that *mod*-1 arithmetic applies in each coordinate: if $\theta, \theta' \in \mathbb{T}^n$ and $s, t \in \mathbb{R}$, then $s\theta + t\theta' = \theta'' \in \mathbb{T}^n$, where θ'' has coordinates $\theta_k'' \equiv s\theta_k + t\theta_k'$ (mod 1). (See *mod, modulo*.) Using these conventions, and given a *frequency vector* $\omega \in \mathbb{R}^n$, we define the *conditionally periodic flow* $F_t^\omega : \mathbb{T}^n \to \mathbb{T}^n$ with frequency ω by $F_t^\omega(\theta) = \theta + \omega t$.

More generally, we speak of conditionally periodic flow (or 'motion') $G_t^\omega : \mathcal{N} \to \mathcal{N}$ on an *embedded* or 'distorted' n-torus $\mathcal{N}$ if there is a *smooth embedding* $g : \mathbb{T}^n \to \mathcal{N}$ and a conditionally periodic flow $F_t^\omega : \mathbb{T}^n \to \mathbb{T}^n$ such that $F_t^\omega = g^{-1} \circ G_t^\omega \circ g$ (i.e., the embedding g smoothly *conjugates* F_t^ω with G_t^ω).

The terminology 'conditionally periodic' comes from the Russian literature, and makes sense when one realizes that conditionally periodic flow on $\mathbb{T}^n$ is either *periodic* or *quasiperiodic*. If ω satisfies $n - 1$ *independent resonance conditions*, then the flow is periodic (i.e., all its *orbits* are periodic); if ω satisfies r independent resonance conditions, with $0 \leq r \leq n - 2$, then the flow is quasiperiodic with $m = n - r$ *fundamental frequencies* (more precisely, each orbit of F_t^ω *densely* fills an m-dimensional subtorus of $\mathbb{T}^n$, and the flow may be 'reduced' (i.e., smoothly conjugated) to *quasiperiodic flow on* $\mathbb{T}^m$ with *nonresonant* frequency vector $\alpha \in \mathbb{R}^m$).

Conditionally periodic function. The *function* $f : \mathbb{R} \to \mathbb{R}$ is *conditionally periodic* if there is a *continuous function* $G : \mathbb{T}^n \to \mathbb{R}$ and a nonzero *frequency vector* $\omega \in \mathbb{R}^n$ such that $f(t) = G(F_t^\omega(\theta))$, where F_t^ω is *conditionally periodic flow* on $\mathbb{T}^n$ with frequency ω, and θ is an IC in $\mathbb{T}^n$. By 'shifting' G, one can always choose the IC $\theta = 0$ and write f in the less precise but simpler form $f(t) = G(\omega t)$.

Note that some authors do not require G to be continuous; note also that f is periodic if $n = 1$ or if ω is *resonant* with *multiplicity* $n - 1$. Conditionally periodic functions arise naturally in *Hamiltonian systems*, since when an *integrable system* is expressed in *action-angle variables*, any continuous function of the angles is conditionally periodic.

Configuration space. In *classical mechanics*, the *configuration space* of a mechanical *system* is the set of all positions the parts of the system can assume, taking constraints into account. For most systems, the configuration space is naturally viewed as a *smooth manifold*, with *local coordinates* called the *generalized coordinates* of the system. The *dimension* of the configuration space is called the number of *degrees of freedom* of the system.

Conjugacy (also *equivalence*). Given C^k-*smooth manifolds* A and B, and C^k maps $f : A \to A$ and $g : B \to B$, a C^j conjugacy of f and g $(0 \le j \le k)$ is a C^j *diffeomorphism* $h : A \to B$ such that $f = h^{-1} \circ g \circ h$. (Here by a C^0 diffeomorphism we mean a *homeomorphism*.) When $j = 0$, the homeomorphism h is called a *topological conjugacy* (or *topological equivalence*), and we say f is *topologically conjugate* (or *topologically equivalent*) to g. When $j > 0$, the diffeomorphism h is called a C^j-*smooth conjugacy* (or *smooth equivalence*), and we say f is C^j-*smoothly conjugate* (or *smoothly equivalent*) to g. (As usual, when 'smooth' is used without qualification, it means 'of class C^∞.') The notion of conjugacy for *flows* is very similar. Note that conjugacy is an *equivalence relation*, and may often be viewed as a change of coordinates.

Conjugate. [1] For *conjugate maps* or *flows*, see *conjugacy*.
[2] In *Hamiltonian* mechanics, one speaks of *conjugate* variables to denote a corresponding pair (q_k, p_k) of *canonical coordinates* (i.e., the members q_k and p_k are conjugate to one another).

Connected. A *topological space* $(X, \mathcal{T})$ is *connected* if there do not exist nonempty $O_1, O_2 \in \mathcal{T}$ with $O_1 \cap O_2 = \emptyset$ and $O_1 \cup O_2 = X$. (In words, there is no *cover* of X consisting of two nonempty disjoint *open sets*.)

Constant of motion (also called *integral of motion* or *first integral*). If (M, φ) is a *continuous dynamical system*, a nonconstant C^k-smooth function $f : M \to \mathbb{R}$ is a C^k-*constant of motion* for φ if, for each $x \in M$, $f(\varphi_t(x))$ is constant for all t. In other words, f is constant along each *orbit* of φ. (A similar definition applies to *discrete systems*.)

Continued fraction. A continued fraction (CF) is an expression of the form $a_0 + 1/(a_1 + 1/(a_2 + 1/(a_3 + \cdots)))$ (which may or may not terminate), where a_0 is any integer and the other a_k $(k \geq 1)$ are positive integers. A standard abbreviation for a CF is $[a_0; a_1, a_2, a_3, \ldots]$, in which the a_k are called the 'quotients.' One also uses the truncation $[a_0; a_1, a_2, \ldots, a_n] = a_0 + 1/(a_1 + 1/(a_2 + \cdots + 1/a_n)) = p_n/q_n$, called the nth *convergent* of the CF (which coincides with the CF if the CF terminates at the nth quotient). CFs have a long history and a number of interesting properties. Every finite (i.e., terminating) CF represents a rational number, and every rational number may be written as a finite CF in precisely two ways $([a_0; a_1, a_2, \ldots, a_{n-1}, a_n] = [a_0; a_1, a_2, \ldots, a_n - 1, 1])$. The CF expansion of an irrational number r is unique and infinite (i.e., does not terminate), with successive convergents that are alternately larger and smaller than r (and yes, the *sequence* of convergents does *converge* to r). Each convergent p_k/q_k is a 'best rational approximation' to r in the sense that there is no other rational number with denominator less than q_k that is closer to r. Finally, the kth convergent satisfies the beautiful estimate $|r - p_k/q_k| < 1/(q_k q_{k+1})$.

The fact that low-dimensional *discrete dynamics* (e.g. circle maps) is a much more completely developed subject than its higher-dimensional analogs is due in large part to the use of CFs. Both number theorists and dynamicists have long sought substitutes for CFs in higher dimensions, with only partial success (see, e.g. [Ar98b]).

Continuity, continuous. If X, Y are *topological spaces*, the map $f : X \to Y$ is *continuous* if given any *open* set $O \in Y$, its *preimage* $f^{-1}(O)$ is an open set in X. Though it may seem somewhat abstract at first, it is one of the wonders of *topology* that this definition perfectly captures the intuitive notion of *continuity*, and agrees with the usual δ-ε definition when X and Y are *metric spaces*. (The usual δ-ε definition is this: If (X, d_X) and (Y, d_Y) are metric spaces, $f : X \to Y$ is *continuous* at $x \in X$ if given any $\varepsilon > 0$, there is a $\delta > 0$ (possibly depending on x) such that $d_X(w, x) < \delta \implies d_Y(f(w), f(x)) < \varepsilon$. To say that $f : X \to Y$ is continuous means f is continuous at every $x \in X$. Note also that a compact way of defining such f to be continuous at x is to require that $\lim_{z \to x} f(z) = f(x)$.)

Continuous dynamical system. A *dynamical system* in which the time *domain* is the real numbers $\mathbb{R}$ (or a subinterval of $\mathbb{R}$, or of some other 'continuous monoid'). Such systems commonly occur as the solutions of a system of ODEs on an appropriate *phase space*. Compare *discrete dynamical system*.

Converge, convergence. To *converge* is to approach a (finite) *limit*. (One uses *diverge* when speaking of the approach toward an infinite limit.) *Convergence* may occur in many different ways (or 'modes'); perhaps the most basic kind is *convergence of a sequence*. See also *absolute convergence*, *conditional convergence*, *pointwise convergence*, *uniform convergence*, *convergence in norm* (and there are many other modes of convergence).

Convergence in norm. If $\{v_k\}_{k=0}^{\infty}$ is a *sequence* of vectors in a *normed vector space* with *norm* $\| \ \|$, we say that $\{v_k\}$ *converges in norm* to w if the sequence of numbers $\|v_k - w\|$ converges to zero.

Convergence of a function. See *limit of a function*.

Convergence of a sequence. If $\{s_k\}_{k=0}^{\infty}$ is a *sequence* in a *metric space* (M, d), we say that it *converges* to $L \in M$ (or has *limit* L, denoted $\lim_{k \to \infty} s_k = L$ or $s_k \to L$) if given any $\varepsilon > 0$, there is a $K \in \mathbb{N}^*$ such that $d(s_k, L) < \varepsilon$ whenever $k \geq K$.

Convergent. [1] (adj.) Said of a *sequence* or *series* or *function* that *converges* to a (finite) *limit*.
[2] (n.) A *term* p_k/q_k in the *sequence* $\{p_k/q_k\}_{k=0}^{\infty}$ of rational approximations to the *continued fraction* $[a_0; a_1, a_2, a_3, \ldots]$ defined by $p_k/q_k := [a_0; a_1, a_2, \ldots, a_k]$ (cf. *continued fraction* for notation).

Converse KAM theory. The collective name of results showing that *invariant tori* do not exist (in a given *system* or *class* of systems, over a specified range of parameter values, etc.).

Cotangent bundle. An even-dimensional *smooth manifold*, denoted T^*M, consisting of the *collection* of all *cotangent spaces* T_p^*M of a smooth manifold M together with their *base points* $p \in M$. In classical mechanics, it is natural to view the *phase space* of a *Hamiltonian system* as the *cotangent bundle* of the *configuration space* Q.

Cotangent space. Given a point p on a *smooth manifold* M of dimension n, the *cotangent space* of M at the *base point* p, denoted T_p^*M, is the

n-dimensional *dual space* of the *tangent space* T_pM. Vectors in T_p^*M are called 'cotangent vectors,' and are familiar to dynamicists, because the momentum variables are naturally identified as cotangent vectors in the *symplectic* picture of *Hamiltonian dynamics*.

Countable (also *denumerable*). A set B is *countable* if there is a *surjection* $f : \mathbb{N}^* \to B$ (in other words, if there is a *sequence* with *range* B).

Countably infinite. A set B is *countably infinite* if it is both *countable* and *infinite*; or equivalently, if there exists a *bijection* $f : \mathbb{N}^* \to B$.

Cover. A *collection* of sets whose union contains a given set (one says that the collection *covers* the given set).

Darboux coordinates (after Gaston Darboux). See *Darboux's theorem*.

Darboux's theorem. Given a $2n$-dimensional *symplectic manifold* (M, ω) and a point $x \in M$, there are smooth *local coordinates* $h : U \to \mathbb{R}^{2n}$ at x (for some open $U \subset M$, $x \in U$) such that for each $y \in U$, in the coordinates induced by h on the *tangent space* T_yM, the *symplectic form* ω is represented by the *standard symplectic matrix* J. (The conclusion may also be stated as follows: If the coordinate functions of h are $(q_1, \ldots, q_n, p_1, \ldots, p_n)$, then $\omega = dq_1 \wedge dp_1 + \cdots + dq_n \wedge dp_n$, where $\wedge$ is the 'wedge product' of differential forms.) The coordinates h are called *Darboux coordinates*. Note that the theorem is local; it does not say that there is always a single *chart* of Darboux coordinates covering all of M.

Degrees of freedom. In *classical mechanics*, the first step in analyzing a 'holonomic' mechanical *system* is to identify and properly coordinatize its *configuration space*. The (unique, minimal) number n of real *local coordinates* needed for the configuration space is the 'number of *degrees of freedom*' of the system (and each coordinate is loosely referred to as 'a degree of freedom,' though it is not generally unique, since alternate coordinates may be used). One then constructs the $2n$-dimensional *phase space* of the system by including, for each degree of freedom, either a *generalized velocity* (in *Lagrangian mechanics*) or a *conjugate* momentum variable (in *Hamiltonian dynamics*).

   More generally, when the phase space of a *Hamiltonian system* is a *symplectic manifold* viewed as the *cotangent bundle* T^*Q of a *smooth manifold* Q, then Q may be seen as the *configuration space*, and its real *local coordinates* as degrees of freedom. But if the phase space is simply given abstractly as a $2n$-dimensional symplectic manifold M, without reference to an underlying configuration space, then it is customary to say that any Hamiltonian system on M has n degrees of freedom, without specifying 'which half' of M constitutes those degrees of freedom.

The reader should be aware that in many applied sciences, 'degrees of freedom' has a meaning related to, but different from, what is discussed here.

Dense. Given a *subspace* B of the *topological space* X, the subset A of B is *dense* in B if the *closure* of A contains B.

Denumerable. See *countable*.

Derivative. If D is an *open subset* of $\mathbb{R}$ and $f : D \to \mathbb{R}$, the *derivative* of f at $x \in D$ is defined as $f'(x) := \lim_{h \to 0}(f(x + h) - f(x))/h$ whenever the *limit* exists. When $f'(x)$ exists for every $x \in D$, the derivative function $f' : D \to \mathbb{R}$ is well defined, so we can ask whether f' is *continuous*, and we can (try to) take its derivative to obtain the second derivative f''. When possible, we continue in this way to obtain $f^{(k)}$, the kth order derivative of f. (See *smoothness*.)

More generally, if A, B are *Banach spaces*, $D \subset A$ is *open*, and $f : D \to B$, we define the *Fréchet derivative* of f at $x \in D$ to be the unique *bounded linear map* Df_x (when it exists) such that $\lim_{h \to 0} \| f(x + h) - f(x) - Df_x(h) \| / \| h \| = 0$.

For a *smooth map* $f : M \to N$ between *smooth manifolds* M and N, the derivative map Df at a point $p \in M$ is the *linear map* $Df(p) : T_pM \to T_{f(p)}N$ from the *tangent space* of M at p to the tangent space of N at $f(p)$ such that the above definition applies in *local coordinates* (i.e., in appropriate *charts*).

Deterministic. This adjective is redundant when applied to *dynamical systems* (as defined here), since they are automatically *deterministic* in the sense that each future state is a *function* of the initial state (i.e., if $\varphi_t : M \to M$ is a *flow* and $x \in M$ is an IC, then for each $t > 0$ the future state $\varphi_t(x)$ is a function of x). But the redundancy is sometimes useful to distinguish such systems from 'stochastic' or 'quantum' systems that are not deterministic. The expression 'deterministic *chaos*' is similarly redundant, but may be used either for emphasis or to exclude so-called 'quantum chaos.' The emphasis reminds us that however chaotic a system may seem, it is still deterministic in the strict sense, and retains a measure of predictability. And of course, the words 'deterministic chaos' have a catchy ring by reason of seeming slightly oxymoronic together.

Diffeomorphic, diffeomorphism. Given two *smooth manifolds* M, N and $r \in \mathbb{N}^*$, the *map* $f : M \to N$ is a C^r *diffeomorphism* if it is a *homeomorphism* of *class* C^r with *inverse* f^{-1} of class C^r. In this case one says that M, N are C^r-*diffeomorphic* (or simply 'diffeomorphic' when r is understood).

Difference equation. The discrete analog of an ordinary *differential equation* (ODE). In other words, just as one solves an ODE $dx/dt = f(t, x)$ to find solutions in terms of continuous time t and IC x_0 (i.e., a *flow*), one solves a *difference equation* $x_{n+1} - x_n = f(n, x_n)$ to find solutions in terms of discrete time n and IC x_0 (i.e., a *map*).

Differentiability class. A synonym of *smoothness class*.

Differentiable manifold, differential manifold. See *smooth manifold*.

Differential equation. No definition of *differential equation* is given here (scientists and mathematicians "know one when they see one"). Differential equations occupy a hallowed place in science, as they are the means by which most of the basic laws of physics are expressed. It's important to remember that differential equations occur as two basic types: ordinary differential equations (ODEs), and partial differential equations (PDEs). ODEs have a single independent variable (usually but not always denoted t, for 'time'), involve only ordinary derivatives, and may be identified with *vector fields* from the *dynamical systems* point of view. PDEs have more than one independent variable, involve partial derivatives, and have a broader, messier, less complete theory than ODEs.

Differential structure. See *smooth manifold*.

Differential topology. This is the part of *topology* that deals with *smooth manifolds* and *smooth maps* between them. It is now a large subject, overlapping both *analysis* and geometry, and is essential background to much of *dynamical systems*.

Dimension. See *Hausdorff dimension* or *topological dimension*; but note that when one speaks without qualification of the 'dimension' of a topological object Z, one means the topological dimension, denoted $\dim Z$.

Diophantine condition, Diophantine set (after Diophantus of Alexandria). In *dynamical systems*, to say that $\omega \in \mathbb{R}^n$ is *Diophantine* (or satisfies a 'Diophantine condition,' or 'Diophantine conditions') means that ω belongs to a *Diophantine set* $\mathcal{D}(\gamma, \tau) := \{\omega \in \mathbb{R}^n \,|\, \text{for each } k \in \mathbb{Z}^n, \ |k \cdot \omega| \geq$

$\gamma|k|^{-\tau}\}$, with $\gamma > 0$ and $\tau > n - 1$. Such sets are nonempty for sufficiently small γ (they approach *full measure* in $\mathbb{R}^n$ as $\gamma \to 0^+$; in fact, it's not hard to show that the *complement* in the *closed ball* $\overline{B}_1(0)$ of such a set has *Lebesgue measure* $O(\gamma)$ as $\gamma \to 0^+$). Diophantine sets are in fact 'Cantor-like' sets in $\mathbb{R}^n$, with empty *interior*. Such sets occur naturally in KAM theory as sets of *frequency* vectors for *invariant KAM tori*.

Note that the adjective Diophantine is most commonly encountered in algebraic geometry and the number theory of polynomials over integers (where 'Diophantine set' means something different).

Direct product. See *Cartesian product.*

Direct sum (of *vector subspaces*). Suppose V, W are vector subspaces (of an ambient vector space G over the field $\mathbb{F}$) such that $V \cap W = \{0\}$. The *direct sum* of V and W, denoted $V \oplus W$, is defined as $span(V \cup W)$. Note that $V \oplus W$ is not defined if one subspace is contained in the other (unless at least one of them is $\{0\}$); and note that the *dimension* $\dim(V \oplus W) = \dim V + \dim W$.

Discrete dynamical system. A *dynamical system* in which the time *domain* is the set of integers $\mathbb{Z}$ (or a subsegment of $\mathbb{Z}$, or of some other 'discrete monoid'). Such *systems* commonly occur as the iterations of a *map* on the *phase space* (each iteration representing one integer increment of the time), and are often derived from *difference equations*. Compare *continuous dynamical system.*

Dissipative system. See *wandering set.*

Distance, distance function. See *metric.*

Diverge. A *limit* that fails to exist finitely is said to *diverge*. In general, a real-valued limit may fail to exist because it 'jumps' or 'oscillates' (i.e., 'never settles down'), or because it 'diverges to infinity,' in which case we write $\lim_{x \to a} f(x) = +\infty$ (or $= -\infty$). This latter notation may seem odd, as it seems to indicate that the limit exists, but when we recall that $\pm\infty$ are not (ordinary) real numbers, we see that it is simply a way of indicating how *divergence* occurs.

Divergence. [1] The process by which a limit *diverges*, or fails to exist. [2] Given a C^1 *vector field* $f : \mathbb{R}^n \to \mathbb{R}^n$, the *divergence* of f, denoted $\mathrm{div} f$ or $\nabla \cdot f$, is the function $\mathrm{div} f : \mathbb{R}^n \to \mathbb{R}$ defined by $\mathrm{div} f(x) = \sum_{k=1}^{n} \frac{\partial f_k}{\partial x_k}(x)$ in standard Cartesian coordinates. If $\mathrm{div} f = 0$ for all x, then the flow of the ODE $\dot{x} = f(x)$ is volume preserving, or 'incompressible.'

Domain. [1] The set on which the *rule* of a *function* acts (i.e., the collection of 'independent variables' or 'inputs'); or the collection of possible first entries in the set of ordered pairs comprising a *relation*. When the notation $f : A \to B$ is used to display all parts of a function, the *domain* is A. See also *relation* for the more general notion of domain of a relation.

[2] A 'nice' set, especially an *open*, *connected* region in the plane $\mathbb{R}^2$ (and sometimes an open, connected region in $\mathbb{R}^n$).

Note that *domain* is often used in senses [1] and [2] simultaneously.

Dual space. This is yet another term with several meanings in mathematics. Most commonly, it refers to the *dual space* V^* of a given *vector space* V, defined as the set of all *linear functionals* $f : V \to \mathbb{R}$. It turns out that V^* thus defined is also a vector space of the same dimension as V. The sense in which V^* is 'dual' to V is perhaps most simply appreciated by thinking of the relationship between row and column vector spaces of the same dimension in linear algebra.

Dynamical system. In its basic form, a *dynamical system* (DS) consists of a time domain, a *phase space* M, and an *evolution law* that determines how points in phase space evolve over time. If the time domain is $\mathbb{Z}$, the DS is *discrete* and its evolution law is a *map*; if the time domain is $\mathbb{R}$, the DS is *continuous* and its evolution law is a *flow*. (More general time domains, called 'monoids,' are sometimes used.) To indicate a DS with phase space M and evolution law φ, we write (M, φ) (other features of the DS may be indicated as necessary; most are understood in context).

Here we mainly consider continuous DSs (M, φ) with *smooth* flows of the form $\varphi : \mathbb{R} \times M \to M$ satisfying (i) $\varphi(0, x) = x$, and (ii) $\varphi(s+t, x) = \varphi(s, \varphi(t, x))$ for all $s, t \in \mathbb{R}$ and $x \in M$ ((i) and (ii) are called the 'group action' properties of φ). It is also common to represent a flow as a one-parameter family of *maps* $\varphi_t : M \to M$, in which case the group action properties become (i) $\varphi_0 = \text{id}$ and (ii) $\varphi_{s+t} = \varphi_s \circ \varphi_t$. We may also fix an *initial condition* (or IC) $x \in M$ to obtain the *orbit* $\varphi(\cdot, x) : \mathbb{R} \to M$. It is often more convenient to indicate the IC as a superscript of φ; in this notation the orbit of x is $\varphi^x : \mathbb{R} \to M$. (Note that 'orbit' is also often used to mean the *image* $\varphi^x(\mathbb{R})$.) The type of flow just described is 'global' (defined at all points of $\mathbb{R} \times M$). Global flows are very nice (e.g., by the group-action properties, each map φ_t is *invertible* with *inverse* φ_{-t}), and for simplicity, in this glossary we assume that flows are global unless otherwise stated. It's important to note, however, that this is not always true: for many DSs of interest, the time domain may consist only of nonnegative times (for so-called 'irreversible' systems), or may be a subinterval of $\mathbb{R}$ depending on the IC (as is generally the case for flows arising from vector fields, i.e., as the solutions of ODEs). The flow may also fail to be defined at some points of M.

Historically, continuous DSs (flows) were the first to be intensively studied, since they arise naturally as the solutions of ODEs (when solutions can be found, that is). But discrete DSs are more fundamental, as they are naturally generated by invertible maps $f : M \to M$, so the evolution of a phase point $x \in M$ after time $k \in \mathbb{Z}$ is simply $f^k(x)$, where f^k indicates k successive applications of f (or successive inversions, if $k < 0$). The group-action properties clearly hold for discrete DSs (subject to caveats similar to those above for non-global flows).

There are many special types of DSs: 'topological,' 'smooth,' 'measurable,' 'Hamiltonian,' etc., indicating respectively that the phase space is a *topological manifold, smooth manifold, measure space, symplectic manifold*, etc., and that the evolution law preserves the appropriate structure(s). The study of each of these types (and others) makes up the separate branches of *dynamical systems*, with their characteristic problems and methods.

Dynamical systems. This is the name given by G.D. Birkhoff both to his 1927 book [Bir27] and to the mathematical domain created by Poincaré out of what were previously the separate subjects of ODEs, *classical mechanics*, and *topology* (which Poincaré also largely created in its algebraic form). The subject has two main branches: *discrete* and *continuous*, with results in one branch usually mirrored in the other. It also has a number of subdisciplines (Hamiltonian, measurable, and topological dynamics; ergodic theory, etc.) and overlaps with other subjects in both mathematics and physics.

Dynamics. Used informally, either as an abbreviation of *dynamical systems*, or to refer to the behavior of a specific *dynamical system* (or some of its *orbits*, etc.).

On p. 11 of [New99], we learn that it was Leibniz who introduced the name 'dynamick' for what Newton had previously called 'rational mechanics.' Newton objected to this name, not because of its "inadequacy to describe the subject matter," but rather because Leibniz had "set his mark upon this whole science of forces calling it Dynamick, as if he had invented it himself & is frequently setting his mark upon things by new names & new Notations."

École Normale Supérieure (often abbreviated *ENS*). Originally conceived during the French revolution as institutions for training teachers, France's ENSs are now among its most prestigious *grandes écoles* (elite institutions of higher learning outside the state university system). There are currently two in Lyon, one each in Cachan and Paris; there is also one in Pisa, Italy, founded by Napoléon.

The ENS in Paris (which has a number of nicknames) has played a vital role in mathematics during the last two centuries, and presently boasts more *Fields medalist* alumni than any other institution in the world (its closest rival being University of Cambridge, U.K.). The ENS-Paris is also home of the *Bourbaki*, with everything that it entails, yet it is certainly not synonymous with Bourbaki, and even occasionally counts opponents of that group within its walls.

It is difficult to describe the culture of mathematics inside the ENS, but much of it is explained by the two or three years of intensive preparatory classes which precede the competitive examinations for entry into the ENS mathematics program. The abstract and rigorous style of these preparatory classes leaves a life-long impression on everyone who passes through them, and those who finish toward the top of the competitive exams often emerge as mathematical innovators later.

Elementary function. A *function* ($f : D \subset \mathbb{R} \to \mathbb{R}$ or $f : D \subset \mathbb{C} \to \mathbb{C}$) with *rule* constructed by combining exponentials, logarithms, constants, the identity, and roots of equations through finitely many compositions and applications of the four arithmetic operations $+ - \cdot \div$. Introduced by J. Liouville during his investigations of *integrable functions* (sense [1]), the concept has endured, though it is not always easy to decide whether or not a function is *elementary*.

Elliptic equilibrium (or *elliptic fixed point*, or *center*).   For the *flow* arising from an *autonomous vector field* f, the *fixed point* p is *elliptic* provided that all eigenvalues of the *linearization* $Df(p)$ are nonzero and lie on the imaginary axis in the complex plane. For the *discrete system* arising from an autonomous *map* g, the fixed point p is elliptic provided all eigenvalues of the linearization $Dg(p)$ lie on the unit circle in the complex plane.

Embed, embedded, embedding (also spelled *imbed*, etc.). Given C^k-smooth manifolds M and N, a C^k embedding $f : M \to N$ is a C^k immersion which is a *homeomorphism onto* its *range* $f(M)$, and such that the *preimage* of every *compact* subset of N is compact in M. (Note that some authors don't require an embedding to satisfy the compactness condition; if it does, they call it a 'proper' embedding.) Perhaps the nicest feature of an embedding $f : M \to N$ is that its *image* $f(M)$ is automatically a smooth submanifold of N (one says $f(M)$ is an *embedded* submanifold of N, or that f *embeds* $f(M)$ in N).

ENS. An abbreviation of *École Normale Supérieure*.

Entropy. Usually denoted S, *entropy* is a measure of the 'disorder' of a *system* (originally starting with thermodynamical systems, later expanded to include more general systems). The story of its introduction and development beginning in the mid 19th century is intricate, but credit goes to R. Clausius—then later to L. Boltzmann—for the first precise and useful definitions. For Clausius, who worked in classical thermodynamics (which deals with macroscopic *observables*), entropy is a measure of a system's nearness to *thermodynamic equilibrium*, where entropy reaches a maximum. From Boltzmann, who worked in statistical thermodynamics (which deals with both macroscopic and microscopic quantities), we have the famous equation $S = k \log W$, now inscribed on his tombstone in Vienna. It states that the entropy S of a given thermodynamical system (at equilibrium and in a particular 'macrostate') is proportional (by Boltzmann's constant k) to the logarithm of W, the number of 'microstates' accessible to the system (or 'consistent with' its current macrostate).

A more complete definition would make clear the meaning of macrostates and microstates, and how to count the latter (but these are not provided in this glossary). There are also now mathematical definitions of entropy, especially in *ergodic theory*; the most important are 'topological entropy' and 'Kolmogorov-Sinai entropy.'

Equilibrium. [1] In *dynamics*, another name for *fixed point* (*equilibrium* is commonly used for *flows*; fixed point is more common for *maps*).
[2] In thermodynamics, an abbreviation of *thermal equilibrium*.

Equivalence class. In mathematics, it frequently happens that one has a large *collection* of objects, and one wants to say that some of them are 'the same' for the purposes at hand. One then divides (or 'partitions') the collection into an appropriate disjoint union of subcollections called *equivalence classes*. (See *equivalence relation*.) Thereafter, one need not distinguish between members of the same equivalence class, and can use any member of the class to 'represent' it.

Equivalence relation. A *binary relation* R on a set A that is (i) reflexive ($a \in A \implies (a,a) \in R$), (ii) symmetric ($(a,b) \in R \iff (b,a) \in R$), and (iii) transitive ($(a,b) \in R$ and $(b,c) \in R \implies (a,c) \in R$). When $(a,b) \in R$, one says that a and b are *equivalent* (or 'congruent'). An *equivalence relation* automatically partitions A into *equivalence classes*.

Equivalent. See *equivalence relation*, or *conjugacy*.

Ergode (plural *Ergoden*). (n.) This is the German word used by L. Boltzmann [Bolt84] to describe what is now called (in statistical mechanics) a 'stationary microcanonical ensemble' with a single *integral of motion*.

Ergodensatz. The term used by Paul and Tatiana Ehrenfest in describing L. Boltzmann's ideas, later translated from German into English as *ergodic hypothesis*.

Ergodic (from German *ergodisch*). In modern *dynamical systems*, if (M, Σ, μ) is a *probability space* and $\varphi_t : M \to M$ is a *measure preserving flow* on M (i.e., μ is an *invariant measure* of φ_t), then we say φ_t is *ergodic* with respect to μ (or μ is ergodic with respect to φ_t) if every $A \in \Sigma$ which is *invariant* under φ_t satisfies $\mu(A) = 0$ or $\mu(A) = 1$. (The definition for *discrete systems* is essentially the same.) In words, all *measurable sets* that are *invariant* under the flow are either *negligible* or of *full measure*. Intuitively, an ergodic flow 'moves almost all sets all over the place.' The only exceptions are full sets (M or *almost all* of M, with nowhere else to go) and some μ-negligible sets. However, the flow may move sets 'rigidly,' without *mixing*.

The *Oxford English Dictionary* gives the etymology of ergodic as Greek ἔργον (work) + ὁδός (way) + -ic, and cites the first appearance of *Ergoden* in Boltzmann's 1887 paper [Bolt87]. Others cite an earlier use of *Ergoden* by Boltzmann in [Bolt84], or even [Bolt71]. (G. Gallavotti, whose father was a philologist, has written a paper [Gall95b] on the etymology of ergodic, and how its misunderstanding delayed progress in *ergodic theory*.)

Ergodicity. The state or quality of being *ergodic*; the *statistical property* possessed by an ergodic *dynamical system*.

Ergodic component. Given a *measure space* (M, Σ, μ) and $\varphi_t : M \to M$ a *measure preserving flow* on M, a set $A \in \Sigma$ is an *ergodic component* of φ_t provided $\mu(A) > 0$, A is *invariant* under φ_t, and any *measurable* $B \subset A$ which is invariant under φ_t satisfies $\mu(B) = 0$ or $\mu(A \backslash B) = 0$. (In other words, the *restriction* $\varphi_t\big|_A$ is *ergodic* with respect to the *probability measure* $\rho = \mu/\mu(A)$.)

Ergodic hypothesis (from German *Ergodenhypothese* or *Ergodensatz*). A key assumption used in the late 19th century by L. Boltzmann and others in dealing with the foundations of statistical mechanics and thermodynamics. At certain points in his discussion, Boltzmann needed to say that the *time* and *space averages* of a relevant (thermo)*dynamical system* were equal. He drew special attention to this assumption, and justified it with a blend of probabilistic and physical intuition, saying that the *trajectories* of the system should, over time, pass through all or *almost all* points of *phase space* that are energetically accessible to them. The success of Boltzmann's theory focused attention on his *ergodic hypothesis*, and mathematicians realized that his justifications were not quite right. The possibility of trajectories passing through all points of phase space was shown to be untenable, and the remaining possibility (passing through almost all points) became known as the 'quasi-ergodic hypothesis.' Even this proved to be difficult, and it was not until the early 1930s (with the advent of *measure* theory and *Hilbert space* techniques) that mathematicians proved *ergodic theorems*, partially vindicating Boltzmann's intuition in a *rigorous* way and establishing *ergodic theory* as a new field of research in mathematics. Thereafter, mathematicians reformulated Boltzmann's original ergodic hypothesis to assert that the *flow* of a *generic Hamiltonian system* is ergodic on each of its energy surfaces (or perhaps on *almost all* of them). But this conjecture was invalidated in the strict mathematical sense by KAM theory. The present state of affairs is thus somewhat paradoxical, since mathematical physicists continue to use assumptions closely related to the (updated) ergodic hypothesis in their successful formulations of statistical mechanics.

Ergodic theorem. There are many *ergodic theorems*, but most of them recall or imply L. Boltzmann's original assumption 'space averages equal time averages' in some way. The first 'pointwise' or 'individual' ergodic theorem was published by G.D. Birkhoff [Bir31], and still serves nicely to illustrate what such theorems look like. In modern language, it reads as follows.

Suppose (M, Σ, μ) is a *probability space* and $\varphi_t : M \to M$ is a *measure preserving flow* on M. Then for any $f \in \mathcal{L}^1(M, \mu)$, and for almost every $x \in M$, the *time average* $\langle f \rangle$ of f exists in $\mathcal{L}^1(M, \mu)$ and $\overline{\langle f \rangle} = \overline{f}$ (i.e., $\langle f \rangle$ and f have the same *space average*).

An immediate corollary of Birkhoff's ergodic theorem is that, if φ_t is *ergodic*, then $\langle f \rangle = \overline{f}$ almost everywhere (i.e., the time and space averages of f are the same constant function in $\mathcal{L}^1(M, \mu)$).

Shortly afterward, J. von Neumann published a different version of the ergodic theorem [Neu32] (sometimes called the 'mean ergodic theorem') in the context of the *Hilbert space* $\mathcal{L}^2(M, \mu)$. Although von Neumann's theorem is slightly *weaker* than Birkhoff's, his proof is considerably simpler, as it takes full advantage of Hilbert space techniques. Despite the publication dates, it's widely acknowledged that von Neumann finished his proof before Birkhoff, and that Birkhoff drew inspiration for his own proof from a manuscript version of von Neumann's paper.

Ergodic theory. This is the branch of *dynamical systems* descended from mathematicians' (mostly successful) attempts to make sense of L. Boltzmann's *ergodic hypothesis*. It deals with both *discrete* and *continuous* systems on *measure spaces*, and has by now established a large body of results on the *statistical properties* of such systems. The subject was expanded at the end of the 1950s by Kolmogorov's introduction of a kind of *entropy* that bears his name; it now overlaps extensively with both probability and information theory. Unfortunately, the foundations of statistical mechanics have not benefited as much as might have been hoped from the growth and success of ergodic theory.

Estimate. In mathematics, and particularly in *analysis*, the word *estimate* may carry its ordinary meaning (v. 'to make a good guess'; n. 'a good guess'), but it usually has the more specialized meaning of finding a *rigorous bound* for a *function*; it is also used as a noun to mean such a bound.

Euclidean space. This is the name of any of the n-dimensional spaces $\mathbb{R}^n$ (the n-fold *Cartesian product* of $\mathbb{R}$ with itself) with 'Cartesian coordinates' $(x_1, \ldots, x_n)$ in which one can do 'Euclidean geometry' based on the usual *inner product* (the dot product $x \cdot y = x_1 y_1 + \cdots + x_n y_n$) and the Euclidean *norm* $\|x\| = \sqrt{x \cdot x}$ it induces. Euclidean spaces are said to be 'flat,' because their 'local curvature' is everywhere zero.

Evolution law. This is the *rule* that determines the motion in a *dynamical system*. In an *autonomous discrete system*, it is a *map* on the *phase space*, with time corresponding to iterates of the map. In an *autonomous continuous system*, it is a 1-parameter family of maps, with time corresponding to values of the parameter.

Extended real numbers. This set (denoted $[-\infty, +\infty]$ or $\overline{\mathbb{R}}$) consists of the 'ordinary' real numbers $\mathbb{R} = (-\infty, +\infty)$ together with the symbols $-\infty$ and $+\infty$ 'attached at the left and right ends, respectively.' Some care must be exercised in using the attached infinities. (Note that one may also use 'half-extensions' of $\mathbb{R}$, such as $[0, \infty]$.)

Extension (of a *function*). If $f : A \to B$ and $g : C \to D$ are *functions*, we say that g is an *extension* of f if $A \subset C$, $B \subset D$, and $f(x) = g(x)$ for all $x \in A$ (i.e., f is a *restriction* of g).

Field. In mathematics, a *field* is a set of objects that satisfy a list of properties called the 'field axioms' (these basically allow arithmetic to be performed, and are not given here). The two most important (and in some sense the only) fields are the real numbers $\mathbb{R}$ and the complex numbers $\mathbb{C}$, with their usual arithmetic operations $+ - \cdot \div$.

Fields medal (after John Charles Fields). The short name of the 'International Medal for Outstanding Discoveries in Mathematics,' first awarded in 1936 at the *ICM* in Oslo, and also at subsequent ICMs (every four years since 1950). Generally regarded as the highest award in mathematics, and often also called the 'Nobel prize of mathematics,' it may eventually be supplanted in this respect by the Abel prize (established in 2003), which is more Nobel-like in several ways.

Finite. A set B is *finite* if there is no *injection* $f : \mathbb{N}^* \to B$. Note also that 'finite' is often used to emphasize that a number is an ordinary real number (rather than one of the endpoints of the *extended real numbers* $\overline{\mathbb{R}} = [-\infty, \infty]$).

Finite measure space. See *measure space*.

First category. A synonym of *meager*. See also *Baire category theory*.

First integral. See *constant of motion*.

Fixed point (also called *equilibrium*, 'critical point,' 'rest point,' etc.). In a *dynamical system*, a *fixed point* is an *invariant set* consisting of one point p in *phase space*. For the *flow* of an *autonomous vector field* f, the fixed points are the points p such that $f(p) = 0$; for the *discrete system* defined by a *map* g, they are the points p with $g(p) = p$.

Flow. A *flow* is the *evolution law* of a *continuous dynamical system*. If M is the *phase space* of the system, depending on context, a flow may be viewed either as a (usually *smooth*) map $\varphi : \mathbb{R} \times M \to M$, or as $\varphi_t : M \to M$ (a one-parameter group of maps, with parameter t). A flow satisfies the fundamental group-action properties (i) $\varphi_0 = \mathrm{id}$, and (ii) $\varphi_{s+t} = \varphi_s \circ \varphi_t$ at all values of s, t for which both sides of (ii) are well defined. In *dynamics*, a flow commonly arises as the set of solutions of an ODE (and, by identifying an ODE with its *vector field*, one also speaks of the 'flow of a vector field').

Foliate, foliation. In *differential topology*, a *foliation* of a *smooth manifold* M is an 'integrable sub-bundle' of the *tangent bundle* TM. Roughly speaking, a foliation is a decomposition of M into a disjoint union of smooth submanifolds (called 'leaves') of smaller *dimension*, with the property that nearby leaves are 'locally parallel' to one another. For example, the *integral curves* of a (nonvanishing) *vector field* on a smooth manifold *foliate* (i.e., form a foliation) of the manifold; and the *invariant tori* of a *completely integrable Hamiltonian system* foliate (at least a portion of) the system's *phase space*.

Formal, formally. [1] Most commonly in mathematics, these terms are used in a loose way to mean "We are writing down this expression without carefully considering whether it makes sense" (or "We're only looking at the computational aspects without considering other subtleties," etc.). [2] In mathematical logic, a *formal* statement means a statement of the utmost *rigor*. Similarly, to speak of a formal proof is to highlight the proof's rigor and completeness.

Although senses [1] and [2] here are nearly opposite, context usually prevents their confusion.

Formal series. This is a special case of sense [1] of the previous entry: to speak of a *formal series* means that we know the terms of the series, but we claim nothing about its *convergence* (perhaps it *diverges*, or we haven't yet checked—or cannot check—whether it *converges*).

Fourier series (after Joseph Fourier). A special kind of *series* commonly used to represent periodic or multiply periodic *functions* of one or more variables. If $f : \mathbb{R} \to \mathbb{R}$ has period p in θ, its (*formal*) Fourier series is $\sum_{k \in \mathbb{Z}} \hat{f}_k\, e^{2\pi i k\theta/p}$ where the kth Fourier coefficient $\hat{f}_k :=$ $(1/p) \int_0^p f(\theta)\, e^{-2\pi i k\theta/p}\, d\theta$, whenever the right-hand side makes sense. More generally, if f is a multiply periodic function $f : \mathbb{T}^n \to \mathbb{R}$ (i.e., f has period 1 in each component of $\theta = (\theta_1, \ldots, \theta_n)$), its (formal) multiple Fourier series is $\sum_{k \in \mathbb{Z}^n} \hat{f}_k\, e^{2\pi i k \cdot \theta}$ with coefficients $\hat{f}_k := \int_{\mathbb{T}^n} f(\theta)\, e^{-2\pi i k \cdot \theta}\, d\theta$, when defined. (Note that if f depends on variables in addition to θ, the Fourier coefficients $\hat{f}_k$ will also depend on those variables.) The conditions under which such series *converge* and equal f make up a very extensive part of *analysis*; in fact the development of those results often drove analysis and integration theory in the 19th and 20th centuries.

Fractal. Although there is no universally accepted definition of this term, in [Man75–89] B.B. Mandelbrot defined a *fractal* set to be a point set with non-integer *Hausdorff dimension* (but, alas, some sets with integer Hausdorff dimension are also commonly called fractals). Fractals are typically 'self-similar,' with structures repeated on finer and finer scales. In nature, many objects exhibit approximate fractal structure (i.e., near self-similarity over a wide range of scales): snowflakes, clouds, coastlines, mountains, trees, rivers, vascular systems, etc. In mathematics, typical fractals are *Cantor sets*, 'Sierpinsky carpets' (or 'sponges'), 'solenoids,' '*Mandelbrot sets*,' 'Koch snowflakes,' and the *collection* of *invariant KAM tori* in *generic Hamiltonian systems*.

Fréchet derivative (after Maurice Fréchet). See *derivative*.

Fréchet space. A 'locally convex topological *vector space*' equipped with a (translation invariant) *metric* with respect to which the *space* is *complete*. (No definition of 'locally convex topological vector space' is given in this glossary; see e.g. [KatH95], A.2.) A *Fréchet space* resembles a *Banach space*, but may fail to have a norm; instead, it has a 'family of *seminorms*' which serve to give it an appropriate *topology*.

Frequencies. The components $\omega_1, \ldots, \omega_n$ of a *frequency vector* $\omega \in \mathbb{R}^n$.

Frequency map, frequency vector. A *frequency vector* is an n-dimensional vector $\omega = (\omega_1, \ldots, \omega_n)$ used to define a *linear* or *conditionally periodic* flow $F_t^\omega : \mathbb{T}^n \to \mathbb{T}^n$ on the *n-torus* $\mathbb{T}^n$ via the formula $F_t^\omega(\theta) = \theta + \omega t$ (here the addition of components is understood as $\theta_k + \omega_k t$ (mod 1), $k = 1, \ldots, n$). Such flows arise naturally in *integrable Hamiltonian systems* expressed in *action-angle variables* (θ, I). Since the *Hamiltonian* $H = h(I)$ in these variables depends only on I, and since I remains constant on any *invariant torus* of the integrable system, the frequency vector on the *torus* defined by $I = I^0$ is simply $\omega = \operatorname{grad} h(I^0)$, and this formula defines the *frequency map* $I^0 \mapsto \omega$ of the integrable system. See also *nondegeneracy*.

Full measure. In a *measure space* (M, Σ, μ), the set $A \in \Sigma$ is of (or has) *full measure* in $B \in \Sigma$ provided $\mu(B \setminus A) = 0$. The reference set B is often understood (to be some *nice* set, such as a ball, or the *base set* M), in which case one simply says 'A is of full measure.'

Function. One of the central objects of study in mathematics (perhaps even more so than 'set' or 'number'). The concise definition is this:

[1] (a) A *function* is a *relation* (sense [1] (a)) from a set A to a set B in which each element of A occurs exactly once. (This definition, in which A and B are not explicitly part of the function, is usually preferred by set theorists and logicians.)

A more elaborate definition, which includes the sets A and B, is as follows.

[1] (b) A function is an ordered triple of sets $(A, B; S)$ where A is called the *domain* and B is called the *codomain* (or *target set*). The third set S, called the *graph* of the function, is a subset of the *Cartesian product* $A \times B$ in which each element of A occurs precisely once.

[1] (c) A third definition, almost the same as the last, replaces the graph S by an unambiguous *rule* f which assigns, to each element of the domain, precisely one element of the codomain. (This definition encourages the view of a function as a *map* or *mapping* which 'acts' on the domain.)

Definitions [1] (b) or especially [1] (c) are usually preferred by working mathematicians. When the domain and codomain are understood, then using [1] (c), a function may be specified simply by giving its rule (usually but not always by a formula, such as $f(x) = x^2$). For clarity (and often by necessity) one displays all three parts of a function using the notation $f : A \to B$, then one specifies the action of the rule on an arbitrary domain element x, again usually by $f(x) = $ [formula], or by $x \mapsto$ [formula].

Functional. In most mathematical contexts, a *functional* is a *map* from a *vector space* into the underlying *field*. The term has its origins in variational methods applied to mechanics, where it applies to quantities (such as energy, or *action*) that are to be extremized (or at least made stationary) by an appropriate choice of *trajectory*.

Function space. A *vector space* in which the vectors are *functions*. Most function spaces (e.g. the L^p spaces) are infinite dimensional and come equipped with a *norm*, or at least a *topology*.

Fundamental frequency. See *quasiperiodic flow on* $\mathbb{T}^m$, or *quasiperiodic function*.

G delta set (or G_δ *set*, or simply G_δ). In a *topological space*, a G_δ set is a *countable* intersection of *open sets*. G_δ sets are important in a number of ways; here are two: First, if X is a topological space, M is a *metric space*, and $f : X \to M$ is a *function*, then the set of points at which f is *continuous* is a G_δ in X. Second, if a property holds on a *dense* G_δ set in a topological space, then that property is *generic* in the *space*.

General, generalize. In mathematics, to say that result B *generalizes* (or 'is more *general* than') result A means that result B includes result A as a special case.

General position. This term from the Russian literature means something like 'the general case' (as opposed to 'particular' or 'exceptional cases'). To make the meaning precise, one introduces a *topology*, then says that the general position is *generic*. For example, to say that two lines in the plane are in general position means that they intersect in one point (the other two cases—coincidence or being parallel—are exceptional), and this is also the generic case in any appropriate topology. (See also *transversal, transverse.*)

Generalized coordinates. These are *local coordinates* of the *configuration space* of a mechanical *system*. *Generalized coordinates* were first introduced by J.-L. Lagrange in his *Mécanique analytique* (1788). Loosely speaking, each generalized coordinate of a mechanical system may be identified with one of its *degrees of freedom*. Apart from their theoretical interest, which is considerable, generalized coordinates can be very practical in solving problems, since judicious choice of coordinates allows constraints and symmetries to appear in especially simple form.

Generalized momenta. In the traditional picture of *Lagrangian* or *Hamiltonian* mechanics, once the *generalized coordinates* q of a mechanical system are chosen and the *Lagrangian* $L = L(q, \dot{q}, t)$ is constructed, one defines the *generalized momenta* p_k by $p_k := \partial L / \partial \dot{q}_k$.

Generalized velocity. If q_k is a *generalized coordinate* of a mechanical *system*, the corresponding *generalized velocity* is the time derivative $\dot{q}_k$.

Generic, genericity. In a *topological space*, a property is *generic* if the set on which it holds is *residual*, i.e., contains a *countable* intersection of *open dense* sets (in other words, if the set on which it holds is *comeager*, i.e., the complement of a *meager* set in the sense of *Baire category theory*).

  Genericity is important in *dynamical systems* as a way of identifying what is 'typical' versus 'exceptional.' When one says, for example, that P is a generic property of *Hamiltonian systems*, it means two things: (i) a *topology* has been placed on the set of Hamiltonian systems (for example the C^ω or C^∞ topology), making that set into a topological space, and (ii) P holds on a subset of Hamiltonian systems that is *residual* in that topological space. A property that is generic with respect to one topology may or may not be generic with respect to another, so to be precise in making statements about genericity, it should be clear what topology is used.

Genericity also became important in attempts to understand the modeling of physical systems by *dynamical systems*; see *structural stability dogma* below.

Note that in some contexts, 'generic' means typical in some other sense, such as occurring on a set with *full measure* (but this is rare in *dynamical systems*).

Germ. If M, N are *smooth manifolds* and $f : M \to N$ is a *smooth map* with $y = f(x)$, then the *germ* of f at x is the *equivalence class* $[f]$ of all smooth maps $g : M \to N$ which agree with f on some open neighborhood of x. One speaks of germs of various *regularity* (continuous, C^k, C^∞, C^ω, etc.) depending on the regularity of maps in the equivalence class.

Gevrey smoothness (after Maurice Gevrey). Let D be a *bounded connected open domain* in $\mathbb{R}^n$ and $f : D \to \mathbb{R}$ a C^∞ function. For $\lambda \geq 1$ and $L > 0$, we say f is of *Gevrey smoothness class* $G_L^\lambda(D)$ provided $\|f\|_L :=$ $\sup_{\beta \in \mathbb{N}_0^n} \sup_{x \in D} |\partial_x^\beta f(x)| L^{-|\beta|} (\beta!)^{-\lambda} < \infty$. Here, $\beta = (\beta_1, \ldots, \beta_n) \in \mathbb{N}_0^n$ is a *multiindex* with $|\beta| = \beta_1 + \ldots + \beta_n$ and $\beta! = \beta_1! \cdots \beta_n!$. The parameters λ, L are the Gevrey index and Gevrey constant, respectively, and the Gevrey constant is sometimes omitted. It isn't hard to see that $G^1(D) = C^\omega(D)$ (i.e., the functions of Gevrey index 1 are precisely the *analytic* functions), and in fact $G^1(D) \subset G^\lambda(D) \subset G^\infty(D)$ for $1 < \lambda < \infty$. In this way the Gevrey smoothness classes provide an interpolation between the smoothness classes $C^\omega(D)$ and $C^\infty(D)$.

Graph. The *graph* of a *function* $f : A \to B$ is the *collection* of all ordered pairs $\{(a, f(a))\}$ with $a \in A$. The graph of a *relation* is defined similarly. (In fact, for many logicians and set theorists, a function and its graph [or a relation and its graph] are the same.)

Habilitation. In many European (and some West Asian) countries, this is the certification needed to become a university professor. It is not usually considered an academic degree, though it involves writing and defending a lengthy thesis (or assembling a thesis from one's published articles). Preparation of the habilitation may take five to ten years time beyond the PhD or its equivalent.

Hamiltonian (after William R. Hamilton). [1] (n.) As a noun, this is an abbreviation of *Hamiltonian function*, a *smooth* (or at least C^2) function $H : M \to \mathbb{R}$ on an even-dimensional *phase space* M used to generate

Hamilton's equations for a *Hamiltonian system*. For simple (e.g. time-independent) physical systems, the Hamiltonian may often be written as the sum of kinetic and potential energy: $H = T + V$. Such Hamiltonians are sometimes called 'natural.'

[2] (adj.) As an adjective, Hamiltonian may modify *system, dynamics, flow*, 'mechanics,' etc., indicating the appropriate association with a *Hamiltonian system*.

Hamiltonian perturbation theory (HPT). The study of *nearly integrable Hamiltonian systems*, i.e., those with *Hamiltonian* of the form $H(\theta, I, \varepsilon) = h(I) + \varepsilon f(\theta, I, \varepsilon)$, where $(\theta, I) = (\theta_1, \ldots, \theta_n; I_1, \ldots, I_n)$ are *action-angle variables* for the *completely integrable* 'unperturbed system' (with Hamiltonian $h = h(I)$), $\varepsilon \in \mathbb{R}$ is the 'perturbation parameter,' and $\varepsilon f(\theta, I, \varepsilon)$ is the *perturbation*. Poincaré famously called HPT the 'fundamental problem of dynamics.'

Hamiltonian system. One version of *Newton's second law of motion* states that a particle of mass m in a potential $V = V(q)$ (defined at each configuration point $q = (q_1, q_2, q_3) \in \mathbb{R}^3$) moves so that its location $q(t)$ at time t obeys $m\ddot{q} = -\operatorname{grad} V(q)$. Defining the kth component of momentum by $p_k := m\dot{q}_k$, and the total energy (or *Hamiltonian*) by $H(q, p) := \|p\|^2/(2m) + V(q)$, we see that the second law is equivalent to *Hamilton's equations* $\dot{q}_k = H_{p_k}$, $\dot{p}_k = -H_{q_k}$, $k = 1, 2, 3$. These (or the *dynamical system* defined by them) constitute the *Hamiltonian system* for a particle moving in the potential V. In the modern generalization, we replace the phase space $\mathbb{R}^6$ by a *symplectic manifold* (or 'Poisson manifold'), but the basic structure of Hamilton's equations remains unchanged in so-called *Darboux coordinates*. See §2.2 above for more detail.

Hamilton's equations (also called *canonical equations of motion*). If M is the $2n$-*dimensional phase space* of a *Hamiltonian system* with Hamiltonian $H : M \to \mathbb{R}$, and $(q, p) = (q_1, \ldots, q_n; p_1, \ldots, p_n)$ are canonical (or Darboux) coordinates, then Hamilton's equations are $\dot{q}_k = H_{p_k}$, $\dot{p}_k = -H_{q_k}$, $k = 1, \ldots, n$.

Hausdorff dimension (or Hausdorff-Besicovitch dimension, after Felix Hausdorff and Abram Besicovitch). A generalization of the concept of *dimension* (e.g. from *topological manifolds*, which have integer dimension) to include irregular sets (such as *fractals*) that have non-integer dimension. If X is a *metric space* and $F \subset X$, the *Hausdorff dimension* of F is defined as $\dim_H(F) := \inf\{a \geq 0 \mid C_H^a(F) = 0\}$, where $C_H^a(F)$ is the

'Hausdorff content' of F, defined as $C_H^a(F) := \inf\{\, \delta \geq 0 \mid$ there is a *cover* $\{B_{r_i}\}_{i=1}^\infty$ of F such that $\sum_i r_i^a < \delta \,\}$. Here $\{B_{r_i}\}_{i=1}^\infty$ is a *countable collection* of balls in X with radii $r_i > 0$ such that $F \subset \bigcup_{i=1}^\infty B_{r_i}$. There are other ways of measuring dimension that admit non-integer values (such as 'box-counting dimension' and 'packing dimension'). For many sets, these other methods agree with the Hausdorff dimension, but there are exceptions.

Hausdorff space. A *topological space* $(X, \mathcal{T})$ such that whenever $x_1, x_2 \in X$, there are *neighborhoods* $O_1, O_2 \in \mathcal{T}$ with $x_i \in O_i$ and $O_1 \cap O_2 = \emptyset$. (A *Hausdorff space* is also called a 'T2 space,' using the language of the 'separation axioms' in *topology*.)

Heteroclinic orbit, heteroclinic point. In a *dynamical system*, a point p is *heteroclinic* to two *fixed points* if it lies in the intersection of the *stable manifold* of one fixed point and the *unstable manifold* of the other. The *orbit* of such p is also called heteroclinic, since all of its points are heteroclinic. (This notion also generalizes to points in the stable and unstable manifolds of other *invariant* sets, such as invariant *tori*.)

Hilbert space (after David Hilbert). A *Banach space* equipped with an *inner product* that is compatible with the *norm*. (In other words, if $\mathcal{H}$ is a Hilbert space with inner product $(\cdot, \cdot)$ and $v \in \mathcal{H}$, then $\|v\| = \sqrt{(v, v)}$.) The inner product and norm permit one to do 'geometry' in a Hilbert space.

Hölder condition, Hölder continuity, Hölder smoothness (after Otto Hölder). If (A, d_A) and (B, d_B) are *metric spaces*, $D \subset A$ is an *open connected domain*, and $f : D \to B$, we say f satisfies a *Hölder condition*, or f is *Hölder continuous* with Hölder exponent $\alpha \geq 0$ (or simply 'f is α-Hölder') if there is a nonnegative constant C such that for all $x, y \in D$, $d_B(f(x), f(y)) \leq C(d_A(x, y))^\alpha$. (Note that the special case $\alpha = 1$ means that f is *Lipschitz*.) Hölder continuity has many uses in *analysis*, *PDE*, and *dynamical systems*; one of the most basic is to provide a means of interpolating between the usual *smoothness classes* $C^k(D)$ defined when k is a nonnegative integer. If $D \subset A = \mathbb{R}^n$ and $0 < \alpha \leq 1$, we say that $f : D \to \mathbb{R}$ is of *Hölder smoothness class* $C^{k,\alpha}(D)$ if f is of class $C^k(D)$ and all kth-order partial derivatives $\partial^\beta f$ are α-Hölder (here β is a *multiindex* with $|\beta| = k$). Note that larger α correspond to 'smoother' f.

Homeomorphic, homeomorphism. If X, Y are *topological spaces*, a *homeomorphism* $f : X \to Y$ is a *continuous bijection* with *continuous inverse*. To say that X, Y are *homeomorphic* means that there exists a homeomorphism $f : X \to Y$. Homeomorphic spaces share the same topological properties.

Homoclinic orbit, homoclinic point. In a *dynamical system*, a point p is *homoclinic* to a *fixed point* if it lies in the intersection of the *stable* and the *unstable manifolds* of the fixed point. The *orbit* of such p is also called homoclinic, since all of its points are homoclinic. (This notion also generalizes to points in the stable and unstable manifolds of other *invariant* sets, such as *invariant tori*.)

Horseshoe map. See *Smale horseshoe map*.

HPT. An abbreviation of *Hamiltonian perturbation theory* (in this book).

Hyperbolic, hyperbolicity. In *dynamics*, a *hyperbolic fixed point* of a *flow* is a fixed point at which all eigenvalues of the *linearization* have nonzero real part. For a *map*, a fixed point is hyperbolic if its linearization has no eigenvalues with *modulus* 0 or 1.

A *hyperbolic system* is, roughly speaking, a system that exhibits either separation or convergence of nearby *orbits* along certain directions with increasing time. Often a hyperbolic system will display simultaneously contraction along some directions and expansion along others. *Hyperbolicity* is important because it—and *invariant sets* on which it occurs—are ordinarily *robust*, or *structurally stable*.

Hyperplane. In most contexts, a *hyperplane* is a 'linear set' of *codimension* 1 in *Euclidean space* $\mathbb{R}^n$. In other words, in Cartesian coordinates $x = (x_1, \ldots, x_n)$ it is a set of points $\{x \in \mathbb{R}^n \mid x \cdot a = c\}$, where $c \in \mathbb{R}$ and $a = (a_1, \ldots, a_n) \in \mathbb{R}^n$ is a fixed vector. It is thus a (planar) *hypersurface* in $\mathbb{R}^n$, perpendicular (or 'orthogonal') to the vector a, and passing through the origin if and only if $c = 0$.

Note however, that in some contexts—e.g. the definition of *steep* in this glossary—a hyperplane in $\mathbb{R}^n$ means a 'linear set' of any positive codimension d ($1 \le d \le n$), in other words any intersection of finitely many hyperplanes of codimension 1.

Hypersurface. If M is a *smooth manifold* of *topological dimension* n, a *hypersurface* in M is a smooth submanifold of M of topological dimension $n - 1$; i.e., of *codimension* 1. In $\mathbb{R}$, the hypersurfaces are points; in $\mathbb{R}^2$ they are curves; in $\mathbb{R}^3$ they are (ordinary) surfaces; and in $\mathbb{R}^n$ they are commonly defined by a single equation in the n variables $x_1, x_2, \ldots, x_n$ (since each such relation 'cuts down' the dimension by 1).

As with *hyperplanes*, in certain contexts a hypersurface means a smooth submanifold of any positive codimension.

IC. An abbreviation of *initial condition*.

ICM. An abbreviation of *International Congress of Mathematicians*.

Image. If $f : A \to B$ is a *map* and $C \subset A$, the *image* of C under f is the set of points $f(C) := \{y \in B \,|\, y = f(x)$ for some $x \in C\}$. When one says 'the image of f' (without reference to a set) one means $f(A)$ (in other words the *range* of f).

Imbed, Imbedded, imbedding. Alternate spellings of *embed, embedded, embedding*.

Immersion. Given C^k-smooth manifolds M and N, a C^k-smooth map $f : M \to N$ is called a C^k immersion if its *derivative map* Df is *injective* at all points. An immersion $f : M \to N$ is a *local embedding*: at each point $p \in M$ there is a *neighborhood* $U \subset M$ of p such that the *restriction* $f|_U : U \to N$ is an embedding. However, because of self-intersections in $f(M)$, an immersion may fail to be an embedding 'globally.'

Implicit function theorem. It often happens that one has a *relation* (described, say, by an equation in x and y) from which one wants to extract a *function* (i.e., a *rule* $y = f(x)$, valid on an appropriate x-*domain*). An *implicit function theorem* gives conditions under which this is possible, at least *locally*. Implicit function theorems are closely related to *inverse function theorems*, and important versions of them have been discovered and proved using KAM techniques.

Inclination lemma. See *lambda lemma*.

Incommensurable. Not *commensurable*. In other words, numbers are *incommensurable* if they do not satisfy a *commensurability relation*. A synonymous term is *rationally independent*.

Independent. This term occurs throughout mathematics with various meanings, but it often means—or reduces to—a certain set being *linearly independent* (cf. *completely integrable, resonance condition*, etc.).

Infimum (also called *greatest lower bound*). If A is a set of real numbers, the *infimum* of A (denoted inf A or glb A) is defined to be $-\infty$ if A has no lower bound, and defined to be the largest lower bound of A if A is *bounded from below*. The fact that every set of real numbers that is bounded from below has a greatest lower bound in $\mathbb{R}$ is called the *completeness* property of $\mathbb{R}$.

Infinite. A set B is *infinite* if it is not *finite*; or equivalently, if there is an *injection* $f : \mathbb{N}^* \to B$.

Initial condition (often abbreviated IC). A 'starting point' in *phase space* for the *flow* (or *map*) of a *dynamical system*. By allowing the system to evolve from an IC, one generates the *orbit* (or *trajectory*) corresponding to the IC. See *dynamical system* for more details and notation.

Injection, injective. A *function* $f : A \to B$ is an *injection* if $x_1, x_2 \in A$ and $x_1 \neq x_2 \implies f(x_1) \neq f(x_2)$. In words, f never takes different elements of the *domain* to the same element of the *codomain*. The adjective is *injective* or *one-to-one* (and sometimes one even sees 'univalent' used synonymously, though this fancy word—courtesy of *Bourbaki*—is ordinarily reserved for injective *analytic* functions on an *open domain* in the complex plane).

Inner product. An *inner product* (on the *vector space* V over the *field* $\mathbb{F}$) is a *map* $(\cdot, \cdot) : V \times V \to \mathbb{F}$ satisfying, for all $x, y, z \in V$ and $\alpha \in \mathbb{F}$: (i) $(x, y) = \overline{(y, x)}$ (ii) $(\alpha x, y) = \alpha(x, y)$ (iii) $(x, y + z) = (x, y) + (x, z)$.

Integrability, integrable. [1] In *analysis*, one says that a *function* $f : \mathbb{R} \to \mathbb{R}$ is '*integrable* in elementary terms' if the indefinite integral $\int f(x)\, dx$ can be expressed as an *elementary function*. This notion of *integrability* was first investigated by J. Liouville in the 19th century [Liou35]. [2] One says that $f : M \to \mathbb{R}$ is *Lebesgue integrable* (written $f \in L^1(M)$ or simply $f \in L^1$ when M is understood) if f is *measurable* and the *Lebesgue integral* $\int_M f\, d\lambda$ exists and is *finite*. (Similar kinds of integrability are defined for other sorts of integrals, e.g. Riemann integrals.) [3] In *Hamiltonian dynamics*, 'integrability' and 'integrable' are usually abbreviations of *complete integrability* and *completely integrable*. The reader should be aware of many other sorts of integrability in mathematics, with subtle distinctions and relations among them.

Integral curve. An *orbit* (in either sense) of a *continuous dynamical system* derived from an ODE. (The word 'integral' refers to 'integrating the *vector field*' of the ODE.)

Integral of motion. See *constant of motion*.

Interior. If $(X, \mathcal{T})$ is a *topological space* and $A \subset X$, the *interior* of A is the *largest open set* contained in A. The interior of A is usually denoted $\mathring{A}$, or int(A). Sets with empty interior are topologically 'thin.'

International Congress of Mathematicians (abbreviated ICM). First held in Zürich in 1897 (and at various locations every four years since 1900, except for a 14-year hiatus covering WWII), this is the world's largest and most prestigious mathematics conference. Several prizes (including the *Fields medal*) are awarded during the opening ceremony, but simply being invited to speak at an ICM is itself a considerable honor.

Invariant manifold. An *invariant set* (of a *dynamical system*) which is also a *manifold*.

Invariant measure. Suppose (M, φ) is a *continuous dynamical system* on the *measurable space* (M, Σ). A measure $\rho : \Sigma \to [0, \infty]$ is *invariant* for the system if for every $t \in \mathbb{R}$ the map $\varphi_t : M \to M$ is a *measure preserving transformation* for ρ. (Note that when one thinks of φ as fixed and ρ as varying, one speaks of '*invariant measures*'; when one thinks of ρ as fixed and φ as varying, one speaks of '*measure preserving transformations*.' Both points of view are useful in *ergodic theory*.)

Invariant set. If (M, φ) is a *continuous dynamical system* and $A \subset M$, we say that A is *invariant* under the *flow* φ (or simply 'A is invariant') if for every t, $x \in A \implies \varphi_t(x) \in A$. (A similar definition applies to *discrete systems*.) In other words, a *trajectory* of the flow that begins in the set A remains in A for all time (forward or backward).

Inverse (of a function). When a function $f : A \to B$ is *injective*, it is possible to define a new function $f^{-1} : f(A) \to A$ called the *inverse* of f, with rule $f^{-1}(y) = x$, where x is the unique element of A such that $f(x) = y$. (Note that uniqueness follows from injectivity of f.) In many cases when f is not injective, it is still desirable to find a *restriction* $f|_C$ of f to a smaller *subdomain* $C \subset A$ so that $f|_C$ is injective (and thus invertible). An inverse $(f|_C)^{-1}$ found in this way is called a *local inverse* of f.

Inverse function theorem. Roughly speaking, the *inverse function theorem* says that a C^1-*smooth map* $f : D \subset A \to A$ has a *local inverse* $(f|_C)^{-1}$ for some *open neighborhood* C of any point $x \in D$ at which the linear map $f'(x)$ (the *derivative* of f at x) is invertible. Elementary versions of the inverse function theorem (with $A = \mathbb{R}$ or $\mathbb{R}^n$) are often included in undergraduate mathematics courses, but there are also much more general versions where A is e.g. a *smooth manifold*, *Banach space*, or *Fréchet space*.

Note that inverse function theorems are closely related to *implicit function theorems*.

Inverse image. See *preimage*.

Invertible. [adj.] Having an *inverse*.

Involution. In *Hamiltonian mechanics* and *symplectic geometry*, two functions $f, g : M \to \mathbb{R}$ (where M is a suitable *domain*) are said to be in *involution* if $\{f, g\} \equiv 0$ (i.e. their *Poisson bracket* vanishes everywhere on M). More generally, in mathematics, an involution is a *map* h such that $h^2 = id$.

Isolated point. If A is a subset of a *topological space*, a point $a \in A$ is an *isolated point* of A if there is a *neighborhood* of a containing no other points of A.

Isotropic. Given a *symplectic vector space* (V, ω), a *vector subspace* U of V is *isotropic* if $U \subset U^{\perp\omega}$, where $U^{\perp\omega}$ is the *symplectic complement* of U. (And U is isotropic if and only if ω vanishes when restricted to U.) A *submanifold* N of a *symplectic manifold* (M, ω) is isotropic if for each $p \in N$, the *tangent space* T_pN is an isotropic subspace of T_pM.

Jacobian (after Carl Gustav Jacob Jacobi). An abbreviation of *Jacobian determinant* (and, more rarely, an abbreviation of *Jacobian matrix*).

Jacobian determinant (often abbreviated *Jacobian*). The determinant of the *Jacobian matrix* (whenever that matrix is square).

Jacobian matrix. This is the matrix representation of the (*Fréchet*) derivative Df_p of a *function* $f : D \to \mathbb{R}^m$ (at the point $p \in D \subset \mathbb{R}^n$ and with respect to particular coordinate systems on D and $\mathbb{R}^n$). If $(x_1, \ldots, x_n) = x$ and $(y_1, \ldots, y_m) = y$ are coordinates on D and $\mathbb{R}^n$, respectively, with $y = f(x)$, a common notation for the *Jacobian matrix* (at p) is $\frac{\partial(y_1, \ldots, y_m)}{\partial(x_1, \ldots, x_n)}(p)$, which reminds us that the i–jth entry of the matrix is $[\partial y_i / \partial x_j](p)$. (In this glossary, the Jacobian matrix is denoted $[\partial y / \partial x]_p$ or $[\partial f / \partial x]_p$.)

Note that, even when working with *maps* between *smooth manifolds*, in order to carry out computations, one ultimately represents the derivative map as a Jacobian matrix after choosing *local coordinates*.

KAM theory. Because it is always growing and changing, it isn't easy to characterize *KAM theory* in a short space. In the opening paragraph of [Pös01], J. Pöschel writes, "KAM theory is not only a collection of specific theorems, but rather a methodology, a collection of ideas of how to approach certain problems in perturbation theory connected with 'small divisors'." A more succinct yet broader characterization due to M.B. Sevryuk (privately communicated) is this: "KAM theory is the theory of quasiperiodic motions in nonintegrable dynamical systems."

KAM torus (plural *tori*). This is the central object of study in KAM theory, first discovered by A.N. Kolmogorov and announced in his 1954 ICM lecture. A *KAM torus* is an *invariant torus* of a *nearly integrable Hamiltonian system* that has survived *perturbation* (of the original integrable system) and survived the accompanying breakdown of classical *integrability*. Because the KAM torus continues to support *quasiperiodic flow*, its existence may be seen as a (partial) continuation of integrability in a nonclassical sense that was not foreseen before Kolmogorov's announcement.

Note that for *twist maps*, the analogous object is called a KAM circle or KAM curve.

Kronecker flow on $\mathbb{T}^n$ (after Leopold Kronecker). See *conditionally periodic flow on* $\mathbb{T}^n$.

Lagrangian (after Joseph-Louis Lagrange). [1] (n.) As a noun, this is an abbreviation of *Lagrangian function*.
[2] (adj.) As an adjective, Lagrangian may modify *dynamics*, *manifold*, *torus*, etc., indicating the appropriate concept or object in *dynamics*.

Lagrangian function. A *smooth* (or at least C^2) function $L : M \to \mathbb{R}$ (or $L : M \times \mathbb{R} \to \mathbb{R}$ when L is time dependent) on an even-dimensional *phase space* M used to generate the equations of motion in *Lagrangian mechanics*. For 'natural' time-independent mechanical systems, the Lagrangian may be written as the difference between kinetic and potential energy: $L = T - V$.

Lagrangian mechanics. In his seminal treatise *Mécanique analytique* (1788), J.-L. Lagrange reformulated and extended *classical mechanics* in a particularly elegant and mathematical way, introducing *generalized coordinates* and the *Lagrangian function*, and making use of variational methods (e.g., calculus of variations, stationary *action* principle). His methodology, since known as *Lagrangian mechanics*, is closely related to *Hamiltonian* mechanics. Together, Lagrange's and Hamilton's methods form the basis of modern mechanics, in both its classical and quantum forms.

Lagrangian (sub)manifold. An *isotropic* n-dimensional *(sub)manifold* L of an ambient $2n$-dimensional *symplectic manifold* M. A prime example of a *Lagrangian submanifold* is the n-dimensional *configuration space* of a mechanical system, viewed as a submanifold of the $2n$-dimensional *phase space*.

Lagrangian torus. A *Lagrangian submanifold* that is also a *torus*. The 'standard' *KAM tori* of classical KAM theory are Lagrangian tori. (For a *system* with n degrees of freedom, these tori are n-dimensional, but one may also investigate the existence of lower- and higher-dimensional invariant tori in KAM theory.)

Lambda lemma, λ-lemma (also called *inclination lemma*). The original version of this lemma (first stated for *maps* by J. Palis on p. 387 of [Pal69]) may be roughly paraphrased as follows: Given a *hyperbolic fixed point* of a C^2 *diffeomorphism* with *stable* and *unstable manifolds*, any C^1 manifold *transverse* to the unstable manifold *converges* under iteration to the stable manifold in the C^1 *topology*.

The λ-lemma turned out to be a very useful tool in *dynamical systems*; Palis first used it to get important results for so-called Morse-Smale systems, and it reappeared in the book [PalM82] in versions for both maps (p. 82) and flows (p. 87).

In the subject of *Arnold diffusion*, versions of the λ-lemma adapted to partially or *normally hyperbolic invariant tori* have been used to justify Arnold's 'obstruction property' in constructing transition chains.

Landau notation, Landau symbol (after Edmund G.H. Landau). See *big O*, or *little o*.

Lebesgue integral (after Henri Lebesgue). Let (M, Σ, μ) be a *measure space*, $A \in \Sigma$ a *measurable set*, and $f : M \to \mathbb{R}$ a *measurable function*. In this case one can find $\int_A f \, d\mu$, the Lebesgue integral of f over A with respect to μ (the integral is a certain extended real number which won't be defined here; see e.g. [Rud86] for a definition and careful treatment). When $(M, \Sigma, \mu) = (\mathbb{R}, \Sigma, \lambda)$ (i.e., the real line with Lebesgue measure λ), the Lebesgue integral coincides with the Riemann integral whenever the latter makes sense; in fact the Lebesgue integral generalizes and 'completes' the Riemann integral in a precise sense. The introduction of the Lebesgue integral at the beginning of the 20th century was a big step, not simply because it incorporated recent advances in set theory, but especially because of the powerful *convergence* theorems that came with it (e.g. the 'dominated' and 'monotone' convergence theorems).

Lebesgue measure. This is the *measure* λ on $\mathbb{R}$ which *generalizes* the notion of length of an interval (i.e., for any interval $[a, b] \subset \mathbb{R}$, $\lambda([a, b]) =$

$b - a$; but the *class* Σ of Lebesgue-measurable sets contains much more than just the intervals). Similarly, the higher-dimensional version of *Lebesgue measure* on $\mathbb{R}^n$ generalizes the notion of volume of a rectangular box in $\mathbb{R}^n$.

Lebesgue space. See L^p *space.*

Limit of a function. If (M, d_M), (N, d_N) are *metric spaces* and $f : M \to N$ a *function*, we say that f has (or approaches) *limit* $L \in N$ as x approaches $a \in M$ (denoted $\lim_{x \to a} f(x) = L$, or $f(x) \to L$ as $x \to a$) if given any $\varepsilon > 0$, there is a $\delta > 0$ such that $d(f(x), L) < \varepsilon$ whenever $0 < d(x, a) < \delta$. (We also say that "f *converges* to L as $x \to a$.")

Limit of a sequence. See *convergence of a sequence.*

Limit point. In a *topological space* $(X, \mathcal{T})$, a point p is a *limit point* of $A \subset X$ if every *open neighborhood* of p contains a point of A different from p. (Compare *isolated point.*) Note that α- or ω-limit point means something more specific in *dynamical systems*; see e.g. [KatH95], §1.6.

Lindstedt series (after Anders Lindstedt). A Lindstedt series is a type of perturbation expansion in a small parameter typically called ε. More precisely, it is a certain 'incomplete' *formal series* expansion in ε of the solution of a *differential equation* depending on ε (described in more detail below in the context of *HPT*). Such series are often found by the 'Poincaré-Lindstedt method' (also called the 'method of strained coordinates'; see Chapter 4 of [Mur91] for an introduction). This method is designed to eliminate the so-called secular terms arising in more naïve perturbation expansion methods.

In HPT and KAM theory, Lindstedt series hold a special place, as it was Poincaré's inconclusive efforts to show their *convergence* that led to his discovery of Hamiltonian *chaos*. Much later, the discovery of *KAM tori* showed indirectly that the Lindstedt series for solutions of perturbed Hamiltonian systems do converge for a large set of initial conditions, against Poincaré's intuition. Still later, convergence of these Lindstedt series was shown directly.

Let us describe the sort of Lindstedt series that arise in HPT and KAM theory. Given a nearly integrable Hamiltonian system $H = H(\theta, I, \varepsilon)$ with n degrees of freedom, we consider a single component of one of its solutions (say, $\theta_j = \theta_j(t, \varepsilon)$) and we seek to expand it in a formal series $a_0(t, \varepsilon) + \varepsilon a_1(t, \varepsilon) + \varepsilon^2 a_2(t, \varepsilon) + \cdots$ in such a way that each a_k is a *conditionally periodic function* of t with (the same) ε-dependent *frequency vector* $\omega = \omega(\varepsilon) \in \mathbb{R}^n$. This means that each a_k may be written as

$a_k(t,\varepsilon) = G_k(\omega(\varepsilon)\,t)$ for some 'nice' function $G_k : \mathbb{T}^n \to \mathbb{R}$. Thus, for fixed ε, each a_k is either *periodic* in t, or *quasiperiodic* in t (with *fundamental frequencies* $\alpha_1, \ldots, \alpha_m$; $2 \le m \le n$). Note that this series is an 'incomplete expansion in ε,' since ε-dependence appears not only in the powers ε^k, but also in the coefficients $a_k = a_k(t,\varepsilon)$ (by way of the frequency vector $\omega = \omega(\varepsilon)$).

Linear combination. In a *vector space*, this is a finite sum of *scalar* multiples of vectors such as $\alpha_1 v_1 + \cdots + \alpha_m v_m$ (each α_k is a scalar, each v_k a vector).

Linear flow on $\mathbb{T}^n$. See *conditionally periodic flow on $\mathbb{T}^n$*.

Linear map. If V, W are *vector spaces* over the same *field* $\mathbb{F}$, the *map* $L : V \to W$ is *linear* if $L(\alpha x + \beta y) = \alpha L(x) + \beta L(y)$ whenever $\alpha, \beta \in \mathbb{F}$ and $x, y \in V$.

Linear part (sometimes used interchangeably with *linearization*). The *linear part* of a *function* at a point a denotes the $k = 1$ term in the *Taylor series* or *Taylor polynomial* of the function. Thus, for $D \subset \mathbb{R}$ and $f : D \to \mathbb{R}$ of class C^1, the linear part of f at a is $f'(a)(x - a)$. For $D \subset \mathbb{R}^n$ and a C^1 *vector field* $g : D \to \mathbb{R}^n$, the linear part of g at a is $Dg_a(x - a)$, where Dg_a is the *Fréchet derivative* of g at a, represented in coordinates by the *Jacobian matrix* $[\partial g / \partial x]_a$ of g at a.

Note that, using the notation above, it's not uncommon to see Dg_a referred to as the linear part of g (in the same way *linearization* is sometimes used at *fixed points*).

Linear space. Another name for *vector space*.

Linearly independent. In a *vector space*, the finite *collection* of vectors $\{v_1, \ldots, v_m\}$ is *linearly independent* if the only collection of *scalars* $\{\alpha_1, \ldots, \alpha_m\}$ for which $\alpha_1 v_1 + \ldots + \alpha_m v_m = 0$ is the collection $\alpha_1 = \ldots = \alpha_m = 0$ (all zeros). This means that no single vector in the collection can be expressed as a *linear combination* of the others.

Linearization (sometimes used interchangeably with *linear part*). In *dynamics*, if a *flow* is given in *local coordinates* $B \subset \mathbb{R}^n$ near a point p by the *vector field* $f : B \to \mathbb{R}^n$ (i.e., by the *autonomous* ODE $\dot{x} = f(x)$), the *linearization* at p is the ODE $\dot{x} = f(p) + A(x - p)$, where $A = [\partial f / \partial x]_p$ is the *Jacobian matrix* of f at p (also denoted more abstractly as the 'derivative operator' or *Fréchet derivative* Df_p). Commonly, one linearizes at an *equilibrium* p, where the linearization reduces to $\dot{x} = A(x - p)$ (or simply

$\dot{u} = Au$, in the shifted coordinates $u = x - p$). In this case, one often speaks loosely of A as the linearization. Similarly, for a *discrete system* defined in local coordinates by the *map* $g : B \to \mathbb{R}^n$, the linearization at p is the 'affine map' $m(x) = g(p) + A(x - p)$, where again $A = [\partial g / \partial x]_p$. (And when p is a *fixed point*, the linearization reduces to $m(p + u) = p + Au$; again in this case A is often called the linearization.)

Lipschitz condition, Lipschitz continuity (after Rudolf Lipschitz). If (A, d_A) and (B, d_B) are *metric spaces*, $D \subset A$ is an *open connected domain*, and $f : D \to B$, we say f satisfies a *Lipschitz condition* (or f is *Lipschitz continuous*, or simply "f is Lipschitz") if there is a non-negative constant L (the Lipschitz constant) such that for all $x, y \in D$, $d_B(f(x), f(y)) \leq L \, d_A(x, y)$. On an *open* interval in $\mathbb{R}$, *local Lipschitz continuity* is *stronger* than *continuity* and *weaker* than differentiability (i.e., differentiability implies local Lipschitz continuity which implies continuity, but the reverse implications do not hold). Lipschitz continuity is a special case of *Hölder continuity*, and local Lipschitz continuity is one of the *weakest* conditions on a *vector field* under which the classical properties of existence and uniqueness hold for the solutions of the ODE defined by that vector field.

Liouville integrability, Liouville integrable (after Joseph Liouville). For *Liouville integrability* in the context of *Hamiltonian systems*, see *complete integrability* above. There is also a more general notion of Liouville integrability in *dynamical systems*, but that is not discussed here.

Liouville measure. This is the natural *invariant measure* for a *Hamiltonian system*, defined in terms of the *symplectic form* ω on *phase space*. Roughly speaking, for a *Hamiltonian system* with n degrees of freedom, the *Liouville measure* is the natural means of computing the volume of a portion of the $2n$-dimensional *phase space* (in fact *Liouville measure* is proportional to the standard 'volume form' ω^n), and this volume is preserved as the portion is transported along the *flow* of the *system*. The Liouville measure is especially useful in investigating the *ergodic* properties of Hamiltonian systems.

Liouville's theorem. More than one mathematical theorem bears this name, but in *Hamiltonian* mechanics, it refers to the invariance of the standard 'volume form' ω^n (or to invariance of the induced *Liouville measure*) under Hamiltonian *flow*. The proof of this basic result is immediate since the Hamiltonian *vector field* has vanishing *divergence*: in canonical coordinates (q, p), the divergence is $\operatorname{div}(H_p, -H_q) = \partial_q H_p - \partial_p H_q = 0$.

Little o. Given *functions* $f : \mathbb{R} \to V$ and $g : \mathbb{R} \to V$, where V is a *normed vector space* and $g(x) \neq 0$ (except possibly $g(a) = 0$), we say $f(x) = o(g(x))$ as $x \to a$ (read '$f(x)$ is *little o* of $g(x)$ as x approaches a') provided $\lim_{x \to a} \|f(x)\| / \|g(x)\| = 0$. A similar definition applies for infinite *limits* $a = \pm\infty$, or for one-sided limits $x \to a^+$ or $x \to a^-$. Note that the limiting value a is often omitted when understood in context. Examples: (i) If $n > m$, $\varepsilon^n = o(\varepsilon^m)$ as $\varepsilon \to 0$. (ii) For $n \in \mathbb{N}^*$, $e^{-1/\varepsilon} = o(\varepsilon^n)$ as $\varepsilon \to 0^+$. (iii) For $n \in \mathbb{N}^*$, $x^n = o(e^x)$ as $x \to \infty$.

Local, locally. In mathematics the precise meaning of this term varies according to context, but a good idea of what is meant is usually obtained by replacing *local* (or *locally*) by "in a sufficiently small *neighborhood* (of the point considered)." See *local Lipschitz continuity* below for an example.

Local coordinates (or 'system of local coordinates'). Another name for *chart* (see *topological manifold*). When used in the singular, a 'local coordinate' refers to one of the real-valued coordinate *functions* $h_k : U \to \mathbb{R}$ of a chart $h : U \to \mathbb{R}^n$. Note also that a system of local coordinates $h : U \to \mathbb{R}^n$ may be referred to loosely by its domain U (thus indicating the part of the manifold 'covered' or 'coordinatized' by the system), without explicitly specifying the coordinate function h.

Local inverse. See *inverse*.

Local Lipschitz continuity, locally Lipschitz. If A and B are *metric spaces*, the *function* $f : A \to B$ is *locally Lipschitz continuous* (or simply *locally Lipschitz*) provided every point $x \in A$ has a *neighborhood* $N \subset A$ such that the *restriction* $f|_N$ is *Lipschitz continuous*.

Logical positivism. In philosophy, a stricter, more scientific form of *positivism* that flourished in the so-called 'Vienna Circle' and 'Berlin Circle' between World Wars I and II. After tangling with such heavyweight thinkers as Ludwig Wittgenstein and Karl Popper, members of the circles dispersed and saw their doctrines take root in the U.K. and U.S., where they evolved into 'logical empiricism.' Though no longer as fashionable, the influence of various positivist tendencies is still undeniable, especially in the philosophy of science and in the rift between what are sometimes called the 'continental' and 'Anglo-Saxon' views of philosophy.

Lower bound. If A is a set of real numbers, a *lower bound* for A is a real number that is less than or equal to all numbers in A. (Note that a lower bound need not belong to A.)

L^p-space. If (M, Σ, μ) is a *measure space* and $p \in [1, \infty)$ is a real number, the *space* $L^p(M, \mu)$ is the *vector space* of all *measurable functions* $f : M \to \mathbb{R}$ such that the *Lebesgue integral* $\int_M |f|^p d\mu < \infty$. Any such space is a *Banach space* (with so-called L^p-norm $\|f\|_p := (\int_M |f|^p d\mu)^{1/p}$) and $L^2(M)$ is also a *Hilbert space* (perhaps the most important Hilbert space in *analysis*). Note that to be precise, one should distinguish between the spaces $L^p(M, \mu)$ just described and the closely related spaces $\mathcal{L}^p(M, \mu)$ consisting of *equivalence classes* defined so that $f, g \in L^p(M, \mu)$ are equivalent whenever $f(x) = g(x)$ for *almost all* $x \in M$.

Lyapunov exponent (after Aleksandr Lyapunov, also transliterated in other ways, e.g. Liapunov). *Lyapunov exponents* (also called 'Lyapunov numbers') provide a quantitative measure of *hyperbolicity* and *chaos* in a *dynamical system*. To say that a system has Lyapunov exponent $\lambda \in \mathbb{R}$ means roughly that if two *trajectories* start a short distance $\delta(0)$ apart, then their distance apart at time $t > 0$ will be $\delta(t) \approx \delta(0)e^{\lambda t}$. Note however that λ may depend strongly on the starting point(s), and on their relative orientation (in fact, for a *system* on an n-dimensional *phase space*, there is a 'spectrum' of n Lyapunov exponents $\{\lambda_1, \ldots, \lambda_n\}$). Positive λ is an indication of *sensitive dependence on initial conditions*, and of positive 'Kolmogorov-Sinai entropy.'

Mandelbrot set (after Benoît B. Mandelbrot). This subset of the complex plane $\mathbb{C}$ is defined as all points c such that the (forward) *orbit* of the *map* $z \mapsto z^2 + c$ with *initial condition* 0 is *bounded*. Thanks to the striking color images of it first produced by H.-O. Peitgen and colleagues, as well as its intensive study by Mandelbrot, A. Douady, J.H. Hubbard, J.-C. Yoccoz, and others, the *Mandelbrot set* (or more precisely its boundary) is perhaps the best-known *fractal* set.

Manifold. See *topological manifold* or *smooth manifold*.

Map, mapping. In mathematics these terms may be used interchangeably with *function*. *Map* may also be used as a verb, emphasizing the 'action' of a function on elements (or subsets) of its *domain*. In *dynamical systems*, the word *map* is often used (in opposition to *flow*) to distinguish *discrete* from *continuous systems*.

Meager (also *nongeneric*). To say that a (sub)set in a *topological space* is meager means that it is 'small' or 'thin' in the sense of *Baire category* *theory*. More precisely, it is 'of *first category*'; i.e., it may be expressed as a *countable* union of *nowhere dense* sets.

Measurable flow. If (M, Σ, μ) is a *measure space*, the flow $\varphi : M \times \mathbb{R} \to M$ is *measurable* if, for every t, the map $\varphi_t : M \to M$ is a *measurable transformation*.

Measurable function. If (M, Σ, μ) is a *measure space*, the real-valued function $f : M \to \mathbb{R}$ is *measurable* if, for every $a \in \mathbb{R}$, the *preimage* $f^{-1}((-\infty, a])$ is a *measurable set*.

Measurable set. A set belonging to the σ-*algebra* on which a given *measure* is defined. (And yes, if one accepts the 'Axiom of Choice,' then lurking inside most *measure spaces* one wants to use, there are non-measurable sets—nonempty subsets of the *base set* that don't belong to the σ-algebra, though these may be difficult to describe.)

Measurable space. A *measurable space* (M, Σ) consists of a *base set* M together with a σ-*algebra* Σ of subsets of M. In other words, a measurable space is a (potential) *measure space* that doesn't (yet) have a *measure*; it is useful when one wants to keep (M, Σ) fixed and consider different measures on Σ.

Measurable transformation. If (M, Σ, μ) is a *measure space*, the map $f : M \to M$ is a *measurable transformation* if $A \in \Sigma \implies f^{-1}(A) \in \Sigma$ (i.e., the *preimage* of every *measurable set* is measurable).

Measure. A *function* $\mu : \Sigma \to [0, \infty]$ (where Σ is the σ-*algebra* of some *base set* M) satisfying (i) $\mu(\emptyset) = 0$, and (ii) $\mu(\bigcup_{k=1}^{\infty} A_k) = \sum_{k=1}^{\infty} \mu(A_k)$ for every *countable collection* $\{A_k\}$ of pairwise disjoint sets $A_k \in \Sigma$. Sets $A \in \Sigma$ with $\mu(A) = 0$ are called *null sets*, or simply 'sets of measure zero.' A measure μ provides a means of comparing the sizes of sets in Σ, and underpins the modern theory of integration (see *Lebesgue integral*).

Measure preserving. Let (M, Σ) be a *measurable space* and $\mu : \Sigma \to [0, \infty]$ a *measure*. The map (or transformation) $\varphi : M \to M$ is *measure preserving* for μ (or simply 'preserves μ') if $A \in \Sigma \implies \mu(\varphi^{-1}(A)) = \mu(A)$. (Note that when one thinks of φ as fixed and μ as varying, one speaks of '*invariant measures*'; when one thinks of μ as fixed and φ as varying, one speaks of 'measure preserving transformations.' Both points of view are useful in *ergodic theory*.)

Measure space. A *measure space* consists of three parts (often listed (M, Σ, μ)): the universal set or *base set* M, a σ-*algebra* Σ of subsets of M

(the collection of *measurable subsets* of M), and the *measure* μ defined on each set in Σ. To say the measure space is *finite* means $\mu(M) < \infty$ (see also *probability space*).

Metric, metric space. Given a set M, the *function* $d : M \times M \to [0, \infty)$ is called a *metric* (or *distance function*) on M provided that, for all $x, y, z \in M$, we have (i) $d(x, y) = 0 \iff x = y$, (ii) $d(x, y) = d(y, x)$, and (iii) $d(x, y) \leq d(x, z) + d(z, y)$. The set M together with the distance function d is called a *metric space*, written (M, d) (or simply M, if d is understood). A metric space is automatically a *topological space*, since the set of *open balls* can be used as a *base* for the *topology*.

Metric transitivity, metrically transitive. *Metric transitivity* is an older term for *ergodicity*; to say that a *system* is *metrically transitive* means that it is *ergodic*. (The term was introduced by G.D. Birkhoff; evidently the word 'metric' in this context refers to *measure*.)

Minimal. If X is a *topological space* and (X, φ) a *continuous dynamical system*, the flow φ is said to be *minimal* if the *orbit* of every $x \in X$ is *dense* in X. (A similar definition applies to *discrete systems*.)

Mixing. A *measure-preserving flow* $\varphi_t : M \to M$ on the finite *measure space* (M, Σ, μ) is *mixing* if for any $A, B \in \Sigma$, we have $\lim_{t \to \infty} \mu(\varphi_{-t}(A) \cap B) = \mu(A) \cdot \mu(B)$. This coincides with intuitive notions of mixing: it says that with increasing time, the portion of A moved inside B approaches the product of the relative sizes of A and B. Mixing is sometimes called *strong mixing* to distinguish it from *weak mixing*, which it implies.

Mod, modular, modulo. Fix $a > 0$. Then given real numbers x and y, to say that $x \equiv y \pmod{a}$ (read "x is congruent to y *modulo* a") means that $x - y$ is an integer multiple of a.

This *relation* between x and y partitions $\mathbb{R}$ into *equivalence classes* with each class represented by a number in the interval $[0, a)$. This can be visualized by imagining first joining the endpoints of $[0, a]$ together to form a circle of circumference a, then wrapping the real line around the circle so that each real number sits atop its representative in the circle. An alternate notation for this *collection* of equivalence classes is $\mathbb{R}/a\mathbb{Z}$, and arithmetic performed in these classes is called *modular* arithmetic. Note that the 1-*torus* (or circle) $\mathbb{T} = \mathbb{R}/\mathbb{Z}$ is obtained from the relation $x \equiv y \pmod{1}$ (some authors use $x \equiv y \pmod{2\pi}$ instead).

Modulus. The *modulus* of a complex number z is its absolute value $|z|$ (i.e., its distance from 0 in the complex plane).

Most. A loose term, usually employed when one wants to avoid use of the more precise term *generic* (thus avoiding a technical explanation).

Multiindex. A notational convenience for partial derivatives. The *multi-index* $\beta = (\beta_1, \ldots, \beta_n) \in \mathbb{N}_0^n$ (i.e., each β_k is a nonnegative integer) has *norm* $|\beta| = \beta_1 + \cdots + \beta_n$ and is used to indicate the partial differential operator $\partial^{|\beta|} / \partial_{x_1}^{\beta_1} \cdots \partial_{x_n}^{\beta_n}$ in the abbreviated form ∂^β (or sometimes D^β). For example, $\partial^{(2,0,1)}$ means $\partial^3 / \partial_{x_1}^2 \partial_{x_3}^1$.

Multiple resonance, multiplicity of resonance. (See *resonance condition*)

Nearly integrable. A *nearly integrable system* is a *system* that is a (small) *perturbation* of an *integrable* system. (See also *Hamiltonian perturbation theory*.)

Negligible set. In a *measure space*, a set is *negligible* if it is contained in a *null set* (i.e., a set of measure zero). Note that in a *complete measure space*, negligible and null sets are the same.

Neighborhood. If $(X, \mathcal{T})$ is a *topological space* and $p \in X$, to say that "U is a *neighborhood* of p" (usually) means that $p \in U \in \mathcal{T}$; in other words, U is an *open set* containing p. However, many writers use 'neighborhood of p' to mean any set which contains an open set containing p. For clarity it may be best to write 'open neighborhood,' but one gets lazy.

Newton's second law of motion (after Isaac Newton). This is the fundamental physical law of classical mechanics. Known to students as "$F = ma$ (force equals mass times acceleration)," it asserts that when a force is applied to a free physical body in an inertial reference frame, the body undergoes acceleration proportional to, and in the direction of, the applied force (the proportionality constant being [the inverse of] the body's mass). See also *Hamiltonian system*.

NHIM. See *normally hyperbolic invariant manifold*.

Nice. In mathematics, this adjective is often used in an informal way (and inside quotes) to indicate that the object it modifies (e.g. a set, function, or other mathematical object) has the properties needed for the purposes at hand, without specifying what those properties are. This is useful for quickly conveying the content of a theorem or result; the precise meaning of *nice* can be provided elsewhere in a more detailed statement if needed (but it can often be guessed).

Nondegeneracy, nondegenerate. [1] *Nondegeneracy* is a necessary hypothesis of most KAM theorems. Consider a *nearly integrable Hamiltonian* system $H(\theta, I, \varepsilon) = h(I) + \varepsilon f(\theta, I, \varepsilon)$ expressed in variables $(\theta, I) = (\theta_1, \ldots, \theta_n; I_1, \ldots, I_n)$ of the 'unperturbed system' with Hamiltonian $h = h(I)$. To say that this system is *nondegenerate* is to place a condition on its frequency map $I \mapsto \omega$ (given as $\omega = \operatorname{grad} h(I)$ on $D \subset \mathbb{R}^n$, an appropriate action-*domain*). The simplest of these conditions is probably Kolmogorov's condition: $\det \partial \omega / \partial I = \det \partial^2 h / \partial I^2 \neq 0$ (i.e., the *Jacobian* of the frequency map is nonzero on the domain D, which [by an appropriate *inverse function theorem*] is equivalent to saying that $I \mapsto \omega$ is a *local diffeomorphism*). A more modern nondegeneracy condition, known to be optimal in certain settings, is Rüssmann's condition that the image $\omega(D)$ of the frequency map not lie in any *hyperplane* passing through the origin in $\mathbb{R}^n$. Such conditions ensure that *invariant tori* with 'highly nonresonant flow' are plentiful in the unperturbed system; it is these tori that persist upon perturbation.
[2] For a different meaning of nondegenerate, see *symplectic form*.

Nongeneric (also *meager*). Not *generic*; see also *Baire category theory*.

Nonintegrable. This is of course the negation of *integrable* (in any of its senses), but in the context of *Hamiltonian systems*, it usually means 'not completely integrable.'

Nonresonant. The *frequency vector* $\omega \in \mathbb{R}^n$ is *nonresonant* if the only integer vector $k \in \mathbb{Z}^n$ such that $k \cdot \omega = 0$ is $k = 0$ (with all components equal to 0). In other words, ω satisfies no *resonance conditions*.

Norm. A *norm* is a way of measuring the 'length' of vectors in a *vector space*. To be precise, if V is a vector space (over the *field* $\mathbb{F}$), a norm on V is a *function* $\| \cdot \| : V \to [0, \infty)$ such that for every $v, w \in V$ and every $\alpha \in \mathbb{F}$, (i) $\|\alpha v\| = |\alpha| \|v\|$, (ii) $\|v + w\| \leq \|v\| + \|w\|$ and (iii) $\|v\| = 0 \implies v = 0$. (Note that if $\| \ \|$ satisfies only (i) and (ii), it is called a *seminorm*.)

Normal form. A simplified form (of a matrix, a *Hamiltonian*, a *vector field*, a *map*, etc.) that displays the essential or desired features. One ordinarily uses a change of coordinates to bring an object into *normal form* on a given *domain*. In KAM theory, one often uses a sequence of coordinate changes to bring the Hamiltonian into a normal form showing both the existence of *invariant tori* and the domain in *phase space* where they exist. See also *Birkhoff normal form*.

Normally hyperbolic invariant manifold (NHIM). Roughly speaking, an *invariant manifold* M of a *continuous dynamical system* is *normally* *hyperbolic* if, in the linear approximation, the 'rate of flow' of *trajectories transverse* to M is greater than the rate of flow of trajectories tangent to M, so the *dynamics* on M is slow compared to the dynamics off M. NHIMs are important: first, they and many of their properties are *robust* (they are *structurally stable*); second, one of the most common mechanisms by which *chaos* arises in dynamical systems is through the transverse intersection of the *stable* and *unstable manifolds* emanating from NHIMs.

Normed. Adjective applied to a *vector space* to indicate that it has (or 'is equipped with') a *norm*.

Nowhere dense. A subset A of a *topological space* X is *nowhere dense* in X if A is not *dense* in any nonempty *open subset* of X; or equivalently, if its *closure* $\overline{A}$ has empty *interior*.

Null set. A set of *measure* zero in a *measure space*.

O or **o** notation (also called *Landau notation*). See *big O* or *little o*.

Observable. In classical physics an *observable* is a real-valued (or vector-valued) *function* on the *phase space* of a *system* whose value may be determined by physical measurement. Examples include the macroscopic variables of thermodynamics (such as temperature, pressure, volume). In quantum physics, an observable is a more elaborate object; it is a 'self-adjoint operator' on the *Hilbert space* representing the system in question.

ODE. An abbreviation of *ordinary differential equation*.

One-to-one. See *injective*.

One-to-one correspondence. See *bijection*.

Onto. See *surjective*.

Open ball. In a *metric space* (M, d), the *open ball* $B_r(x)$ (with center x and radius $r > 0$) is defined as the set $B_r(x) := \{z \in M \mid d(x, z) < r\}$.

Open cover. A *cover* all members of which are *open sets*.

Open set. In a *topological space* $(X, \mathcal{T})$ a set $G \subset X$ is open if $G \in \mathcal{T}$. Another useful characterization of an open set G is the following: Given any point $x \in G$, there is a *neighborhood* N of x such that $N \subset G$.

Orbit (also called *trajectory*). [1] If (M, φ) is a *continuous dynamical system* and $x \in M$, the *orbit* starting at x (or with IC x) is the curve $\varphi(\cdot, x) : \mathbb{R} \to M$, sometimes written $\varphi^x : \mathbb{R} \to M$. (For a *discrete system*

$f : M \to M$, the orbit of x is the *sequence* $\{f^k(x)\}_{k \in \mathbb{Z}}$.)
[2] The words 'orbit' and 'trajectory' are also used to refer to the *image* $\varphi^x(\mathbb{R})$, i.e., to the set of points traced out by the orbit in sense [1] (this is sometimes written $\bigcup_{t \in \mathbb{R}} \varphi_t(x)$, or $\bigcup_{k \in \mathbb{Z}} f^k(x)$ for discrete systems).

Although some authors use 'orbit' in one sense and '*trajectory*' in the other, there is no standard convention; in fact it is natural to use these words in both senses [1] and [2], depending on context.

Order symbols (also called *Landau symbols*). See *big O*, or *little o*.

Ordinary differential equation (often abbreviated ODE). See *differential equation*; see also *vector field*.

Orrery. A mechanical 'clockwork' device that models the solar system. Elaborate orreries were constructed in 18th century Europe (especially at royal courts) following widespread acceptance of the heliocentric solar system.

Paradigm shift. This is the term coined by T.S. Kuhn in his influential book [Kuh62] to describe the 'change in worldview' that occurs in scientific revolutions, both large and small. The term proved to be immensely popular over time, moving outside its original context and eventually becoming a 'buzzword' used even by market researchers. See [Sard00] for an interesting discussion.

Partial differential equation (often abbreviated PDE). See *differential equation*.

Periodic orbit. An *orbit* (in sense [1]) which is periodic. In more detail, if $\varphi : \mathbb{R} \times M \to M$ is a *flow* and $\varphi^x : \mathbb{R} \to M$ is an orbit with initial condition $x \in M$, it is *periodic* if there is a real number $T > 0$ (the smallest of which is called the *period*) such that for all $t \in \mathbb{R}$, $\varphi^x(t + T) = \varphi^x(t)$. In other words, at the end of every time interval of length T, the orbit returns to where it was in *phase space* at the beginning of the interval. (A similar definition applies to *discrete systems*; note that a *fixed point* may be regarded as a special limiting case of a periodic orbit with period 0.)

Periodic point. An *initial condition* for a *periodic orbit* (especially in a *discrete system*).

Perturb, perturbation. To *perturb* a *system* is to change it, usually only slightly. The noun *perturbation* may refer either to the change, or to the system including the change. In 'perturbation theory,' one commonly uses a 'perturbation parameter' (traditionally denoted ε) to both measure and control the size of the perturbation, and one usually seeks to show that some feature of the unperturbed system (the system with $\varepsilon = 0$) is retained for all sufficiently small perturbations (i.e., for all systems with $0 < \varepsilon < \varepsilon_c$, where $\varepsilon_c > 0$ is the 'critical value'). KAM theory is a perturbation result, since it shows the persistence of *invariant tori* under small perturbations of *integrable systems*.

Phase curve. An *orbit* (in either sense) of a *continuous dynamical system*.

Phase space (also called *state space*). The set of all possible 'states' of a *dynamical system*. Each point in *phase space* uniquely represents or encodes all that is relevant about the *system*; if any relevant attribute of the system changes, the system is represented by a different point in phase space. For the special case of a mechanical system, each point in phase space specifies the 'spatial configurations' and 'corresponding velocities' (or 'momenta') of all constituents of the system. Since the configuration and velocity variables occur in pairs, the phase space of a mechanical system has even *dimension*. (Other systems may have any dimension, though *infinite-dimensional* systems are not discussed much in this book.) The phase space of a dynamical system is often a *smooth manifold* (or *symplectic manifold*, for a *Hamiltonian system*).

Phase space is one of those remarkable inventions that looks obvious and natural once it is pointed out, yet a significant part of the success of dynamical systems in modeling physical phenomena rests on it.

Poincaré map (after Henri Poincaré). If M is a finite-dimensional *phase space* and (M, φ) is a *continuous dynamical system* with *Poincaré section* S, it may happen that *initial conditions* in S give rise to *orbits* that 'return' to S (more precisely, for $x \in S$, the so-called 'first return time' $T(x) := \inf\{t > 0 \mid \varphi_t(x) \in S\}$ is finite). If $U \subset S$ consists of those points $x \in S$ with finite first return time $T(x)$, then the *Poincaré map* $P : U \to S$ is defined by $P(x) = \varphi_{T(x)}(x)$. Most commonly, one defines a Poincaré map by choosing S *transverse* to a *periodic orbit*; this guarantees the existence of returning orbits near the periodic orbit, and one can use the map to study the stability properties of the periodic orbit.

Poincaré recurrence theorem. Here is a modern statement (for *discrete systems*) of this famous theorem. Let (M, Σ, μ) be a *finite measure space* and $\varphi : M \to M$ a *measure preserving transformation*. Then for any measurable set $A \subset M$, we have $\mu(\{x \in A \mid \varphi^k(x) \in A$ for infinitely many $k \in$

$\mathbb{N}^*\}) = \mu(A)$. In other words, *almost every initial condition* in A generates an *orbit* that returns to A infinitely often.

Poincaré section (also called *surface-of-section*). If M is a finite-dimensional *phase space* and (M, φ) is a *continuous dynamical system*, a *Poincaré section* is a *hypersurface* (or a 'piece' of a hypersurface) S that is transverse to the *orbits* of φ. (See also *Poincaré map.*)

Pointwise convergence. If (M, d_M), (N, d_N) are *metric spaces* and $\{f_k\}_{k=0}^\infty$ is a *sequence of functions* $f_k : M \to N$, we say that $\{f_k\}$ *converges pointwise* at $x \in M$ if the sequence of points $\{f_k(x)\}$ converges (see *convergence of a sequence*). If $\{f_k\}$ converges pointwise at x for all $x \in M$, we simply say $\{f_k\}$ *converges pointwise*. Note that pointwise convergence is (considerably) weaker than *uniform convergence*.

Poisson bracket (after Siméon Poisson). In *Hamiltonian* mechanics, given a *phase space* M with *canonical coordinates* $(q_1, \ldots, q_n; p_1, \ldots, p_n)$, and *observables* $f, g : M \to \mathbb{R}$ of class C^1, the *Poisson bracket* of f and g is the function $\{f, g\} : M \to \mathbb{R}$ given by $\sum_{k=1}^n \left(\frac{\partial f}{\partial p_k} \frac{\partial g}{\partial q_k} - \frac{\partial g}{\partial p_k} \frac{\partial f}{\partial q_k} \right)$.

Poisson brackets also exist in more general settings (e.g. in so-called Poisson spaces). In mechanics, they are especially useful, e.g., in identifying *constants of motion*, and in generating *canonical transformations* via the Lie method.

Poisson matrix. See *standard symplectic matrix*.

Positivism. In philosophy, the idea that true knowledge is gained only through the senses; metaphysics has little or no meaning. It is no great leap to see that *positivism* supports experimental results in science, and devalues 'theory.' With roots in the Enlightenment, positivism was first explicitly formulated by Auguste Comte in the 19th century, and inspired *logical positivism* in the early 20th century. Though positivists helped refine the scientific method and its interpretation, they also distorted or delayed the advance of theory, as when L. Boltzmann's theory of the atom was repeatedly ridiculed and rejected by E. Mach and others.

Power series. A special kind of *series* of the form $\sum_{k=0}^\infty a_k (x - x_0)^k$, where a_k, x_0 and x are numbers (real or complex). The a_k are called the 'coefficients,' x_0 is called the 'center,' and x is the 'variable.' Every power series has a so-called 'radius of convergence' $R \in [0, \infty]$ such that if $R > 0$, the series *converges pointwise* in the *open ball* $B_R(x_0)$, converges *absolutely* and *uniformly* on any *closed ball* $\overline{B}_\rho(x_0)$ of positive radius $\rho < R$, and *diverges* at every x outside the closed ball $\overline{B}_R(x_0)$ (there are of course no such x when $R = \infty$). When $R = 0$, the series converges only at the center $x = x_0$. On the boundary $|x - x_0| = R$, convergence must be determined point-by-point. Power series are one of the main tools in *analysis*, and have a number of special properties (most known already to Newton, as V.I. Arnold is fond of pointing out; cf. [Ar90]).

Preimage (also called *inverse image*). If $f : A \rightarrow B$ is a *function* and $C \subset B$, the *preimage* of C under f, denoted $f^{-1}(C)$, is the set $\{x \in A \,|\, f(x) \in C\}$. In other words, it is the set of all points in the *domain* A that are *mapped* to the set C by f. Note that the preimage of any set $C \subset B$ always exists, even though the *inverse* f^{-1} of f may not exist (but when the inverse does exist, its *images* coincide with the preimages of f, so there are no serious ambiguities of notation).

Probability space. A *probability space* (M, Σ, ρ) is a *measure space* with $\rho(M) = 1$. Note that any measure space (M, Σ, μ) with *finite* measure $(\mu(M) < \infty)$ can be made into a probability space by normalizing the measure (i.e., redefining it as $\rho = (1/\mu(M))\mu$). The name derives of course from probability theory, where members of Σ are events, and $\rho(A) \in [0, 1]$ is interpreted as the probability that A occurs. A.N. Kolmogorov played a major role in the development of measure-theoretic probability.

Proper degeneracy, properly degenerate. In KAM theory, a nearly integrable *Hamiltonian system* $H(\theta, I, \varepsilon) = h(I) + \varepsilon f(\theta, I, \varepsilon)$ is *properly degenerate* if the integrable part h does not depend on all the action variables $(I_1, I_2, \ldots, I_n) = I$. (Clearly this violates the standard *nondegeneracy* conditions of KAM theory, such as Kolmogorov's and Rüssmann's.) A *proper degeneracy* is a mild degeneracy, because it is often still possible to show a KAM result in spite of it.

Quadrature. This word may be familiar from the Latin expression *quadratura circuli* ('squaring the circle'). More generally, it may refer to the geometric process of constructing a square with area equal to that of a given figure, or to the process of numerical integration. However, in *classical mechanics* and ODE, it has (or had) a related meaning that is now difficult to find in dictionaries: one says that a *system* is 'reduced to *quadrature*' if the system's solution is given by explicit formulas, including perhaps 'indicated integrals' (integrals that have not been—or cannot be—evaluated in terms of *elementary functions*). On p. 86 of [Tay96], we have the following (slightly enigmatic) definition: "Given a collection of functions $\{u_j\}$, a map is said to be constructed by quadrature if it is produced by a composition of the following operations: (i) elementary algebraic manipulation, (ii) differentiation, (iii) integration, (iv) constructing inverses of maps."

Quasiperiodic flow on $\mathbb{T}^m$. This refers to *conditionally periodic flow on* $\mathbb{T}^m$ with *nonresonant frequency vector* $\alpha \in \mathbb{R}^m$ $(m \geq 2)$. The m components of α are called the *fundamental frequencies* of the flow. (Note also that *quasiperiodic flow*—without reference to the dimension of the *torus* on which it occurs—means 'conditionally periodic flow that is not periodic,' since such flow is necessarily quasiperiodic on some torus of dimension $m \geq 2$; cf. *conditionally periodic flow on* $\mathbb{T}^n$.)

Quasiperiodic function. A *conditionally periodic function* that is not periodic. Every quasiperiodic function $q : \mathbb{R} \to \mathbb{R}$ may be written as $q(t) = G(F_t^\alpha(\theta))$ for some *continuous* function $G : \mathbb{T}^m \to \mathbb{R}$, and some quasiperiodic flow F_t^α starting at θ on $\mathbb{T}^m$ with *nonresonant frequency vector* $\alpha \in \mathbb{R}^m$ $(m \geq 2)$. As with conditionally periodic functions, it is often convenient to take the IC $\theta = 0$ and write q in the less precise but simpler form $q(t) = G(\alpha t)$. The components $\alpha_1, \dots, \alpha_m$ of α are unique (up to rearrangement) and are called the *fundamental frequencies* of q.

Note that some authors define quasiperiodic functions in the way conditionally periodic functions are defined in this glossary (so that periodic is a special case of quasiperiodic). Also, there is another meaning of *quasiperiodic function* sometimes used in complex analysis (involving a so-called scalar 'quasiperiod'), but this doesn't ordinarily arise in *dynamical systems*.

Range. The *range* of the *function* $f : A \to B$ is $f(A)$ (i.e., it is the *image* of the *domain*, or in other words, the set of all points in the *target set* that are 'hit' by f).

Rationally dependent (or **independent**). See *commensurable* (or *incommensurable*).

Recurrence, recurrent. Consider the *discrete dynamical system* (X, f), where X is a *Hausdorff space* and $f : X \to X$ is a map. The point $p \in X$ is *recurrent* in (X, f) if for every *neighborhood* U of p there is an $n \in \mathbb{N}^*$ such that $f^n(p) \in U$. (A similar definition applies to *continuous systems*.) Every recurrent point of a system is 'nonwandering' (i.e., is not *wandering*), but a system may also have nonwandering points that are not recurrent. *Recurrence* (or recurrence that is not simply *periodicity*) may be viewed as a weak form of *chaos*. See also *Poincaré recurrence theorem*.

Regular, regularity. A way of referring to the *continuity*, *smoothness*, or *analyticity* properties of *functions*. If $D \subset \mathbb{R}^n$ is open, and $1 < \lambda < \infty$, $1 \le k < \infty$, then $C^\omega(D) \subset G^\lambda(D) \subset C^\infty(D) \subset C^k(D) \subset C^0(D)$ shows *smoothness classes* arranged in a hierarchy of decreasing *regularity*: analytic functions are the 'most regular,' followed by *Gevrey-smooth* functions, then C^∞ functions, C^k functions, and finally the continuous functions in the 'least regular' position. (Of course interpolations and extensions of the list are possible.)

Relation. Here are two closely related definitions of this fundamental notion:

[1] (a) Given arbitrary sets A and B, a *relation* between them (or between their elements) is a subset S of the *Cartesian product* $A \times B$. (Since the order of the sets matters, one often says 'a relation from A to B' to emphasize that the first factor is A and the second is B.)

[1] (b) A relation is an ordered triple of sets $(A, B; S)$ in which A is called the *domain*, B the *codomain*, and $S \subset A \times B$ the *graph* of the relation.

Definition **[1]** (a) is usually preferred by logicians and set theorists, while **[1]** (b) is usually preferred by working mathematicians.

Note that, to be more precise, what is defined here should be called a '2-place relation' (or *binary relation*), to distinguish it from the obvious generalization to an 'n-place relation,' defined (analogously to **[1]** (a)) as a subset of the Cartesian product $A_1 \times \cdots \times A_n$ of n sets $A_1, \ldots, A_n$.

Relative measure. If (M, Σ, μ) is a *measure space* and $A, B \in \Sigma$, with $\mu(B) > 0$, the *relative measure* of A in B is $\mu(A \cap B)/\mu(B)$. (One often speaks of 'the relative measure of A,' without mentioning B when B is understood in context.) Note that relative measure is a number in the interval $[0, 1]$.

Residual. A synonym for *generic* (see *Baire category theory*).

Resonance, resonance condition. For the *frequency vector* $\omega \in \mathbb{R}^n$, a *resonance condition* is a *relation* $k \cdot \omega = 0$, where $k \in \mathbb{Z}^n \setminus \{0\}$ (i.e., k is a nonzero integer vector of dimension n). The r resonance conditions $k^{(1)} \cdot \omega = 0, \ldots, k^{(r)} \cdot \omega = 0$ are *independent* if the set $\{k^{(1)}, \ldots, k^{(r)}\}$ of integer vectors is *linearly independent* in $\mathbb{R}^n$. An ω satisfying precisely r independent resonance conditions is also said to have r (*multiple*) resonances, or to satisfy a resonance condition of *multiplicity* r.

Restriction (of a *function*). Given a function $f : A \to B$ and any (proper) subset $C \subset A$, we can define a new function $f|_C : C \to B$ called the *restriction* of f to C and defined by $f|_C(x) = f(x)$. In other words $f|_C$ has the same *rule* as f, but its *domain* is smaller. (Note also that a restriction of f may be defined as any function having f as an *extension*.)

Reversible, reversibility. The idea of *reversibility* arises naturally from mechanical systems in which time may be 'run forward or backward' without revealing a 'preferred' time direction. As a prime example, note that many mechanical systems have a *Hamiltonian* of the form $H(q,p) = \frac{1}{2}p^2 + V(q)$, $((q,p) \in \mathbb{R}^n \times \mathbb{R}^n, p^2 = p \cdot p)$ which clearly satisfies $H(q,p) = H(q,-p)$. From *Hamilton's equations*, it follows that whenever $(q(t), p(t))$ is a *trajectory* of the system with IC (q_0, p_0), then $(q(-t), -p(-t))$ is also a trajectory with IC $(q_0, -p_0)$. (Note the 'reversal' of time and momentum; this is the sort of reversibility often discussed in thermodynamics.)

But such Hamiltonian systems are not the only systems enjoying the reversibility property. Given a 'nice' *autonomous vector field* $f : D \to \mathbb{R}^n$, $(D \subset \mathbb{R}^n)$, the *flow* arising from the ODE $\dot{x} = f(x)$ is *reversible* if there is a so-called 'reversing symmetry' $R : D \to D$ (*invertible* and sufficiently *smooth*) such that $\frac{d}{dt}R(x) = -f(R(x))$ along the flow. (In practice, the symmetry R is also often an involution; i.e., $R^2 = id$.) This more general reversibility property is readily extended to *maps* and non-autonomous systems, and there are also other related or overlapping notions such as 'weak reversibility' and 'equivariance.' See §4.8 (d) above for references to KAM theory for reversible systems.

Rigor, rigorous. In mathematics, *rigor* refers to the certainty or logical persuasiveness of an argument. To say that an argument (or an *estimate*, or line of reasoning, etc.) is *rigorous* means that it carries the persuasiveness of a mathematical proof (as opposed to, say, the persuasiveness of a physical argument, or a numerical calculation that may have round-off errors, etc.).

Robust. See *structurally stable*.

Rotation number. If $\phi : \mathbb{T} \to \mathbb{T}$ is an (orientation preserving) *homeomorphism* of the circle $\mathbb{T}$, its *rotation number* $\rho = \rho(\phi)$ is, roughly speaking, the average amount by which it rotates points of $\mathbb{T}$. More precisely, it is (the fractional part of) $\lim_{n\to\infty} (\phi_L^n(x) - x)/n$, where ϕ_L is any 'lift' of ϕ and x is any point of $\mathbb{T}$. For more detail (including definitions of 'orientation preserving' and 'lift'), see Section B.2 of Appendix B, above. Note that rotation numbers may also be defined in other settings (e.g. for *mappings* on simple closed curves, or for *maps* of the plane near *fixed points*, etc.).

Rule. The essential part of a *function*, describing precisely how to get the output from an input.

Scalar. In mathematics, *scalar* simply means an element of (or pertaining to) a *field*, and is usually used to distinguish the elements of a field from the vectors of a *vector space*. In physics, nearly the same meaning holds, but emphasis is placed on the fact that "scalars appear the same in any coordinate system" (which is not true for vectors or tensors).

Second category. Not of *first category* (i.e., not *meager*) in the sense of *Baire category theory*. (Note that *second category* is not a synonym of *comeager*.)

Second law of thermodynamics. Here is a rough statement of this famous law: The *entropy* of an isolated, macroscopic thermodynamical system not in *equilibrium* increases until the system reaches equilibrium and the entropy is maximized. Many other statements have been formulated in different contexts and at different levels of precision.

Seminorm. See *norm*.

Sensitive dependence on initial conditions (abbreviated SDOIC in this book). To say that a *dynamical system* has SDOIC means roughly that small uncertainties in *initial conditions* are magnified over time, so that it becomes impractical to use the *system* as a model of a physical system.

A basic type of SDOIC may be characterized more precisely as follows. Suppose (M, d) is a *metric space* and $\varphi_t : M \to M$ is a *flow*. The flow has *sensitive dependence on initial conditions* on $S \subset M$ provided there is a $K > 0$ such that given any $x \in S$, for every $\varepsilon > 0$ there is a y in the *open ball* $B_\varepsilon(x)$ and a $T > 0$ with $d(\varphi_T(x), \varphi_T(y)) \geq K$. (Here K is sometimes called the 'sensitivity constant.' A similar definition applies to *maps*; note also that some authors require S to be *compact* and *invariant* under φ_t.)

A stronger form of SDOIC on S requires the existence of a *Lyapunov exponent* $L > 0$ such that given any $x \in S$, for every $\varepsilon > 0$ there is a $y \in B_\varepsilon(x)$ such that for all $t \geq 0$, $d(\varphi_t(x), \varphi_t(y)) \geq e^{Lt} d(x, y)$.

SDOIC is the primary ingredient of *deterministic chaos*, and was discussed more than a century ago by J. Hadamard, P. Duhem, and Poincaré (see, e.g. Quote **Q**[Poi6] in Appendix E). It is often called the *butterfly effect* in popular *chaos* literature.

Sequence. A *sequence* is a *function* $s : \mathbb{N}_0 \to B$, where B is any set. (More generally, the *domain* may be $\mathbb{N}^*$ or some other 'segment' of $\mathbb{Z}$; the *codomain*

B is commonly $\mathbb{R}$, $\mathbb{C}$, or a *function space*.) One usually indicates the independent variables of a sequence by subscripts, called 'indices'; hence, when the codomain is understood, one writes $\{s_0, s_1, s_2, \ldots\}$ or more compactly $\{s_k\}_{k=0}^{\infty}$ (or simply $\{s_k\}$), and one indicates the *rule* for the sequence by a formula involving the indices. The dependent variables of a sequence are called 'terms.' One sometimes indicates the codomain of a sequence by saying "$\{s_k\}$ is a sequence in B."

Series. An expression of the form $\sum_{k=0}^{\infty} a_k$, where the terms a_k may be numbers (real or complex) or *functions* of an independent variable (say x, so that $a_k = a_k(x)$). A series is in fact a special kind of *sequence* whose terms s_n are the so-called 'partial sums' of the series: $s_n = \sum_{k=0}^{n} a_k$. To say that the series $\sum_{k=0}^{\infty} a_k$ converges β, where β is a qualifier (such as *pointwise*, *uniformly*, weakly, etc.), means that s_n converges β.

Shadow, shadowing. In *dynamical systems*, *shadowing* is a loose term for a collection of techniques used to detect 'true' *orbits* of a *system* by first finding 'pseudo-orbits,' then showing that the true orbits closely follow (or 'shadow') them. Shadowing techniques were first introduced by D. Anosov and R. Bowen for *hyperbolic* systems, and have since been extended to more general settings. Such techniques are important in *Arnold diffusion*.

Sharp estimate. The best possible *estimate* (in a given context).

Sigma-algebra (or σ-*algebra*). A collection Σ of subsets of a nonempty universal set (or *base set*) M with the following properties: (i) $M \in \Sigma$, (ii) $A \in \Sigma \implies M \setminus A \in \Sigma$, and (iii) $A_i \in \Sigma$ $(i = 1, 2, \ldots) \implies \bigcup_{i=1}^{\infty} A_i \in \Sigma$. In words, Σ contains M and is closed under *complements* and *countable* unions. When a σ-*algebra* Σ is used in a *measure space* (M, Σ, μ), it is also called the *collection* of *measurable subsets* of M. When *measures* were introduced and studied in the early 20th century, it was quickly discovered that they often could not be consistently defined on the power set of M (the collection of all subsets of M), but rather must be restricted to a σ-algebra of subsets of M.

Simply connected. In *topology*, to say that a *topological (sub)space* is *simply connected* means that it is 'path connected' and the 'fundamental group' at each of its points is trivial (however, definitions of the terms in quotes are not given in this glossary). Intuitively speaking, a *space* is simply connected if it does not have any 'handles.' For a region in the plane, any 'hole' creates a handle and renders the region 'multiply connected' (the negation of simply connected). But in higher dimensions, a simply connected space may have enclosed, bubble-like voids; a 'wormhole' passing all the way through is required to form a handle and make it multiply connected.

Smale horseshoe map (after Stephen Smale). In the early 1960s, S. Smale devised a simple *discrete dynamical system* with paradigmatic *chaotic* behavior ([Sma65]). The system is a *map* of (part of) the unit square in $\mathbb{R}^2$ into itself: one takes the square and stretches it into a long rectangle, bends it into a horseshoe shape, and lays it again on top of the original square (whence the name *horseshoe map*). When this map is iterated (forward and backward), a 2-dimensional *invariant Cantor set* emerges on which the *dynamics* of the map may be conveniently studied using *symbolic dynamics*. One readily shows that the Cantor set has (i) *countably infinitely* many *periodic points* with *orbits* of all periods, (ii) *uncountably infinitely* many nonperiodic points, and (iii) a *dense* orbit. Moreover, the map itself has no *analytic first integral*. See Chapter 4 of [Wig90–03] for a self-contained discussion. The Smale horseshoe is particularly important in *Hamiltonian dynamics*, as it provides a simple model for what happens when *stable* and *unstable manifolds* (of invariant sets) intersect *transversally* to form *homo-* or *heteroclinic* points.

Smooth. [**1**] To say that a *function* $f : D \to B$ is *smooth* means that $f \in C^\infty(D)$; in other words, f has *continuous* derivatives of all orders at every point of its (usually *open*) *domain*.
[**2**] In some cases, smooth is used informally in a way similar to the adjective *nice*; i.e., to say that "f is smooth" means that it is of whatever *smoothness class* C^k is required for the purposes at hand without specifying k (though k may be specified later in a more precise statement). This meaning is perhaps better conveyed by saying "f is sufficiently smooth." (See also *smoothness*.)

Smooth manifold. A *topological manifold* M equipped with a so-called *differential structure* which allows one to do calculus on the manifold to a certain level of *smoothness* (called the *smoothness class* of the structure; one says 'C^k-manifold' to be clear). More precisely, a C^k-differential structure is a (maximal) *atlas* for M in which all *transition maps* are of *class* C^k.

Note that it is artificial to speak of a 'C^k-smooth manifold,' because it turns out that (by results due to H. Whitney) any manifold supporting a C^k differential structure ($0 < k < \infty$) automatically supports a C^∞ differential structure. In other words, all smooth manifolds (of any smoothness $k > 0$) are intrinsically C^∞; a C^k differential structure may be used (or inherited from a particular *embedding*), but isn't 'natural' and doesn't reflect the smoothness of the manifold itself.

Smoothness, smoothness class. *Smoothness* generally refers to the number of times a *function* may be differentiated continuously. To be precise, given an *open set* $D \subset \mathbb{R}^n$, and $k \in \mathbb{N}^*$, we say that $f : D \to \mathbb{R}$ belongs to the *smoothness class* $C^k(D)$ (or the *differentiability class* $C^k(D)$, or simply f is of *class* C^k) if f has continuous derivatives of all orders up to and including k at all points in D. The superscript k may also be 0, ∞, or ω; in these cases, $f \in C^0(D)$ means "f is continuous on D"; $f \in C^\infty(D)$ means "f has continuous derivatives of all orders on D" (often simply "f is smooth"); and $f \in C^\omega(D)$ means "f is *analytic* on D." There are also ways of interpolating between these smoothness classes: for interpolating between the various C^k, see *Hölder smoothness*; for interpolating between C^ω and C^∞, see *Gevrey smoothness*; also see *regularity*.

For functions on or between *smooth manifolds*, these notions generalize as follows. If M is a *smooth manifold* with a C^k differential structure, to say that the function $f : M \to \mathbb{R}$ is of smoothness class $C^r(M)$ ($r \le k$) means that $f \circ h^{-1}$ is of class $C^r(B)$ for any *chart* $h : U \to B$ used as *local coordinates* on M. Similarly, given smooth manifolds M, N with respective C^k *differential structures* $\{(U_\alpha, h_\alpha)\}$, $\{(V_\beta, k_\beta)\}$, to say that $f : M \to N$ is of class $C^r(M; N)$ ($r \le k$) means that the induced map $k_\beta \circ f \circ h_\alpha^{-1} : B_U \to B_V$ is of class C^r for any charts $h_\alpha : U_\alpha \to B_U$ and $k_\beta : V_\beta \to B_V$ such that $f(U_\alpha) \subset V_\beta$.

Space. In mathematics, this word may have its ordinary physical or geometric meaning, but it is also often a short form of *vector space* or *topological space*, as in 'the *space* of continuous functions $C^0(M)$.'

Space average. If (M, Σ, μ) is a *finite measure space* and $f : M \to \mathbb{R}$ is a *measurable function* (sometimes called an *observable*), the *space average* $\overline{f}$ of f is defined as $\overline{f} := \int_M f(x) \, d\mu(x) / \mu(M)$. In statistical mechanics, M is often an 'energy surface'; i.e., the subset of *phase space* where the *system* under consideration has a given energy.

Span. If G is a *vector space* and $S \subset G$, the *span* of S (denoted span S) is defined as the smallest *vector subspace* of G that contains S. (It may also be defined as the set of all (finite) *linear combinations* of vectors in S.)

Stable manifold. Suppose (M, d) is a *metric space*, $\varphi : M \times \mathbb{R} \to M$ a *flow* and $p \in M$ a *fixed point*. The *stable manifold* of p is defined as $W_s(p) := \{x \in M \mid \lim_{t \to \infty} \varphi_t(x) = p\}$ (i.e., the set of phase points that are (forward) asymptotic to p under the flow). This generalizes to any *invariant set A* of the flow (such as an *invariant torus*): its stable manifold is $W_s(A) := \{x \in M \mid \lim_{t\to\infty} d(A, \varphi_t(x)) = 0\}$. (Note that *unstable manifolds* are defined in almost the same way, but with time reversed.)

Standard map. See *Chirikov standard map*.

Standard symplectic matrix (also called the *Poisson matrix*). The $2n \times 2n$ matrix $J := \left[\begin{smallmatrix} 0 & I \\ -I & 0 \end{smallmatrix}\right]$, where I is the $n \times n$ identity matrix. (Of course, if you prefer to write *canonical coordinates* as (p, q) rather than (q, p), you should instead use $J := \left[\begin{smallmatrix} 0 & -I \\ I & 0 \end{smallmatrix}\right]$.)

State space. See *phase space*.

Statistical properties. This is the collective term used by *ergodic theorists* to describe the various sorts of *chaotic* behavior a *dynamical system* may possess. In rough order of increasing *chaos*, a basic list of statistical properties is this: *recurrence, nonintegrability, ergodicity, weak mixing*, (strong) *mixing*, and the Axiom A, Anosov, K, C and Bernoulli properties. (The list may of course be refined or extended.)

Steep, steepness. In the earliest announcement of the theorem now bearing his name (see §6.2), N.N. Nekhoroshev formulated the *steepness* property for a *function* $f : D \to \mathbb{R}$ on an *open domain* $D \subset \mathbb{R}^n$ as follows. Assume that f has nonvanishing gradient on D, and denote by $\{\lambda^r(x)\}$ the set of *hyperplanes* of *dimension* r, passing through $x \in D$ and perpendicular to grad $f(x)$. Then f is *steep* on D if for each $r = 1, \ldots, n-1$, there are positive constants C_r, δ_r (called steepness coefficients), and $\alpha_r \geq 1$ (the steepness indices) such that for all $x \in D$, all $\lambda^r(x) \in \{\lambda^r(x)\}$ and all $\xi \in (0, \delta_r]$, $\max_{0 \leq \eta \leq \xi} \min_{\{y \in \lambda^r(x) \cap D, \|x-y\|=\eta\}} \|\mathrm{grad}(f|_{\lambda^r(x)})(y)\| > C_r \xi^{\alpha_r}$. Conditions similar to steepness may be traced back to J. Glimm's work [Gli64] on the (*formal*) stability of Hamiltonian systems; but it was in [Nek73] that Nekhoroshev first showed steepness to be a strongly *generic* property of *analytic* functions, in the sense that the coefficients of the *Taylor series* of a nonsteep function satisfy an infinite number of algebraic conditions. For detailed discussions of steepness, its generalizations, and its role in the stability of *Hamiltonian systems* (including its relation to convexity and various *nondegeneracy* conditions), see [LocM88], [Nek73], [Nek77], or [Nie06].

Strong mixing. See *mixing*.

Strong solution (also called *classical solution*). One says that a solution of a *differential equation* (DE) is a *strong solution* or is *classical* in order to distinguish it from other solutions which may only be *weak solutions*. More precisely, if k is the order of the highest order derivative occurring in the DE, then a strong solution is of *smoothness class* C^k on its domain (where the independent variable(s) of the DE live), and satisfies the DE at all points of its domain.

Stronger, strongest. If elements of set A satisfy condition α and elements of set B satisfy condition β, to say that α is *stronger* than β means that $A \subset B$. (We may of course also say that β is *weaker* than α.) On the other hand, if Theorems 1 and 2 have the same conclusions, but the hypotheses of Theorem 1 are stronger than those of Theorem 2, we say that Theorem 2 is stronger than Theorem 1 (or Theorem 1 is weaker than Theorem 2), since Theorem 2 has wider applicability. To say that a theorem is the *strongest* possible generally means that the conclusions fail to hold under any weaker hypotheses.

Structuralism. Not quite a branch of philosophy, but rather an 'intellectual trend' with roots in the early 20th century work of Swiss linguist Ferdinand de Saussure, who saw a common 'structure' in languages. The hallmark of *structuralism* is an emphasis on the relations between elements in a system, rather than on the elements themselves. By the mid 20th century, the structuralist approach was widespread in literary theory, anthropology, psychoanalysis, architecture, and elsewhere. In mathematics, its influence was seen especially in the *Bourbaki*, in the work of A. Grothendieck, and in *category theory*.

Structural stability dogma. Following the formulation of *structural stability* for *dynamical systems*, it was often asserted that only *structurally stable systems* could be good models of physical systems. (The idea was that uncertainties in physical measurement *perturb* a given model to a neighboring system.) This is the so-called *structural stability dogma*. Most mathematical physicists now accept a *weaker* version stating that the only properties of (a family of) dynamical systems which are physically relevant are those preserved under small perturbations.

Structurally stable. Roughly speaking, a *system* is *structurally stable* if it is *topologically conjugate* to all nearby systems, in other words, if small *perturbations* of the system do not cause qualitative changes in its solutions' behavior. (To make the definition precise, one needs to specify a *topology* on the *collection* of systems in which *perturbations* occur.) One also says that an *invariant set* of a system is structurally stable, meaning that, if the system is perturbed, the set before perturbation is *homeomorphic* (or *diffeomorphic*, as appropriate) to the set after perturbation. The word *robust* is sometimes used to mean structurally stable.

Sub-. A common prefix in mathematics. If the set X is a Γ, to say that Y is a sub Γ of X typically means that Y is a subset of X and Y is itself a Γ. Additional qualifiers may of course be added; for example, if X is a β *space* (here β stands for *Banach, metric, topological, vector*, etc.) then $Y \subset X$ is a β subspace of X provided Y is by itself a β space.

Superintegrable. [adj.] Said of a *Hamiltonian system* with n degrees of freedom having more than n independent *integrals of motion* (see *completely integrable*).

Supremum (also called *least upper bound*). If A is a set of real numbers, the *supremum* of A (denoted $\sup A$ or $\operatorname{lub} A$) is defined to be $+\infty$ if A has no *upper bound*, and defined to be the smallest upper bound of A if A is *bounded from above*. The fact that every set of real numbers that is bounded from above has a *least upper bound* in $\mathbb{R}$ is called the *completeness* property of $\mathbb{R}$.

Surface-of-section. See *Poincaré section*.

Surjection, surjective. The *function* $f : A \to B$ is a *surjection* if $f(A) = B$ (i.e., the *range* of f is the entire *codomain*). The adjective is *surjective* or *onto*; one also says 'f *maps* A *onto* B.'

Symbolic dynamics. A means of representing the *dynamics* of a *dynamical system* by *sequences* of abstract symbols, with each symbol sequence corresponding to a state of the system, and time iteration represented by 'shifts' in the sequence. This technique was introduced by J. Hadamard in [Had98], and has proved to be extremely fertile, leading to the general theory of 'Markov partitions,' and to a number of classical results in *dynamical systems*, including Smale's results on the *horseshoe map* and O.M. Sharkovsky's theorem for *maps* of the interval.

Symplectic. According to the article [Ig98], this word was coined by Hermann Weyl in 1946 [Wey46]. Weyl sought a word like 'complex,' but without the mathematical meaning already attached to the latter. The term *symplectic geometry* first occurred in a 1953 paper by J.-M. Souriau, although the essential feature of symplectic geometry—the *symplectic form*—goes back to Lagrange.

Symplectic complement. Given a *symplectic vector space* (V, ω) and a *vector subspace* U of V, the *symplectic complement* of U is defined as $U^{\perp\omega} = \{v \in V \mid \omega(u, v) = 0 \text{ for every } u \in U\}$. Note that, unlike ordinary orthogonal complements in inner product spaces, it is not necessarily true that $U \cap U^{\perp\omega} = \{0\}$. If the last equality is true, then U is a *symplectic subspace* of V (in which case ω restricted to U is nondegenerate, so that $(U, \omega|_U)$ is a symplectic vector space in its own right).

Symplectic coordinates. In most contexts, this means *Darboux coordinates* on a $2n$-dimensional *symplectic manifold*. More generally, it may also mean any system of coordinates on a symplectic manifold. In this more general case, the matrix representation of the *symplectic form* may be a fixed, nonsingular, skew symmetric matrix (rather than the *standard symplectic matrix J* for Darboux coordinates).

Symplectic diffeomorphism. See *symplectomorphism*.

Symplectic form. If V is a *vector space*, a *symplectic form* ω on V is a *bilinear form* $\omega : V \times V \to \mathbb{R}$ that is *nondegenerate* ($\omega(u, v) = 0$ for all $v \in V \implies u = 0$) and skew-symmetric ($\omega(u, v) = -\omega(v, u)$). When $\dim V$ is finite, a simple form of *Darboux's theorem* shows that one can always choose special coordinates on V (called *Darboux coordinates*) in which ω has the matrix $J := \begin{bmatrix} 0 & I \\ -I & 0 \end{bmatrix}$ (the *standard symplectic matrix*, where I is the $n \times n$ identity matrix).

If M is a *smooth manifold* of even dimension, a symplectic form ω on M is as just defined on each *tangent space* T_pM of M; and furthermore ω varies smoothly as a function of $p \in M$ and satisfies $d\omega = 0$. The existence of such ω makes M into a *symplectic manifold* (M, ω).

Symplectic geometry. This is the mathematical subject that grew out of *Hamiltonian mechanics* and reached a certain maturity in the mid to late 20th century; thus one way to view it is as the most abstract version of mechanics (though mechanics on 'Poisson manifolds' is perhaps more abstract still). It deals with the geometry of—and *dynamics* on—*symplectic manifolds*. Unfortunately a meaningful description of it is beyond the scope of this book; it is however essential background for the deeper study of KAM theory.

Symplectic linear map. If (V, ω) and (W, ρ) are *symplectic vector spaces*, the *linear map* $L : V \to W$ is *symplectic* if $\rho(L(u), L(v)) = \omega(u, v)$ for all $u, v \in V$. (The left-hand side of the equation is called the 'pullback form' $L^*[\rho]$ at (u, v), so L is symplectic means that $L^*[\rho] = \omega$.)

Symplectic manifold. A *smooth manifold* M of even *dimension* equipped with a closed, smoothly varying *symplectic form* ω on each *tangent space* T_pM (often written as the pair (M, ω); note that here 'closed' means that the 'exterior derivative' $d\omega = 0$). This is the *phase space* of a *Hamiltonian system* in its modern mathematical formulation. For an *n-degree-of-freedom* physical Hamiltonian system, one first identifies the n-dimensional smooth manifold Q representing the *configuration space* of the system; the phase space is then the *cotangent bundle* of Q with its natural symplectic form. A symplectic manifold may also be given abstractly, without reference to any mechanical system (as in *symplectic geometry*).

Symplectic structure. Another name for the *symplectic form* of a *symplectic manifold*.

Symplectic subspace. See *symplectic complement*.

Symplectic vector space. A *vector space* V equipped with a *symplectic form* ω (often written as a pair (V, ω)). Because ω is nondegenerate and skew symmetric, if $\dim V$ is finite, it must be even.

Symplectomorphism (or *symplectic diffeomorphism*). Given *symplectic manifolds* (M, ω) and (N, ρ), a *symplectomorphism* $f : M \to N$ is a *diffeomorphism* such that for each $p \in M$, the *derivative map* $Df : T_pM \to T_{f(p)}N$ is a *symplectic linear map*. A symplectomorphism is the modern version of a *canonical* transformation in *classical mechanics*: it preserves *Hamiltonian vector fields*.

System. Often used as an abbreviated form of *dynamical system*, *Hamiltonian system*, mechanical system, thermodynamical system, etc.

Tangent bundle. An even-dimensional *smooth manifold*, denoted TM, consisting of the *collection* of all *tangent spaces* TM_p of a smooth manifold M together with their *base points* $p \in M$. In *Lagrangian mechanics*, it is natural to view the *phase space* as the *tangent bundle* of the *configuration space* Q.

Tangent space. Given a point p on a *smooth manifold* M of dimension n, the *tangent space* of M at the *base point* p, denoted T_pM, is an n-dimensional *vector space* comprised of so-called 'tangent vectors.' It may be thought of as being 'attached' to M with 'center' at p, and with the same 'orientation' as M at p. By collecting together the tangent spaces based at all points of M, one forms the *tangent bundle* TM of M; this smooth manifold of dimension $2n$ is the natural setting for a *vector field* (hence an ODE).

Target set. Another name for the *codomain* of a *function*.

Taylor polynomial (after Brook Taylor). See *Taylor series*.

Taylor series. Let $D \subset \mathbb{C}$ be an *open domain* and $f : D \to \mathbb{C}$ a *smooth function* (i.e., $f \in C^{\infty}(D)$). Given any $a \in D$, the *Taylor series* of f at a is the series $\sum_0^{\infty}[f^{(k)}(a)/k!](x - a)^k$. (The same definition applies to real functions, i.e., with $D \subset \mathbb{R}$ and $f : D \to \mathbb{R}$.) For $n \in \mathbb{N}_0$, the partial sum $T_{n,a}(x) = \sum_0^n[f^{(k)}(a)/k!](x-a)^k$ is called the nth degree *Taylor polynomial* of f at a. (Note that f need only be of *class* C^n at a to have a well-defined Taylor polynomial $T_{n,a}$ of degree n at a.)

The Taylor series of f at a may or may not *converge* to f in a *neighborhood* of a; if it does, f is said to be *analytic* at a. Taylor series and polynomials are of fundamental importance in analysis. Both objects generalize in a natural way to real or complex functions of several variables (though the generalizations can be somewhat cumbersome to write down).

Taylor's theorem. Here is a simple version of this fundamental theorem, using *little o* notation. Given $n \in \mathbb{N}_0$, suppose $a \in D$, where $D \subset \mathbb{C}$ is an *open domain* and $f : D \to \mathbb{C}$ is of *class* C^{n+1}. Then $f(x) - T_{n,a}(x) = o(|x - a|^n)$ as $x \to a$, where $T_{n,a}$ is the nth degree *Taylor polynomial* of f at a (see preceding entry). The real version of the theorem is the same, with $D \subset \mathbb{R}$ and $f : D \to \mathbb{R}$.

The difference $f(x) - T_{n,a}(x)$ is called the 'remainder,' and there are a number of ways of *estimating* its size as $x \to a$. In other words, in more detailed versions of *Taylor's theorem*, one gives *rigorous estimates* of the rate at which the absolute value of the remainder shrinks to zero as $x \to a$. These estimates make the theorem a powerful tool in analysis.

Thermal equilibrium. In thermodynamics and statistical mechanics, a *system* is in *thermal equilibrium* if its macroscopic *observables* (e.g. its temperature, pressure, volume) are constant in time.

Thermalization, thermalize. In thermodynamics and statistical mechanics, *thermalization* is a loose term for the process by which a *system* approaches *thermal equilibrium*. (The corresponding verb is *thermalize*.)

Time average. Let M be a *phase space*, $\varphi_t : M \to M$ a *flow* on M, and $f : M \to \mathbb{R}$ a *function* (sometimes called an *observable*). The T-*time average* of f along the flow φ is the function $\langle f \rangle_T : M \to \mathbb{R}$ defined by

$\langle f \rangle_T(x) := (1/T) \int_0^T f(\varphi_t(x)) \, dt$ (whenever the integral exists). When one says the *time average* of f without reference to T, one means the function $\langle f \rangle : D \to \mathbb{R}$ defined by $\langle f \rangle(x) := \lim_{T \to \infty} \langle f \rangle_T(x)$, where the *domain* D consists of the $x \in M$ for which the *limit* exists.

Titius-Bode law (often simply *Bode's law*, after Johann Bode and Johann Titius). The hypothesis that the semi-major axes (roughly the mean distances from the Sun) of the planets of our solar system follow a regular mathematical *rule* of the form $a_1 = 0.4$, $a_k = 0.4 + 0.3(2^{k-2})$ where k designates the planet and a_k the planet's semi-major axis in AU (astronomical units, i.e. multiples of the Earth's semi-major axis). Here one starts with $k = 1$ corresponding to Mercury, $k = 2$ to Venus, etc.

When J.D. Titius, then J.E. Bode, first clearly stated this purported law in the 1760s (Titius cited an earlier source; Bode cited Titius), only the visible five planets and Earth were known, and there was a gap in the predicted sequence corresponding to $k = 5$, between Mars and Jupiter. After the sensational discovery of Uranus in 1781, which fit very neatly into the $k = 8$ position, Bode urged astronomers to renew their search of the gap at $k = 5$. Two decades later, G. Piazzi found *Ceres* at almost precisely the predicted $k = 5$ distance from the Sun (and Ceres retained its status as a planet for more than half a century, until the discovery in the 1860s of many other minor bodies nearly in its orbit). In fact the first eight planets (counting Ceres) fit the Titius-Bode law with surprising accuracy. The largest deviation is for Mars, which falls short of its predicted semi-major axis by about 5 percent. But the discovery of Neptune in 1846 largely discredited Titius-Bode, as the giant planet is far short of its predicted location (though Pluto—now like Ceres classified as a 'dwarf planet'—is close to where Neptune should have been). Modern theories of planet formation regard Titius-Bode as more of a statistical likelihood than a law. But during the century in which it was believed to be more than merely statistical, it guided astronomers remarkably well.

Topological conjugacy. See *conjugacy*.

Topological dimension. This is the 'ordinary' (nonnegative integer) dimension of a (*connected*) *topological manifold*, defined as the dimension n of the Euclidean space $\mathbb{R}^n$ to which the manifold is *locally homeomorphic*.

Topological equivalence. See *conjugacy*.

Topological manifold. Intuitively speaking, a *topological manifold* is a *topological space* that looks like a portion of *Euclidean space* when viewed up close, but which may have a global nonlinear structure when viewed from afar. (The introduction of manifolds in *topology* may be seen as a recapitulation of the discovery that the earth is round rather than flat.) More precisely, a topological (sub)space $(X, \mathcal{T})$ is a topological manifold if it is a *Hausdorff space* with a *countable base*, each point p of which is contained in an *open neighborhood* U *homeomorphic* to a nonempty *open ball* B of a Euclidean space $\mathbb{R}^n$. A pair (U, h) of such a neighborhood and a *homeomorphism* $h : U \to B$ is called a *chart*, (or a 'system of *local coordinates* near p'), and a *(countable) collection* $\{(U_k, h_k)\}$ of charts covering X is an *atlas* of X (the colorful terminology is due to H. Whitney). If (U_1, h_1), (U_2, h_2) are charts in an atlas with $h_k : U_k \to B_k$ and the overlap $U_1 \cap U_2$ is nonempty, then the associated *transition map* (or 'coordinate change') is the *restriction* of $h_2 \circ h_1^{-1}$ to the *domain* $h_1(U_1 \cap U_2) \subset B_1$. By a theorem due to L.E.J. Brouwer, if a manifold is *connected*, then n (the dimension of the Euclidean space $\mathbb{R}^n$) is constant, in which case n is also the *(topological) dimension* of the manifold.

Topological space. A *topological space* is a pair $(X, \mathcal{T})$ consisting of a set X and a *collection* $\mathcal{T}$ (called the *topology* of X) of subsets of X such that (i) $\emptyset, X \in \mathcal{T}$ (ii) the union of any *collection* of sets in $\mathcal{T}$ is also in $\mathcal{T}$, and (iii) the intersection of any finite collection of sets in $\mathcal{T}$ is also in $\mathcal{T}$. In words, $\mathcal{T}$ contains the empty set and the whole set, and is closed under unions and finite intersections. When the topology $\mathcal{T}$ is understood, one simply says 'the topological space X.' Members of $\mathcal{T}$ are called *open sets* (or *neighborhoods*, if nonempty). The complement of an open set is called a *closed set*. An amusing truism about topological spaces is this: "A door is either open or closed, but in a topological space, a set may be open, closed, both, or neither."

Topology. [1] In the broad sense, *topology* is the part of mathematics that grew from modest beginnings in the 19th century into one of the major branches of mathematics today. It incorporates and generalizes parts of geometry (e.g. surfaces and their structure) and *analysis* (e.g. continuity of functions) and also makes extensive use of algebra (e.g. in homology and cohomology theory).

[2] In the narrow sense, a *topology* is the collection $\mathcal{T}$ of *open sets* of a *topological space*.

Torus (plural *tori*). A basic object in *topology* (in fact a torus is a *manifold*), and also in *Hamiltonian dynamics* because of its ubiquitous appearance in *integrable systems* and KAM theory. Begin with the so-called 'one-torus,' denoted $\mathbb{T}^1$ (or simply $\mathbb{T}$), which has the topological structure of a circle or one-dimensional loop, and is defined by $\mathbb{T} := \mathbb{R}/\mathbb{Z}$. This latter notation (read '$\mathbb{R}$ mod $\mathbb{Z}$') means that numbers on the real line $\mathbb{R}$ are identified 'mod 1' (see *mod, modular, modulo*); i.e., real numbers that differ by an integer are considered the same. In this way the real line is 'wrapped around' the circle of circumference one (many authors use a circle of circumference 2π). Once the 1-torus is defined, the n-torus is simply its n-fold Cartesian product: $\mathbb{T}^n := \mathbb{T} \times \mathbb{T} \times \cdots \times \mathbb{T}$, which may also be written $\mathbb{T}^n = \mathbb{R}^n/\mathbb{Z}^n$. The 1-torus $\mathbb{T}$ may be visualized as a circle, and the 2-torus $\mathbb{T}^2$ as the surface of a 'doughnut' (or as a square with opposite edges identified). Higher-dimensional tori ($n \geq 3$) can't be visualized directly in 3-dimensional space, but $\mathbb{T}^3$ may be thought of as a cube with opposite faces identified, and tori with $n > 3$ have a similarly simple structure.

(Note that in another part of topology—surface theory—the term 'n-torus' means an 'orientable surface of genus n' or 'sphere with n handles' instead.)

Total variation. See *bounded variation*.

Trajectory. See *orbit* (both senses).

Transformation. Yet another name for *function*, but often used for functions with the same *domain* and *codomain*.

Transition map. See *topological manifold*.

Transitive, transitivity. If X is a *topological space* and (X, φ) a *continuous dynamical system*, the *flow* φ is said to be (topologically) *transitive* if there is an $x \in X$ such that the *orbit* $\bigcup_{t \in \mathbb{R}} \varphi_t(x)$ is *dense* in X. (A similar definition applies to *discrete systems*.) Note that *transitivity* is ingredient (ii) in the provisional definition of *chaos*.

Transversal, transverse. Two *smooth submanifolds* M, N of an ambient finite-dimensional smooth manifold G are said to intersect *transversally* (or to be *transverse*) if, at each point p of their intersection, the union of their *tangent spaces* at p *spans* the tangent space of G at p (in symbols, span $(TM_p \cup TN_p) = TG_p$). When M and N intersect transversally (denoted $M \pitchfork N$) their intersection forms a submanifold of G with *codimension* equal to the sum of the codimensions of M and N. (And if the sum of codimensions of M and N exceeds $\dim G$, then the only way M and N can intersect transversally is to not intersect at all.) Note that when M and N intersect transversally, they are in *general position* in the sense that their transversal intersection is resistant to small *perturbations* of their location or shape (i.e., transversal intersections are *structurally stable*).

Twist condition. See *twist map*.

Twist map. Roughly speaking, a *twist map* is a *continuous map* of the annulus *onto* the annulus (the region between and including two concentric circles in the plane) which leaves the two boundary circles invariant, but rotates them by different (average) amounts (i.e., the map has distinct *rotation numbers* on the boundary circles). This last condition is called the *twist condition*, and it takes the place of the *nondegeneracy conditions* required in higher-dimensional KAM theorems.

Twist maps are important in KAM theory: it was for twist maps (of smoothness *class* C^{333} with the 'self-intersection property') that Moser gave the first detailed proof of a KAM theorem in [Mos62]. This showed, among other things, that KAM results apply to *systems* more general than *Hamiltonian systems* (twist maps naturally arise from the *Poincaré map* of an integrable Hamiltonian system with two *degrees of freedom*, but there are also twist maps that don't arise this way).

The *standard map* may be viewed as a twist map (by mapping a segment of the cylinder to the annulus, or more naturally, by expanding the concept of twist maps to include maps of the cylinder).

Uncountable, uncountably infinite. A set A is *uncountable* or *uncountably infinite* if it is *infinite* and not *countable* (i.e., if there is an *injection* $f : \mathbb{N}^* \to A$ but no *surjection* $g : \mathbb{N}^* \to A$).

Uniform convergence. *Convergence* in the *uniform norm*. More explicitly, if (M, d_M), (N, d_N) are *metric spaces* and $\{f_k\}_{k=0}^{\infty}$ is a *sequence* of functions $f_k : M \to N$, we say that $\{f_k\}$ converges *uniformly* on $A \subset M$ to $f : A \to N$ if the sequence of numbers $\|f_k - f\|_A$ converges to zero, where $\|\ \|_A$ is the *uniform norm* on A (see below). Note that uniform convergence is (considerably) *stronger* than *pointwise convergence*.

Uniform norm. Suppose V is a *normed vector space* with norm $\|\ \|$, and $\mathcal{F}$ is the *function space* consisting of *bounded* functions $f : A \to V$ with some common *domain* A. We define a *norm* on $\mathcal{F}$, called the *uniform norm* on A and denoted $\|\ \|_A$, by $\|f\|_A = \sup_{x \in A} \|f(x)\|$. (See *supremum*.)

Unstable manifold. Suppose (M, d) is a *metric space*, $\varphi : M \times \mathbb{R} \to M$ a *flow* and $A \subset M$ an *invariant set*. The *unstable manifold* of A is $W_u(A) := \{x \in M \mid \lim_{t \to -\infty} d(A, \varphi_t(x)) = 0\}$. (Note that this is almost the definition of a *stable manifold*, but with time reversed.)

Upper bound. If A is a set of real numbers, an *upper bound* for A is a real number that is greater than or equal to all numbers in A. (Note that an upper bound need not belong to A.)

Vector field. Speaking loosely, a *vector field* is an n-dimensional *base space* with an n-dimensional 'vector' attached at each point (each vector indicating both magnitude and direction). This is most simply represented by a *map* $f : D \to \mathbb{R}^n$ where each $x \in D \subset \mathbb{R}^n$ is a point of the base space D and $f(x) \in \mathbb{R}^n$ is the attached vector. If such f is *regular* enough (say, *locally Lipschitz continuous*), we can use it as the right-hand side of the (first-order *autonomous*) ODE $\dot{x} = f(x)$. Solutions of the ODE may be viewed geometrically as curves everywhere tangent to the vectors, and parametrized by time so that their speed at any point is the magnitude of the vector attached to that point. (One can of course also consider non-autonomous vector fields $f(t, x)$, or vector fields depending on other parameters, as the right-hand sides of ODEs.)

In the more abstract setting of *differential topology*, the base space of a vector field is a *smooth manifold* M and the vector attached to an arbitrary point $p \in M$ belongs to the *tangent space* $T_p M$ (such a vector field is also called a 'section of the *tangent bundle* TM'). In this way one defines ODEs—hence *flows*—on smooth manifolds. Yet ultimately, using *charts*, one reduces this situation (at least *locally*) to vector fields on $\mathbb{R}^n$.

Vector space (also called a *linear space*). A *collection* V of things called 'vectors' which may be added together or multiplied by numbers in a *field* $\mathbb{F}$ to produce other vectors. (One says that "V is closed under *linear combinations*.") To specify the field, one says "V is a vector space over $\mathbb{F}$." More commonly, when the associated field $\mathbb{F} = \mathbb{R}$, V is called a *real vector space*; when $\mathbb{F} = \mathbb{C}$, it is called a *complex vector space*. Note that a vector space is commonly *normed*, in which case it is automatically a *topological space*. All vector spaces have a zero vector (or 'center').

Wandering point, wandering set. These concepts were introduced by G.D. Birkhoff in [Bir27]. Consider the *discrete dynamical system* (X, f), where X is a *Hausdorff space* and $f : X \to X$ is a *map*. The point $p \in X$ is *wandering* in (X, f) if there is a *neighborhood* U of p and an $N \in \mathbb{N}^*$ such that for all $n \geq N$, $f^n(U) \cap U = \emptyset$. (A similar definition applies to *continuous systems*.) The set of all wandering points of a dynamical system is its *wandering set*. When a *measure* on X is available, it may be used to characterize *dissipative systems* as those whose wandering sets have positive measure.

Weak KAM theory. This relatively new subject partly extends Aubry-Mather theory (cf. §6.1 above) to higher dimensions, and also studies so-called 'viscosity solutions' of certain Hamilton-Jacobi partial differential equations.

Weak mixing. A *measure-preserving flow* $\varphi : M \times \mathbb{R} \to M$ on a finite *measure space* (M, Σ, μ) is *weakly mixing* if for any $A, B \in \Sigma$, we have $\lim_{t\to\infty}(1/t) \int_0^t |\mu(\varphi_{-s}(A) \cap B) - \mu(A) \cdot \mu(B)| ds = 0$. (A similar definition applies to *maps*.) Weak mixing implies *ergodicity*, and is implied by *mixing*.

Weak solution. A special sort of solution of a *differential equation* (DE) that may not be a *strong solution* (i.e., it may fail to satisfy the DE in the usual, classical sense), and instead satisfies some 'weak formulation' of the DE. The most common weak formulation of a DE involves 'generalized functions' or the 'theory of distributions' (initiated respectively by S. Sobolev and L. Schwartz in the 1930s, after pioneering work of P. Dirac). More recently, other sorts of weak formulations have been used, leading most notably to so-called 'viscosity solutions' for certain DEs. It is this latter type of weak solution that appears in *weak KAM theory*.

Weaker, weakest. If elements of set A satisfy condition α and elements of set B satisfy condition β, to say that α is *weaker* than β means that $B \subset A$. (We may of course also say that β is *stronger* than α.) To say that α is the *weakest* condition under which γ holds means that γ does not hold under any condition β weaker than α. (See also *stronger, strongest*.)

Weierstrass M-test (after Karl Weierstrass). A simple means of establishing the *uniform convergence* of a *series* of *functions*. Suppose $\{f_n\}$ is a *sequence* of real- or complex-valued functions with common *domain A*. The *Weierstrass M-test* says that if $\sum_{n=0}^{\infty} M_n < \infty$, where $M_n = \sup_{x \in A} |f_n(x)|$, then the series $\sum_{n=0}^{\infty} f_n(x)$ converges uniformly on A.

Whitney smoothness (after Hassler Whitney). A means of defining *smoothness* of a *function* whose *domain* is a *closed set* (the usual approach requires the domain to be *open*). If X is a *metric space* and $F \subset X$ a closed set, to say that $f : F \to \mathbb{R}$ is *Whitney smooth* means that there is an *extension* g of f that is *smooth* in the usual sense. In other words, there is an open set $G \subset X$ with $F \subset G$ and a smooth function $g : G \to \mathbb{R}$ such that, for every $x \in F$, $g(x) = f(x)$. (Note that the extension g is not unique, since its behavior on the open set $G \setminus F$ is unspecified.) See §6.1.4 of [BroeHuS96] for a discussion of Whitney smoothness, including definitions of Whitney *smoothness classes* as they are used in KAM theory.

Wolf prize in mathematics (after Ricardo Wolf). A prestigious prize awarded most years since 1978 by the Wolf Foundation in Israel. It was awarded to all the principal founders of KAM: Siegel, Kolmogorov, Moser, and Arnold (Siegel was one of the first two recipients in 1978).

Bibliography

[AbdA01] K. Abdullah and A. Albouy, On a strange resonance noticed by M. Herman, *Regul. Chaotic Dyn.* **6** (4) (2001) 421–432.

[Ab67] R.H. Abraham, *Foundations of Mechanics*, Benjamin, Reading, Mass., 1967.

[Ab94] R.H. Abraham, *Chaos, Gaia, Eros: A chaos pioneer uncovers the three great streams of history*, Harper, San Francisco, 1994.

[AbM78] R.H. Abraham and J.E. Marsden, *Foundations of Mechanics*, 2nd Ed., Benjamin/Cummings, Reading, Mass., 1978.

[AbU00] R.H. Abraham and Y. Ueda (Eds.) *The Chaos Avant-Garde, Memories of the Early Years of Chaos Theory*, World Scientific Series on Nonlinear Science, Series A, Vol. 39, World Scientific Publishers, Singapore/River Edge, NJ, 2000.

[Acz06] A. Aczel, *The Artist and the Mathematician: The Story of Nicolas Bourbaki, the Genius Mathematician Who Never Existed*, Thunder's Mouth Press, New York, 2006.

[Al02] A. Albouy, Lectures on the two-body problem, in *Classical and Celestial Mechanics: The Recife Lectures* (H. Cabral and F. Diacu, Eds.), pp. 63–116, Princeton University Press, Princeton, NJ, 2002.

[Ap25] P. Appell, *Henri Poincaré*, Librairie Plon, Paris, 1925.

[Ar61] V.I. Arnold, Small divisors I: Mappings of the circle onto itself (Russian), *Izvest. Akad. Nauk SSSR*, Ser. Mat. 25 (1) (1961) 21–86; Corrigenda: *ibid.* 28 (2) (1964) 479–480. [English translation: *Amer. Math. Soc. Transl.*, Ser. 2, **46** (1965) 213–284.]

[Ar62] V.I. Arnold, On the classical perturbation theory and the stability problem of planetary systems (Russian), *Dokl. Akad. Nauk SSSR* 145 (1962), 487–490. [English translation: *Soviet Math. Dokl.* 3 (1962), 1008–1012.]

[Ar63a] V.I. Arnold, Proof of a theorem of A.N. Kolmogorov on the preservation of conditionally periodic motions under a small perturbation of the Hamiltonian (Russian), *Uspehi Mat. Nauk* **18** (1963), no. 5 (113), 13–40. [English translation: *Russian Math. Surveys* **18** (1963), no. 5, 9–36.]

[Ar63b] V.I. Arnold, Small denominators and problems of stability of motion in classical and celestial mechanics (Russian), *Uspehi Mat. Nauk* **18** (1963), no.

6 (114), 91–192. [English translation: *Russian Math. Surveys* **18** (1963), no. 6, 85–191.]

[Ar64] V.I. Arnold, Instability of dynamical systems with many degrees of freedom (Russian), *Dokl. Akad. Nauk SSSR* 156 (1964), 9–12. [English translation: *Soviet Math. Dokl.* 5 (1964), 581–585.]

[Ar68] V.I. Arnold, A stability problem and ergodic properties of classical dynamical systems (Russian), in *Proc. Inter. Congr. Math.* (Moscow, 1966), Mir, Moscow, 1968, pp. 387–392 [English translation: *Amer. Math. Soc. Transl.* **70** (2) (1968), 5–11.]

[Ar73] V.I. Arnold, *Ordinary Differential Equations* [translated from Russian], MIT Press, Cambridge, Mass., 1973.

[Ar78–97] V.I. Arnold, *Mathematical Methods of Classical Mechanics* [translated from Russian], Springer-Verlag, Berlin, 1978 (2nd edition 1997).

[Ar83–88] V.I. Arnold, *Geometrical Methods in the Theory of Ordinary Differential Equations* [translated from Russian], Springer-Verlag, New York, 1983 (2nd edition 1988).

[Ar90] V.I. Arnold, *Huygens and Barrow, Newton and Hooke, Pioneers in mathematical analysis and catastrophe theory from evolvents to quasicrystals* [translated from Russian], Birkhäuser Verlag, Basel, 1990.

[Ar93] V.I. Arnold, On A.N. Kolmogorov, in *Reminiscences about Kolmogorov* (Russian), 144–172, Fizmatlit "Nauka," Moscow, 1993. [English translation: in *Golden Years of Moscow Mathematics*, 129–153, Hist. Math. **6**, Amer. Math. Soc., Providence, RI, and London Math. Soc., London, 1993 (2nd Ed. 2007); another English translation: in *Kolmogorov in Perspective*, 89–108, Hist. Math. **20**, Amer. Math. Soc., Providence, RI, 2000.]

[Ar94] V.I. Arnold, Mathematical problems in classical physics, in *Trends and Perspectives in Applied Mathematics*, Appl. Math. Sc. Series 100, Springer-Verlag, New York, 1994, pp. 1–20.

[Ar97] V.I. Arnold, From superpositions to KAM theory (Russian), in *Vladimir Igorevich Arnold. Selected–60*, PHASIS, Moscow, 1997, pp. 727–740 [English translation (with translator's remarks) by M.B. Sevryuk, available online at http://www.ma.utexas.edu/mp_arc-bin/mpa?yn=12-81 with plan to appear in FAOM (*Functional Analysis and Other Mathematics*) to be published by PHASIS.]

[Ar98a] V.I. Arnold, On the teaching of mathematics (Russian) *Uspekhi Mat. Nauk* **53** (1) (1998) 229–234. [English translation: *Russian Math. Surveys* **53** (1) (1998) 229–236; also available online.]

[Ar98b] V.I. Arnold, Higher dimensional continued fractions (Russian) [English translation: *Regul. Chaotic Dyn.* **3** (3) (1998) 10–17.]

[Ar99] V.I. Arnold, From Hilbert's superposition problem to dynamical systems, in *The Arnoldfest, Proceedings of a Conference in Honour of V.I. Arnold for his Sixtieth Birthday* (Toronto, ON, June 15–21, 1997), E. Bierstone, B. Khesin, A. Khovanskii, and J.E. Marsden, Eds., Fields Institute Communications, Vol. 24. American Mathematical Society, Providence, RI, 1999, pp. 1–18.

[Ar02] V.I. Arnold, The mathematical duel over Bourbaki (Russian) *Vestnik*

Ross. Akad. Nauk **72** (3) (2002) 245–250.

[Ar06] V.I. Arnold, The underestimated Poincaré, *Uspekhi Mat. Nauk* **61** (1) (2006) 3–24 (Russian) [English translation: Forgotten and neglected theories of Poincaré, *Russian Math. Surveys* **61** (1) (2006) 1–18.]

[ArA68] V.I. Arnold and A. Avez, *Ergodic Problems of Classical Mechanics* [translated from French], W.A. Benjamin, New York/Amsterdam, 1968.

[ArKN06] V.I. Arnold, V.V. Kozlov, and A.I. Neishtadt, *Mathematical Aspects of Classical and Celestial Mechanics, Third Edition* [translated from Russian] (Encyclopedia of Mathematical Sciences, Vol. 3; Dynamical Systems III) Springer-Verlag, Berlin-Heidelberg, 2006.

[ArV89] V.I. Arnold and V.A. Vasil'ev, Newton's Principia read 300 years later, *Notices Amer. Math. Soc.* **36** (9) (1989) 1148–1154.

[ArrP90] D.K. Arrowsmith and C.M. Place, *An Introduction to Dynamical Systems*, Cambridge University Press, Cambridge, UK, 1990.

[AubiD02] D. Aubin and A. Dahan Dalmedico, Writing the history of dynamical systems and chaos: *longue durée* and revolution, disciplines and cultures, *Historia Mathematica* **29** (2002) 1–67.

[AubrL83] S. Aubry and P.V. LeDaeron, The discrete Frenkel-Kontorova model and its extensions. I. Exact results for the ground-states, *Phys. D* **8** (3) (1983) 381–422.

[Bai99] R.-L. Baire, Sur les fonctions de variables réelles (French) [On functions of real variables] *Annali di Mat.* **3** Ser. 3 (1899) 1–123.

[Bal88] P.R. Baldwin, Soft billiard systems, *Physica* D **29**, no. 3 (1988) 321–342.

[BamP06] D. Bambusi and A. Ponno, On metastability in FPU, *Comm. Math. Phys.* **264** (2006), 539–561.

[Barr66] R. Barrar, A Proof of the convergence of the Poincaré-von Zeipel procedure in celestial mechanics, *Amer. J. Math.* **88** (1966) 206–220.

[Barr86] R. Barrar, KAM today, in *Local and Global Methods of Nonlinear Dynamics* (Silver Spring, Md., 1984), pp. 40–48, Lecture Notes in Phys. Vol. 252, Springer-Verlag, Berlin, 1986.

[Barro97] J. Barrow-Green, *Poincaré and the Three Body Problem* (History of Mathematics, Vol. 11) Amer. Math. Soc., Providence, RI, 1997.

[BartG02] M. Bartuccelli and G. Gentile, Lindstedt series for perturbations of isochronous systems: a review of the general theory, *Rev. Math. Phys.* **14** (2) (2002) 121–171.

[BazMT89] A. Bazzani, S. Marmi, and G. Turchetti, Nekhoroshev estimate for isochronous non resonant symplectic maps, *Celestial Mech. Dynam. Astronom.* **47** (4) (1989/90) 333–359.

[Bel37] E.T. Bell, *Men of Mathematics*, Simon and Schuster, New York, 1937.

[Belli56] A. Bellivier, *Henri Poincaré, ou la vocation souveraine* (French) [Henri Poincaré, or sovereign vocation] Librairie Gallimard, Paris, 1956.

[Ben01] G. Benettin, Moti ordinati e moti caotici (Italian) [Ordered motions and chaotic motions], *Archimede*, no. 4 (2001) 171–185.

[Ben05a] G. Benettin, Physical applications of Nekhoroshev theorem and exponential estimates, in *Hamiltonian Dynamics, Theory and Applications* (Lectures at the C.I.M.E.-E.M.S. Summer School held in Cetraro, Italy, July

1–10, 1999; A. Giorgilli, Ed.) pp. 1–76, Lecture Notes in Mathematics Vol. 1861, Springer-Verlag, Berlin-Heidelberg, 2005.

[Ben05b] G. Benettin, Time scales for energy equipartition in a two-dimensional FPU model, *Chaos* **15** (1) 015108 (2005) 10 pp.

[BenCGG08] G. Benettin, A. Carati, L. Galgani and A. Giorgilli, The Fermi-Pasta-Ulam problem and the metastability perspective, in *The Fermi-Pasta-Ulam Problem: A Status Report*, G. Gallavotti et al., Eds., pp. 152–189, Lecture Notes in Physics, Vol. 728, Springer-Verlag, Berlin, 2008.

[BenG86] G. Benettin and G. Gallavotti, Stability of motions near resonances in quasi-integrable Hamiltonian systems, *J. Statist. Phys.* **44** (3–4) (1986) 293–338.

[BenGG85a] G. Benettin, L. Galgani, and A. Giorgilli, Poincaré's non-existence theorem and classical perturbation theory for nearly integrable Hamiltonian systems, in *Advances in Nonlinear Dynamics and Stochastic Processes* (R. Livi and A. Politi, Eds.) pp. 1–22, World Scientific, Singapore, 1985.

[BenGG85b] G. Benettin, L. Galgani, and A. Giorgilli, A proof of Nekhoroshev's theorem for the stability times in nearly integrable Hamiltonian systems, *Celestial Mech.* **37** (1985) 1–25.

[BenGG87] G. Benettin, L. Galgani, and A. Giorgilli, Exponential law for the equipartition times among translational and vibrational degrees of freedom, *Phys. Lett. A* **120** (1) (1987) 23–27.

[BenGG87–89] G. Benettin, L. Galgani, and A. Giorgilli, Realization of holonomic constraints and freezing of high frequency degrees of freedom in the light of classical perturbation theory, Part I, *Comm. Math. Phys.* **113** (1987) 87–103; *ibid.*, Part II, *Comm. Math. Phys.* **121** (1989) 557–601.

[BenGGS84] G. Benettin, L. Galgani, A. Giorgilli, and J.-M. Strelcyn, A proof of Kolmogorov's theorem on invariant tori using canonical transformations defined by the Lie method, *Il Nuovo Cimento*, **79** B (2) (1984) 201–223.

[BercGG04] L. Berchialla, L. Galgani and A. Giorgilli, Localization of energy in FPU chains, *Disc. Contin. Dynam. Sys.* A **11** (2004), 855–866.

[BermI05] G.P. Berman and F.M. Izrailev, The Fermi-Pasta-Ulam problem: Fifty years of progress *Chaos* **15** (1) 015104 (2005) 18 pp.

[Bernal54] J.D. Bernal, *Science in History*, Vol. 1, Penguin, London, 1954.

[Bernar96] P. Bernard, Perturbation d'un hamiltonien partiellement hyperbolique (French) [Pertubation of a partially hyperbolic Hamiltonian] *C.R. Acad. Sci. Paris*, tome 323, Série I (1996) 189–194.

[Bernar08] P. Bernard, The dynamics of pseudographs in convex Hamiltonian systems, *J. Amer. Math. Soc.* **21**, no. 3 (2008) 615–669.

[Bernar11] P. Bernard, Large normally hyperbolic cylinders in a priori stable Hamiltonian systems, *Ann. Henri Poincaré* **11**, no. 5 (2010) 929–942.

[Berr78] M.V. Berry, Regular and irregular motion, in *AIP Conference Proceedings* **46** (1978) 16–120.

[BertBB03] M. Berti, L. Biasco, and P. Bolle, Drift in phase space: a new variational mechanism with optimal diffusion time, *J. Math. Pures Appl.*, **82** (2003) 613–664.

[Bes96] U. Bessi, An approach to Arnold's diffusion through the calculus of variations, *Nonlinear Analysis, Theory, Methods and Applications*, **26** (6) 1996, 1115–1135.

[Bes97a] U. Bessi, Arnold's example with three rotators, *Nonlinearity* **10** (1997) 763–781.

[Bes97b] U. Bessi, Arnold's diffusion with two resonances, *J. Differential Equations*, **137** (1997), no. 2, 211–239.

[BesCV01] U. Bessi, L. Chierchia, and E. Valdinoci, Upper bounds on Arnold diffusion times via Mather theory. *J. Math. Pures Appl.* (9) **80** (2001), no. 1, 105–129.

[BibP67] Yu.N. Bibikov and V.A. Pliss, The existence of invariant tori in the neighborhood of the zero solution of a system of ordinary differential equations (Russian) *Differentsialnye Uravneniya* **3** (1967), 1864–1881 [English translation: *Diff. Equat.* **3** (1967) 967–976.]

[Bir13] G.D. Birkhoff, Proof of Poincaré's last geometric theorem, *Trans. Amer. Math. Soc.* **14** (1913) 14–22.

[Bir25] G.D. Birkhoff, An extension of Poincaré's last geometric theorem, *Acta. Math.* **47** (1925) 297–311.

[Bir27] G.D. Birkhoff, *Dynamical Systems*, Amer. Math. Soc., Providence, RI, 1927. [Reprinted 1966, and available online]

[Bir31] G.D. Birkhoff, Proof of the ergodic theorem, *Proc. Nat. Acad. Sci. USA* **17** (1931) 656–660.

[Bir32] G.D. Birkhoff, Sur quelques courbes fermées remarquables (French) [On some remarkable closed curves], *Bull. Soc. Math. de France* **60** (1932) 1–26.

[Bir42] G.D. Birkhoff, What is the ergodic theorem? *Amer. Math. Monthly* **49** (1942) 222–226.

[BirK32] G.D. Birkhoff and B.O. Koopman, Recent contributions to ergodic theory, *Proc. Nat. Acad. Sci. USA* **18** (1932) 279–282.

[Bla01] A.D. Blaom, A geometric setting for Hamiltonian perturbation theory, *Mem. Amer. Math. Soc.* **153** (2001), no. 727, xviii+112 pp.

[BoccSBL70] P. Bocchieri, A. Scotti, B. Bearzi, and A. Loinger, Anharmonic chain with Lennard-Jones interaction, *Phys. Rev.* A **2** (1970) 2013–2019.

[Boch66] S. Bochner, *The Role of Mathematics in the Rise of Science*, Princeton University Press, Princeton, NJ, 1966.

[BogM61] N.N. Bogoliubov, Y.A. Mitropolskiy, *Asymptotic Methods in the Theory of Non-Linear Oscillations* [translated from Russian], Gordon and Breach, New York, 1961.

[BoloT99] S. Bolotin and D. Treschev, Unbounded growth of energy in nonautonomous Hamiltonian systems, *Nonlinearity* **12** (2) (1999) 365–388.

[Bolt71] L. Boltzmann, Einige allgemeine Sätze über das Wärmegleichgewicht (German) [Some general results on thermal equilibrium], *Wien. Ber.* **63** (1871) 679–711.

[Bolt84] L. Boltzmann, Über die Eigenschaften monzyklischer und anderer damit verwandter Systeme (German) [On the properties of monocyclic and other related systems], *Wien. Ber.* **90** (1884) 291–321; reprinted in *Wissenschaftliche Abhandlungen* Vol. III, F.P. Hasenhörl, Ed., Chelsea, New York, 1968, pp. 122–152.

[Bolt87] L. Boltzmann, *Jrnl. f. d. reine und angewandte Math.* C. 208 (1887).

[Bolt95] L. Boltzmann, On certain questions of the theory of gases, *Nature* **51** (1895) 413–15.

[Bolt96–98] L. Boltzmann, *Vorlesungen über Gastheorie* (German), J.A. Barth, Leipzig, (Teil I) 1896, (Teil II) 1898 [English translation (Parts I and II combined): *Lectures on Gas Theory*, translated by S. Brush, U. California Press, Berkeley, CA, 1964].

[Boo83] D.J. Boorstin, *The Discoverers*, Random House, New York, 1983.

[Bor98] A. Borel, Twenty-five years with Nicolas Bourbaki (1949–1973), *Notices Amer. Math. Soc.* **45** (3) (1998) 373–380.

[Bos85] J.-B. Bost, Tores invariants des systèmes dynamiques hamiltoniens (d'après Kolmogorov, Arnold, Moser, Rüssmann, Zehnder, Herman, Pöschel, ...) (French) [Invariant tori of Hamiltonian dynamical systems (following Kolmogorov, Arnold, Moser, Rüssmann, Zehnder, Herman, Pöschel, ...] *Sém. Bourbaki*, no. 639 (1984–85) 113–157; also in *Astérisque*, no. 133–134 (1986) 113–157.

[BounF13] A. Bounemoura and S. Fischler, A Diophantine duality applied to the KAM and Nekhoroshev theorems, *Math. Z.* DOI 10.1007/s00209–013–1174–5 (33 pp. electronic, 2013) [also to appear on paper].

[BounM13] A. Bounemoura and J.-P. Marco, Existence de connexions hétéroclines entre tores isotropes contenus dans une chaîne d'anneaux hyperboliques pour les systèmes à trois degrés de liberté (French) [Existence of heteroclinic connections between isotropic tori contained in a chain of hyperbolic annuli for systems with three degrees of freedom], 2013 preprint (to appear).

[BounP12] A. Bounemoura and E. Pennamen, Instability for a priori unstable Hamiltonian systems: a dynamical approach, *Discrete Contin. Dynam. Sys.*, Ser. A, **32**, no. 3 (2012) 753–793.

[Bour95] J. Bourgain, Harmonic analysis and nonlinear partial differential equations, in *Proceedings of the International Congress of Mathematicians* (Zürich, 1994), Vol. 1, pp. 31–44, Birkhäuser, Basel, 1995.

[Bour99a] J. Bourgain, Nonlinear Schrödinger equations, in *Hyperbolic Equations and Frequency Interactions* (Park City, UT, 1995) pp. 3–157, Amer. Math. Soc., Providence, RI, 1999.

[Bour99b] J. Bourgain, Periodic solutions of nonlinear wave equations, in *Harmonic Analysis and Partial Differential Equations* (Chicago, IL, 1996), pp. 69–97, Univ. Chicago Press, Chicago, IL, 1999.

[Boy68] C.B. Boyer, *A History of Mathematics*, John Wiley and Sons, New York, 1968.

[BraaBH90] B.L.J. Braaksma, H.W. Broer, and G.B. Huitema, Toward a quasi-periodic bifurcation theory, *Mem. Amer. Math. Soc.* **83** (1990), no. 421, 83–170.

[Brau80] H. Braun, *Eine Frau und die Mathematik 1933–1949—Der Beginn einer wissenschaftlichen Laufbahn* (German) [A Woman and Mathematics 1933–1949—The Start of a Scientific Journey], Springer-Verlag, Berlin, 1980.

[BricGK99] J. Bricmont, K. Gawędzki, and A. Kupiainen, KAM theorem and quantum field theory, *Comm. Math. Phys.* **201** (1999) no. 3, 699–727.

[Broe04] H.W. Broer, KAM theory: the legacy of A.N. Kolmogorov's 1954 paper, *Bull. Amer. Math. Soc. (N.S.)* **41**, no. 4 (2004) 507–521.

[Broe10] H.W. Broer, Do Diophantine vectors form a Cantor bouquet? *J. Difference Equ. Appl.* **16** (2010), no. 5–6, 433–434.

[BroeHa09] H.W. Broer and H. Hanßmann, Hamiltonian perturbation theory (and transition to chaos), in *Encyclopedia of Complexity and Systems Science* (10 vols.), R.A. Meyers, Ed., Springer-Verlag, New York, Heidelberg, Berlin, 2009.

[BroeHaS10] H.W. Broer, H. Hanßmann, and M. Sevryuk, Preface: KAM theory and its applications, *Discrete Contin. Dyn. Syst.*, Ser. S, **3**, no. 4 (2010) i–iii.

[BroeHu91] H.W. Broer and G.B. Huitema, A proof of the isoenergetic KAM-theorem from the "ordinary" one, *J. Differential Equations* **90** (1991), no. 1, 52–60.

[BroeHu95] H.W. Broer and G.B. Huitema, Unfoldings of quasi-periodic tori in reversible systems, *J. Dynam. Differential Equations* **7** (1995), no. 1, 191–212.

[BroeHuS96] H.W. Broer, G.B. Huitema, and M.B. Sevryuk, *Quasi-Periodic Motions in Families of Dynamical Systems*, Springer-Verlag, Berlin, 1996.

[BroeHuT90] H.W. Broer, G.B. Huitema, and F. Takens, Unfoldings of quasi-periodic tori, *Mem. Amer. Math. Soc.* **83** (1990), no. 421, 1–81.

[BroePS03] H.W. Broer, J. Puig, and C. Simó, Resonance tongues and instability pockets in the quasi-periodic Hill-Schrödinger equation, *Commun. Math. Phys.* **641** (2003) 467–503.

[BroeS10] H.W. Broer and M.B. Sevryuk, KAM theory: quasi-periodicity in dynamical systems, in *Handbook of Dynamical Systems, Vol. 3*, H.W. Broer, B. Hasselblatt, and F. Takens, Eds., North-Holland (Elsevier), Amsterdam, 2010, Chapter 6, pp. 249–344.

[BroeT10] H.W. Broer and F. Takens, *Dynamical Systems and Chaos* (Applied Mathematical Sciences Vol. 172), Springer-Verlag, New York, 2010.

[Bruno65] A.D. Bruno, Convergence of transformations of differential equations to normal form (Russian), *Dokl. Akad. Nauk. SSSR* **165** (1965) 987-989. [English translation: *Soviet Math. Dokl.* **6** (1965), 1536–1538.]

[Bruno67] A.D. Bruno, Divergence of transformations to normal form of differential equations (Russian), *Dokl. Akad. Nauk SSSR* **174** (1967) 1003–1006. [English translation: *Soviet Math. Dokl.* 8 (1967), 692–695.]

[Bruno71] A.D. Bruno, Analytic form of differential equations. I, II (Russian), *Trudy Moskov. Mat. Obšč.* **25** (1971), 119–262; *ibid.* **26** (1972), 199–239. [English translations: *Trans. Moscow Math. Soc.* **25** (1971), 131–288 (1973); *ibid.* **26** (1972), 199–239 (1974).]

[Bruns87] H. Bruns, Über die Integrale des Vielkörper-Problems (German) [On integrals of the many-body problem], *Acta Math.* **11** (1887) 25–96.

[Brus83] S.G. Brush, *Statistical Physics and the Atomic Theory of Matter, from Boyle and Newton to Landau and Onsager*, Princeton University Press, Princeton, NJ, 1983.

[Brus03] S.G. Brush, *The Kinetic Theory of Gases, an anthology of classic papers with historical commentary,* Imperial College Press, London, 2003.

[Cam87] D.K. Campbell, Hamiltonian chaos and statistical mechanics, in *Los Alamos Science* No. 15, Special Issue: Stanislaw Ulam 1909–1984 (1987) 242–245.

[CamRZ05] D.K. Campbell, P. Rosenau, and G.M. Zaslavsky, Introduction: The Fermi-Pasta-Ulam problem—The first fifty years, *Chaos* **15** (1) 015101 (2005) 4 pp.

[Can80–08] M.B. Cantor, *Vorlesungen über Gesichte der Mathematik,* Bände 1–4 (German) [Lectures on the History of Mathematics] Vols. 1–4, Teubner, Leipzig, 1880, 1882, 1894–98, 1908.

[Car07] A. Carati, An averaging theorem for Hamiltonian dynamical systems in the thermodynamic limit, *J. Stat. Phys.* **128** (2007) 1057–1077.

[CarGG05] A. Carati, L. Galgani, and A. Giorgilli, The Fermi-Pasta-Ulam problem as a challenge for the foundations of physics, *Chaos* **15** (1) 015105 (2005) 8 pp.

[Cel09a] A. Celletti, Perturbation theory in celestial mechanics, in *Encyclopedia of Complexity and Systems Science* (10 vols.), R.A. Meyers, Ed., Springer-Verlag, New York, Heidelberg, Berlin, 2009.

[Cel09b] A. Celletti, *Stability and Chaos in Celestial Mechanics,* Springer-Praxis, New York, 2009.

[Cel13] A. Celletti, Celestial mechanics: From antiquity to modern times, to appear in *Celestial Mechanics* (A. Celletti, Ed.) a volume of the *Encyclopedia of Life Support Systems* (EOLSS), Developed under the Auspices of the UNESCO, Eolss Publishers, Oxford, UK, [http://www.eolss.net] (2013).

[CelC88] A. Celletti and L. Chierchia, Construction of analytic KAM surfaces and effective stability bounds, *Comm. Math. Phys.* **118** (1) (1988) 119–161.

[CelC06] A. Celletti and L. Chierchia, KAM tori for N-body problems: a brief history, *Celestial Mech. Dynam. Astronom.* **95** (1–4) (2006) 117–139.

[CelC07] A. Celletti and L. Chierchia, KAM stability and celestial mechanics, *Mem. Amer. Math. Soc.* **187** (2007), no. 878, viii+134 pp.

[CelP07] A. Celletti and E. Perozzi, *Celestial Mechanics: The Waltz of the Planets,* Springer-Praxis, New York, 2007.

[CelG91] A. Celletti and A. Giorgilli, On the stability of the Lagrange points in the spatially restricted problem of three bodies, *Celestial Mech. Dynam. Astronom.* **50** (1991) 31–58.

[Cer98] C. Cercignani, *Ludwig Boltzmann, the Man Who Trusted Atoms,* Oxford University Press, Oxford, UK, 1998.

[Ces98] D. Cesarani, *Arthur Koestler: The Homeless Mind,* W. Heinemann, London, 1998.

[ChaGL10] E. Charpentier, E. Ghys, and A. Lesne (Eds.), *The Scientific Legacy of Poincaré* (History of Math., Vol. 36) American Mathematical Society, Providence, RI, 2010 [Translated from French: *L'héritage scientifique de Poincaré,* Editions Belin, Paris, 2006].

[Chenc02] A. Chenciner, Michel Herman, la mécanique céleste et quelques souvenirs (French) [Michael Herman, celestial mechanics and some recollections] *SMF Gazette* **88** avril (2001) 83–89. [available online]

[Chenc07] A. Chenciner, The three-body problem, *Scholarpedia* 2(10):2111 (2007).

[Chenc12] A. Chenciner, Poincaré and the three-body problem, in *Poincaré, 1912–2012, Séminaire Poincaré XVI*, Ecole Polytechnique, Paris, 2012, pp. 45–133.

[Cheng11] C.-Q. Cheng, Non-existence of KAM torus, *Acta Math. Sin.* (Engl. Ser.) **27** (2011), no. 2, 397–404.

[Cheng13] C.-Q. Cheng, Arnold diffusion in nearly integrable systems, 2013 preprint (to appear).

[ChengS94] C.-Q. Cheng and Y.S. Sun, Existence of KAM tori in degenerate Hamiltonian systems, *J. Differential Equations* **114** (1) (1994), 288–335.

[ChengW13] C.-Q. Cheng and L. Wang, Destruction of Lagrangian torus for positive definite Hamiltonian systems, *Geom. Funct. Anal.* **23** (2013), no. 3, 848–866.

[ChengY04] C.-Q. Cheng and J. Yan, Existence of diffusion orbits in a priori unstable Hamiltonian systems, *J. Differ. Geom.* **67** (3) (2004), 457–517.

[ChengY08] C.-Q. Cheng and J. Yan, Variational construction of diffusion orbits in convex Hamiltonian systems with multiple degrees of freedom, in *Third International Congress of Chinese Mathematicians*, Parts 1, 2, AMS/IP Stud. Adv. Math., 42, pt. 1, 2, Amer. Math. Soc., Providence, RI, 2008, pp. 11–27.

[ChengY09] C.-Q. Cheng and J. Yan, Arnold diffusion in Hamiltonian systems: a priori unstable case, *J. Differential Geom.* **82** (2) (2009) 229–277.

[Cher64] T.M. Cherry, A singular case of iteration of analytic functions: A contribution to the small-divisor problem, in *Nonlinear Problems of Engineering*, Academic Press, New York, 1964, pp. 29–50.

[Chi08] L. Chierchia, Kolmogorov's 1954 paper on nearly-integrable Hamiltonian systems, *Regul. Chaotic Dyn.* **13** (2) (2008) 130–139.

[Chi09a] L. Chierchia, Meeting Jürgen Moser, *Regul. Chaotic Dyn.* **14** (1) (2009) 5–6.

[Chi09b] L. Chierchia, Kolmogorov-Arnol'd-Moser (KAM) theory, in *Encyclopedia of Complexity and Systems Science* (10 vols.), R.A. Meyers, Ed., Springer-Verlag, New York, Heidelberg, Berlin, 2009.

[Chi13] L. Chierchia, The planetary N-body problem, to appear in *Celestial Mechanics* (A. Celletti, Ed.) a volume of the *Encyclopedia of Life Support Systems* (EOLSS), Developed under the Auspices of the UNESCO, Eolss Publishers, Oxford, UK, [http://www.eolss.net] (2013).

[ChiF94] L. Chierchia and C. Falcolini, A direct proof of a theorem by Kolmogorov in Hamiltonian systems, *Ann. Scuola Norm. Sup. Pisa Cl. Sci. Serie IV*, **21**, no. 4 (1994), 541–593.

[ChiF96] L. Chierchia and C. Falcolini, Compensations in small divisor problems, *Commun. Math. Phys.* **175** (1996), 135–160.

[ChiG82] L. Chierchia and G. Gallavotti, Smooth prime integrals for quasi-integrable Hamiltonian systems, *Nuovo Cimento B* (11) 67 (1982), no. 2, 277–295.

[ChiG94] L. Chierchia and G. Gallavotti, Drift and Diffusion in phase space, *Ann. Inst. H. Poincaré Phys. Théor.* **60** (1994) 1–144.

[ChiG98] L. Chierchia and G. Gallavotti, Erratum for "Drift and diffusion in phase space," *Ann. Inst. H. Poincaré* **68** (1998) 135.

[ChiM10] L. Chierchia and J.N. Mather, Kolmogorov-Arnold-Moser theory, *Scholarpedia* 5(9):2123 (2010) [available online].

[ChiPi10] L. Chierchia and G. Pinzari, Properly degenerate KAM theory (following V.I. Arnold), *Discrete Contin. Dyn. Syst.* Ser. S **3**, no. 4 (2010) 545–578.

[ChiPi11a] L. Chierchia and G. Pinzari, Deprit's reduction of the nodes revisited, *Celestial Mech. Dynam. Astronom.* **109** (2011), no. 3, 285–301.

[ChiPi11b] L. Chierchia and G. Pinzari, The planetary N-body problem: symplectic foliation, reductions and invariant tori, *Invent. Math.* **186** (2011), no. 1, 1–77.

[ChiPu09] L. Chierchia and F. Pusateri, Analytic Lagrangian tori for the planetary many-body problem, *Ergodic Theory Dynam. Systems* **29** (2009), no. 3, 849–873.

[Chir59] B.V. Chirikov, Resonance processes in magnetic traps, *Atomnaya Energiya* **6**: 630–637 (1959) (Russian) [English translation: *J. Nucl. Energy, Part C: Plasma Phys.* **1**: 253–260 (1960)].

[Chir69] B.V. Chirikov, Research in the theory of nonlinear resonance and stochasticity (Russian), [Preprint, Novosibirsk Institute for Nuclear Physics of the USSR Academy of Sciences, no. 267, 1969; English translation: CERN Transl. no. 71–40, 1971].

[Chir79] B.V. Chirikov, A universal instability of many-dimensional oscillator systems, *Phys. Rep.* **52** (5) (1979) 264–379.

[ChirS08] B.V. Chirikov and D. Shepelyansky, Chirikov standard map, *Scholarpedia* 3(3):3550 (2008).

[ChirV89] B.V. Chirikov and V.V. Vecheslavov, How fast is the Arnold diffusion? (Preprint, Budker Institute of Nuclear Physics, Novosibirsk, no. 72, 1989).

[ChirV92] B.V. Chirikov and V.V. Vecheslavov, Theory of the fast Arnold diffusion in many-frequency systems (Preprint, Budker Institute of Nuclear Physics, Novosibirsk, no. 25, 1992).

[ClC00] D. Clark and S.P.H. Clark, *Newton's Tyranny, the suppressed scientific discoveries of Stephen Gray and John Flamsteed*, W.H. Freeman, New York, 2000.

[CodL55] E.A. Coddington and N. Levinson, *Theory of Ordinary Differential Equations*, McGraw-Hill, New York, 1955.

[ConteM08] M. Conte and W.W. MacKay, *Introduction to the Physics of Particle Accelerators* (2nd Ed.), World Scientific, Singapore, 2008.

[Conto02] G. Contopoulos, *Order and Chaos in Dynamical Astronomy*, Springer-Verlag, Berlin, 2002.

[Cord08] B. Cordani, Arnold's web and diffusion in the Stark-quadratic-Zeeman problem, *Regul. Chaotic Dyn.* **13** (1) (2008) 46–56.

[CornFS82] I.P. Cornfeld, S.V. Fomin, and Ya.G. Sinai, *Ergodic Theory*, Springer-Verlag, Berlin, 1982.

[Cra00] W. Craig, *Problèmes de petits diviseurs dans les équations aux dérivées partielles* (French) [Small divisor problems in partial differential equations] (Panoramas et Synthèses **9**) Société Mathématique de France, Paris, 2000.

[CraW93] W. Craig and C. E. Wayne, Newton's method and periodic solutions of nonlinear wave equations, *Comm. Pure Appl. Math.*, **46** (11): 1409–1498 (1993).

[CraW94] W. Craig and C. E. Wayne, Periodic solutions of nonlinear Schrödinger equations and the Nash-Moser method, in *Hamiltonian Mechanics* (Torún, 1993), pp. 103–122, Plenum, New York, 1994.

[CranEL84] M.G. Crandall, L.C. Evans, and P.-L. Lions, Some properties of viscosity solutions of Hamilton-Jacobi equations, *Trans. Amer. Math. Soc.* **282** (2): 487–502 (1984).

[CranL83] M.G. Crandall and P.-L. Lions, Viscosity solutions of Hamilton-Jacobi equations, *Trans. Amer. Math. Soc.* **277** (1): 1–42 (1983).

[Craw84] E. Crawford, *The Beginnings of the Nobel Institution, The Science Prizes, 1901-1915*, Cambridge University Press, Cambridge, & Editions de la Maison des Sciences de l'Homme, Paris, 1984.

[Cre28] H. Cremer, Zum Zentrumproblem (German) [On the center problem], *Math. Ann.* **98** (1928) 151–163.

[Cres97] J. Cresson, A λ-lemma for partially hyperbolic tori and the obstruction property, *Lett. Math. Phys.* **42** (4) (1997) 363–377.

[Cres00] J. Cresson, Un λ-lemme pour des tores partiellement hyperboliques (French) [A λ-lemma for partially hyperbolic tori] *C.R. Acad. Sci. Paris Sér. I Math.* **331** (1) (2000) 65–70.

[CuB97] R.H. Cushman and L.M. Bates, *Global Aspects of Classical Integrable Systems*, Birkhäuser Verlag, Berlin/Boston, 1997.

[DaCC92] A. Dahan Dalmedico, J.-L. Chabert, and K. Chemla (Eds.), *Chaos et déterminisme* (French) [Chaos and Determinism], Editions du Seuil, Paris, 1992.

[Dar14] G. Darboux, Eloge historique d'Henri Poincaré (French) [Historical elegy of Henri Poincaré] *Mémoires de l'Académie des Sciences de l'Institut de France* **52**, Paris, Gauthier-Villars, 1914, lxxxi–cxlviii.

[DauR08] T. Dauxois and S. Ruffo, Fermi-Pasta-Ulam nonlinear lattice oscillations, *Scholarpedia* (2008) 3 (8): 5538

[Dav85] H. Davenport, Reminiscences of conversations with Carl Ludwig Siegel, *Math. Intelligencer* **7** (2) (1985) 76–79.

[DavisH81] P.J. Davis and R. Hersh, *The Mathematical Experience*, Birkhäuser, Boston, MA, 1981.

[delaL01] R. de la Llave, A tutorial on KAM theory, in *Smooth Ergodic Theory and its Applications* (Seattle, WA, 1999), Proc. Sympos. Pure Math., **69** (A. Katok et al., Eds.), Amer. Math. Soc., Providence, RI, 2001, pp. 175–292 [also published separately as Vol. 32 of University Lecture Series, Amer. Math. Soc., Providence, RI, 2003, with expanded, updated version available online].

[delaLGJV05] R. de la Llave, A. González, À. Jorba, and J. Villanueva, KAM theory without action-angle variables, *Nonlinearity* **18** (2005) 855–895.

[delaLR90] R. de la Llave and D. Rana, Accurate strategies for small divisor problems, *Bull. Amer. Math. Soc. (N.S.)* **22** (1) (1990) 85–90.

[Delam67] M. Delambre, Notice sur la vie et les œuvrages de M. le comte J.-L. Lagrange (French) [Note on the life and works of Count J.-L. Lagrange], in *Œuvres de Lagrange*, Vol. I, Paris, 1867.

[Delau60–67] C.-E. Delaunay, *La Théorie du mouvement de la lune* (French) [Theory of the Motion of the Moon] (2 volumes) Tome 1er, Mallet-Bachelier, Paris, 1860; Tome 2nd, Guathier-Villars, Paris, 1867.

[DelsG94] A. Delshams and P. Gutierrez, Nekhoroshev and KAM theorems revisited via a unified approach, in *Hamiltonian Mechanics: Integrability and Chaotic Behaviour*, (J. Seiminis, Ed., Proceedings of the NATO ARW on Hamiltonian Mechanics in Torun, Poland, 1993), NATO ASI series B, physics, Vol. 331, Plenum Press, New York, 1994, pp. 299–306.

[DelsG96] A. Delshams and P. Gutierrez, Effective stability and KAM theory, *J. Differential Eqs.* **128** (1996) 415–490.

[DelsH09] A. Delshams and G. Huguet, Geography of resonances and Arnold diffusion in a priori unstable Hamiltonian systems, *Nonlinearity* **22**, no. 8 (2009) 1997–2077.

[DelsH11] A. Delshams and G. Huguet, A geometric mechanism of diffusion: rigorous verification in a priori unstable Hamiltonian systems, *J. Differential Equations* **250**, no. 5 (2011) 2601–2623.

[DelsLS06] A. Delshams, R. de la Llave, and T.M. Seara, A geometric mechanism for diffusion in Hamiltonian systems overcoming the large gap problem: heuristics and rigorous verification on a model, *Memoirs Amer. Math. Soc.* **179** (844) (2006) 1–141. [Results announced earlier in *Electron. Res. Announc. Amer. Math. Soc.* **9** (2003), 125–134 (electronic)]

[Den32] A. Denjoy, Sur les courbes définies par les équations différentielles à la surface du tore (French) [On the curves defined by differential equations on the surface of the torus], *J. Math. Pures Appl.* **11** (9) (1932) 333–375.

[Dep83] A. Deprit, Elimination of the nodes in problems of n bodies, *Celestial Mech.* **30** (1983), no. 2, 181–195.

[Di93] F.N. Diacu, Painlevé's conjecture, *Math. Intelligencer* **15** (2) (1993) 6–12.

[Di96] F.N. Diacu, The solution of the n-body problem, *Math. Intelligencer* **18** (3) (1996) 66–70.

[Di00] F.N. Diacu, A century-long loop, *Math. Intelligencer* **22** (2) (2000) 19–25.

[DiH96] F.N. Diacu and P. Holmes, *Celestial Encounters, the origins of chaos and stability*, Princeton University Press, Princeton, NJ, 1996.

[Din92] S. Diner, Les voies du chaos déterministe dans l'école russe (French) [The paths of deterministic chaos in the Russian school], in *Chaos et déterminisme* (A. Dahan Dalmedico, J.-L. Chabert, et K. Chemla, Eds.), Editions du Seuil, Paris, 1992, pp. 331–370.

[DipL89] R. DiPerna and P.-L. Lions, Ordinary differential equations, transport theory and Sobolev spaces, *Inv. Math.* **98** (1989), 511–548.

[Diz08] P. Dizikes, The meaning of the butterfly; Why pop culture loves the 'butterfly effect,' and gets it totally wrong, *Boston Globe* June 8, 2008.

[Don96] V.J. Donnay, Elliptic islands in generalized Sinai billiards, *Ergodic Theory Dynam. Systems* **16**, no. 6 (1996) 975–1010.

[Don99] V.J. Donnay, Non-ergodicity of two particles interacting via a smooth potential, *J. Statist. Phys.* **96**, no. 5–6 (1999) 1021–1048.

[DonL91] V.J. Donnay and C. Liverani, Potentials on the two-torus for which the Hamiltonian flow is ergodic, *Comm. Math. Phys.* **135**, no. 2 (1991) 267–302.

[Dou88] R. Douady, Stabilité ou instabilité des points fixes elliptiques (French) [Stability or instability of elliptic fixed points] *Ann. Sci. Ecole Norm. Sup.* (4) **21** (1988), no. 1, 1–46.

[Dow07] T. Downarowicz, Entropy, *Scholarpedia* 2(11):3901 (2007).

[Drey53] J.L.E. Dreyer, *A History of Astronomy from Thales to Kepler*, Dover Publications, New York, 1953.

[Dug55] R. Dugas, *Histoire de la mécanique*, Editions du Griffon, Neuchâtel, Sw., 1955 [English translation: *A History of Mechanics*, Dover, New York, 1988].

[Duma93] H.S. Dumas, A Nekhoroshev-like theory of classical particle channeling in perfect crystals, *Dynamics Reported (NS)* **2** (1993) 69–115.

[EarR96] J. Earman and M. Rédei, Why ergodic theory does not explain the success of equilibrium statistical mechanics, *British Journal for Philosophy of Science* **47** (1996) 63–78.

[Eas78] R. Easton, Homoclinic phenomena in Hamiltonian systems with several degrees of freedom, *J. Diff. Equations* **29** (2) (1978) 241–252.

[Eas81] R. Easton, Orbit structure near trajectories biasymptotic to invariant tori, in *Classical Mechanics and Dynamical Systems* (Medford, Mass., 1979) pp. 55–67, Lecture Notes Pure Appl. Math. Vol. 70, Dekker, New York, 1981.

[EdS93] D.A. Edwards and M.J. Syphers, An Introduction to the Physics of High Energy Accelerators, J. Wiley & Sons, New York, 1993.

[EhE11] P. Ehrenfest and T. Ehrenfest, Begriffliche Grundlagen der statistischen Auffassung in der Mechanik (German) [Conceptual foundations of the statistical approach in mechanics], in *Enzyklopädie der Mathematischen Wissenschaften*, Vol. 4, Teil 32, B.G. Teubner, Leipzig, 1911.

[Ei05] A. Einstein, Über einen die Erzeugung und Verwandlung des Lichtes betreffenden heuristischen Gesichtspunkt (German) [On a heuristic point of view concerning the production and transformation of light] *Annalen der Physik* (ser. 4) **17** (1905) 132–148.

[Ei06] A. Einstein, Theorie der Lichterzeugung und Lichtabsorption (German) [The theory of light production and light absorption] *Annalen der Physik* (ser. 4) **20** (1906) 199–206.

[Ei07] A. Einstein, Plancksche Theorie der Strahlung und die Theorie der Spezifischen Wärme (German) [Planck's theory of radiation and the theory of specific heat] *Annalen der Physik* (ser. 4) **22** (1907) 180–190.

[Ek88] I. Ekeland, *Mathematics and the Unexpected*, University of Chicago Press, Chicago and London, 1988 [Author's English translation of the book published in French as: *Le Calcul, L'Imprévu: Les figures du temps de Kepler à Thom*, Editions du Seuil, Paris, 1984].

[Ek11] I. Ekeland, An inverse function theorem in Fréchet spaces, *Ann. Inst. H. Poincaré Anal. Non Linéaire* **28**, no. 1 (2011), 91–105.

[Eli96] L.H. Eliasson, Absolutely convergent series expansions for quasi periodic motions, *Math. Phys. Electron. J.* **2** (1996), Paper 4, 33 pp. (electronic).

[Eli98] L.H. Eliasson, Reducibility and point spectrum for linear quasi-periodic skew-products, in *Proceedings of the International Congress of Mathematicians* (Berlin, 1998), Vol. II, *Doc. Math.* 1998, Extra Vol. II, pp. 779–787 (electronic).

[EliK10] L.H. Eliasson and S.B. Kuksin, KAM for the nonlinear Schrödinger equation, *Ann. of Math.* (2) **172** (2010), no. 1, 371–435.

[EliKMY02] L.H. Eliasson, S.B. Kuksin, S. Marmi, J.-C. Yoccoz, Eds., *Dynamical systems and small divisors* (Lectures from the C.I.M.E. Summer School held in Cetraro, Italy, June 13–20, 1998) Lecture Notes in Math. Vol. 1784, Springer-Verlag, Berlin 2002.

[EscD81] D.F. Escande and F. Doveil, Renormalization method for computing the threshold of the large-scale stochastic instability in two degrees of freedom Hamiltonian systems, *J. Statist. Phys.* **26** (1981) no. 2, 257–284.

[Ev04] L.C. Evans, A survey of partial differential equations methods in weak KAM theory, *Commun. Pure Appl. Math.* **57** (2004) 445–480.

[Ev08] L.C. Evans, Weak KAM theory and partial differential equations, in *Calculus of Variations and Nonlinear Partial Differential Equations* pp. 123–154, (Lecture Notes in Math. Vol. 1927) Springer-Verlag, Berlin, 2008.

[Ev09] L.C. Evans, Further PDE methods for weak KAM theory, *Calc. Var. Partial Differential Equations* **35** (2009), no. 4, 435–462.

[Fa98] F. Fassò, Quasi-periodicity of motions and complete integrability of Hamiltonian systems, *Erg. Th. Dynam. Systems* **18** (6) (1998) 1349–1362.

[FaL01] F. Fassò and D. Lewis, Stability properties of the Riemann ellipsoids, *Arch. Ration. Mech. Anal.* **158** (4) (2001) 259–292.

[Fath97a] A. Fathi, Théorème KAM faible et théorie de Mather sur les systèmes lagrangiens. *C.R. Acad. Sci. Paris* Sér. I Math. **324** (9): 1043–1046 (1997).

[Fath97b] A. Fathi, Solutions KAM faibles conjuguées et barrières de Peierls, *C.R. Acad. Sci. Paris* Sér. I Math. **325** (6): 649–652 (1997).

[Fath10] A. Fathi, *The Weak KAM Theorem in Lagrangian Dynamics*, Cambridge University Press, Cambridge, UK, 2010.

[Fato19–20] P. Fatou, Sur les équations fonctionelles (French) [On functional equations] *Bull. Soc. Math. France* **47** (1919) 161–271; *ibid.* **48** (1920) 33–94; **48** (1920) 208–314.

[Féj04] J. Féjoz, Démonstration du 'théorème d'Arnold' sur la stabilité du système planétaire (d'après Herman) (French) [Proof of 'Arnold's theorem' on the stability of the planetary system (following Herman)], *Erg. Th. Dynam. Sys.* **25** (4) (2004) 1521–1582 [Updated version (and erratum) available online].

[Féj07] J. Féjoz, A summary of M. Herman's proof of 'Arnold's theorem' in celestial mechanics (manuscript, 2007) [available online].

[Féj11] J. Féjoz, Introduction to KAM theory, Ciclo di lezioni presso l'Università degli Studi Milano-Bicocca (2011) [available online].

[Féj12] J. Féjoz, A proof of the invariant torus theorem of Kolmogorov, *Regul. Chaotic Dyn.* **17** (1) (2012) 1–5 [available online].

[Féj13] J. Féjoz, The three and N-body problem, to appear in *Celestial Mechanics* (A. Celletti, Ed.) a volume of the *Encyclopedia of Life Support Systems* (EOLSS), Developed under the Auspices of the UNESCO, Eolss Publishers,

Oxford, UK, [http://www.eolss.net] (2013).

[Fer23a] E. Fermi, Beweis dass ein mechanisches Normalsystem im allgemeinen quasi-ergodisch ist (German) [Proof that a normal mechanical system is quasi-ergodic in general] *Phys. Zeits.* **24** (1923) 261–265.

[Fer23b] E. Fermi, Dimostrazione che in generale un sistema meccanico è quasi ergodico (Italian) [Proof that in general a mechanical system is quasi-ergodic] *Nuovo Cimento* **25** (1923) 267–269.

[Fer23c] E. Fermi, Generalizzazione del teorema di Poincaré sopra la non esistenza di integrali uniformi di un sistema di equazioni canoniche normali (Italian) [Generalizations of Poincaré's theorem on the non-existence of uniform integrals for a system of normal canonical equations] *Nuovo Cimento* **26** (1923) 105-115.

[Fer24] E. Fermi, Über die Existenz quasi-ergodischer Systeme (German) [On the existence of quasi-ergodic systems] *Phys. Zeits.* **25** (1924) 166–167.

[FerPU55] E. Fermi, J. Pasta, and S. Ulam, Studies of nonlinear problems, Los Alamos report LA-1940 (1955).

[For98] E. Forest, *Beam Dynamics: A New Attitude and Framework*, CRC Press, New York, 1998.

[For06] E. Forest, Geometric integration for particle accelerators, *J. Phys. A: Math. Gen.* **39** (2006) 5321–5377.

[Fr89a] J. Franks, Review of *Chaos: Making a New Science*, by James Gleick, *Math. Intelligencer*, **11** (1) (1989) 65–69; John Franks Responds to James Gleick, *ibid.* 70–71.

[Fr89b] J. Franks, Comments on the responses to my review of *Chaos*, *Math. Intelligencer*, **11** (3) (1989) 12–13.

[Fri08] R.P. Frigg, A field guide to recent work on the foundations of statistical mechanics, in *The Ashgate Companion to Contemporary Philosophy of Physics* (D. Rickles, Ed.) Ashgate Publishing, London, 2008, 99–196.

[FriW11] R.P. Frigg and C.S. Werndl, Explaining thermodynamic-like behavior in terms of epsilon-ergodicity, *Philosophy of Science* **78** no. 4 (2011) 328–352.

[FröS83] J. Fröhlich and T. Spencer, Absence of diffusion in the Anderson tight binding model for large disorder or low energy, *Comm. Math. Phys.* **88**, no. 2 (1983) 151–184.

[FuMMPPRV82] E. Fucito, F. Marchesoni, E. Marinari, G. Parisi, L. Peliti, S. Ruffo and A. Vulpiani, Approach to equilibrium in a chain of nonlinear oscillators, *J. Phys.* **43** (1982) 707–713.

[Galg93] L. Galgani, Merging of classical mechanics into quantum mechanics, in *Stochastic Processes in Astrophysics* (J.R. Buchler and H.E. Kandrup, Eds.), Annals of New York Acad. of Sciences, Vol. 706, New York, 1993.

[Gall83–07] G. Gallavotti, *The Elements of Mechanics*, Springer-Verlag, New York, 1983 [2nd ed. G. Gallavotti, Rome, 2007, available online].

[Gall95a] G. Gallavotti, Invariant tori: a field-theoretic point of view of Eliasson's work, in *Advances in Dynamical Systems and Quantum Physics* (Capri, 1993), World Sci. Publ., River Edge, NJ, 1995, pp. 117–132.

[Gall95b] G. Gallavotti, Ergodicity, ensembles, irreversibility in Boltzmann and beyond, *J. Stat. Physics* **78** (1995) 1571–1589.

[Gall99] G. Gallavotti, *Statistical Mechanics, A Short Treatise*, Springer-Verlag, Berlin-Heidelberg, 1999.

[Gall08] G. Gallavotti (Ed.), *The Fermi-Pasta-Ulam Problem: A Status Report* (Lecture Notes in Physics, Vol. 728), Springer-Verlag, Berlin, 2007.

[GallG95] G. Gallavotti and G. Gentile, Majorant series convergence for twistless KAM tori, *Ergodic Theory Dynam. Systems* **15**, no. 5 (1995), 857–869.

[GallGM95] G. Gallavotti, G. Gentile, and V. Mastropietro, Field theory and KAM tori, *Math. Phys. Electron. J.* **1** Paper 5 (1995) 13 pp. (electronic).

[GanT10] W. Gangbo and A. Tudorascu, Lagrangian dynamics on an infinite-dimensional torus; a weak KAM theorem, *Adv. Math.* **224**, no. 1 (2010) 260–292.

[GidL06] M. Gidea and R. de la Llave, Topological methods in the instability problem of Hamiltonian systems, *Discr. Contin. Dyn. Syst.* **14** (2) (2006) 295–328.

[GidR07] M. Gidea and C. Robinson, Shadowing orbits for transition chains of invariant tori alternating with Birkhoff zones of instability, *Nonlinearity* **20** (2007) 1115–1143.

[GidR09] M. Gidea and C. Robinson, Obstruction argument for transition chains of tori interspersed with gaps, *Discrete Contin. Dyn. Syst. Ser. S* **2**, no. 2 (2009) 393–416.

[Gio98] A. Giorgilli, On the problem of stability for near to integrable Hamiltonian systems, *Proceedings of the International Congress of Mathematicians*, Vol. III (Berlin, 1998); *Doc. Math.* 1998, Extra Vol. III, 143–152 (electronic).

[Gio00] A. Giorgilli, A tribute to Jürgen Moser (1928–1999), *Cel. Mech. Dynam. Astron.* **77** (3) (2000) 153–155.

[GioDFGS89] A. Giorgilli, A. Delshams, E. Fontich, L. Galgani, and C. Simó, Effective stability for a Hamiltonian system near an elliptic fixed point, with an application to the restricted three body problem, *J. Diff. Eqs.* **77** (1989) 167–198.

[Glei87] J. Gleick, *Chaos: Making a New Science*, Viking Penguin, New York, 1987.

[Glei89a] J. Gleick, "James Gleick Replies," *Math. Intelligencer*, **11** (1) (1989) 69–70.

[Glei89b] J. Gleick, "James Gleick Replies," *Math. Intelligencer*, **11** (3) (1989) 8–9.

[Glei03] J. Gleick, *Isaac Newton*, Pantheon Books, New York, 2003.

[Gli64] J. Glimm, Formal stability of Hamiltonian systems, *Comm. Pure Appl. Math.* **17** (1964), 509–526.

[Gold59] H. Goldstein, *Classical Mechanics*, Addison-Wesley, Reading, Mass., 1959

[Gold80] H. Goldstein, *Classical Mechanics* (2nd Ed.), Addison-Wesley, Reading, Mass., 1980.

[GoldPS02] H. Goldstein, C. Poole, and J. Safko, *Classical Mechanics* (3rd Ed.), Addison-Wesley, Reading, Mass., 2002.

[Golé92] C. Golé, Ghost tori for monotone maps, in *Twist mappings and their applications* (IMA Vols. Math. Appl. **44**) Springer-Verlag, New York, 1992, pp. 119–133.

[GonJLV05] A. Gonzáles, À. Jorba, R. de la Llave, and J. Villanueva, KAM theory without action-angle variables, *Nonlinearity* **18** (2) (2005) 855–895.

[Grau94] H. Grauert, Gauss und die Göttinger Mathematik (German) [Gauss and mathematics in Göttingen] *Naturwissenschaftliche Rundschau* **47** (6) (1994), 211–219.

[Gray13] J. Gray, *Henri Poincaré: A Scientific Biography*, Princeton U. Press, Princeton, NJ & Oxford, UK, 2013.

[Gre79] J.M. Greene, A method for determining a stochastic transition, *J. Math. Phys.* **20** (6) (1979) 1183–1201.

[Gre93] J.M. Greene, The status of KAM theory from a physicist's point of view, in *Chaos in Australia* (Sydney, 1990), World Sci. Publ., River Edge, NJ, 1993, pp. 8–23.

[Guc78] J. Guckenheimer, The catastrophe controversy, *Math. Intelligencer* **1** (1) (1978) 15–20.

[GucH83–02] J. Guckenheimer and P. Holmes, *Nonlinear Oscillations, Dynamical Systems, and Bifurcations of Vector Fields* (Applied Mathematical Sciences Vol. 42) Springer-Verlag, New York, 1983 [Second Ed., revised & corrected, 2002].

[Gus66] F.G. Gustavson, On constructing formal integrals of a Hamiltonian system near an equilibrium point, *Astron. J.* **71** (1966) 670–686.

[Gut98] M.C. Gutzwiller, Moon-Earth-Sun: the oldest three-body problem, *Rev. Mod. Phys.*, **70** (2) April 1998, 589–639.

[Guz04] M. Guzzo, A direct proof of the Nekhoroshev theorem for nearly integrable symplectic maps, *Ann. Henri Poincaré* **5** (6) (2004) 1013–1039.

[GuzLF02] M. Guzzo, E. Lega, and C. Froeschlé, On the numerical detection of the effective stability of chaotic motions in quasi-integrable systems, *Phys. D*, **163** (1–2) (2002) 1–25.

[Had98] J. Hadamard, Les surfaces à courbures opposées et leurs lignes géodesiques (French) [Surfaces with opposite curvatures and their geodesic lines] *J. Mathématiques* **4** (5) (1898) 27–73. Reprinted in *Œuvres de Jacques Hadamard* **2**, Editions du CNRS, Paris, 1968, pp. 729–780.

[Hale69] J.K. Hale, *Ordinary Differential Equations*, Wiley-Interscience, New York, 1969.

[Hall80] A.R. Hall, *Philosophers at War: The quarrel between Newton and Leibniz*, Cambridge U. Press, Cambridge, U.K., 1980.

[Ham82] R.S. Hamilton, The inverse function theorem of Nash and Moser, *Bull. AMS (new series)* **7** (1) (1982) 65–222.

[Har82] P. Hartman, *Ordinary Differential Equations*, Birkhäuser, Boston, MA, 1982.

[HasK02a] B. Hasselblatt and A. Katok, The development of dynamics in the 20th century and the contribution of Jürgen Moser, *Erg. Th. Dynam. Sys.* **22** (2002) 1343–1364.

[HasK02b] B. Hasselblatt and A. Katok, Eds., *Handbook of Dynamical Systems*, Vol. 1A, North-Holland, Amsterdam, 2002.

[HasP07] B. Hasselblatt and Y. Pesin, Hyperbolic dynamics, *Scholarpedia*

3(6):2208 (2008).

[Hén66] M. Hénon, Exploration numérique du problème restreint IV. Masses égales, orbites non périodiques (French) [Numerical exploration of the restricted problem IV. Equal masses, non-periodic orbits], *Bull. Astronom.* **3** (1:2) (1966) 49–66.

[HenrK08] A. Henrici and T. Kappeler, Results on normal forms for FPU chains, *Comm. Math. Phys.* **278** (2008), 145–177.

[Herm76] M.R. Herman, Conjugaison C^∞ des difféomorphismes du cercle pour presque tout nombre de rotation (French) [C^∞ conjugation of circle diffeomorphisms for almost every rotation number] *C.R. Acad. Sci. Paris*, Sér. A–B, 1976, **283**, no. 8, Aii, A579–A582.

[Herm79] M.R. Herman, Sur la conjugaison différentiable des difféomorphismes du cercle à des rotations (French) [On the differentiable conjugation of circle diffeomorphisms to rotations] *Publ. Math. IHES*, **49** (1979) 5–234.

[Herm83] M.R. Herman, Sur les courbes invariantes par les difféomorphismes de l'anneau (French) [On curves invariant under diffeomorphisms of the annulus] (vol. 1), Astérisque **103–104**, Soc. Math. France, Paris, 1983.

[Herm86] M.R. Herman, Sur les courbes invariantes par les difféomorphismes de l'anneau (French) [On curves invariant under diffeomorphisms of the annulus] (vol. 2), Astérisque **144**, Soc. Math. France, Paris, 1986.

[Hersh13] R. Hersh, *Peter Lax, A Memoir, mathematics in New York in the 20th century*, book manuscript, to appear.

[HershJ10] R. Hersh and V. John-Steiner, *Loving and Hating Mathematics: Challenging the Myths of Mathematical Life*, Princeton University Press, Princeton, NJ, 2010.

[Hir89] M. Hirsch, Chaos, Rigor, and Hype, *Math. Intelligencer*, **11** (3) (1989) 6–8.

[HirSD04] M. Hirsch, S. Smale, and R. Devaney, *Differential Equations, Dynamical Systems, and an Introduction to Chaos*, Elsevier Academic Press, San Diego, CA, 2004.

[Hol90] P. Holmes, Poincaré, celestial mechanics, dynamical-systems theory and "chaos", *Physics Reports* **193** (3) (1990) 137–163.

[Hol05] P. Holmes, Ninety plus thirty years of nonlinear dynamics: More is different and less is more, *Int. J. Bifurcation and Chaos*, **15** (9) (2005) 2703–2716.

[Hol07] P. Holmes, History of dynamical systems, *Scholarpedia*, 2(5):1843 (2007).

[HolM82] P. Holmes and J. Marsden, Melnikov's method and Arnol'd diffusion for perturbations of integrable Hamiltonian systems, *J. Math. Phys.* **23**, no. 4 (1982), 669–675.

[Hop39] E. Hopf, Statistik der geodätischen Linien in Mannigfaltigkeiten negativer Krümmung (German) [Statistics of geodesic curves on manifolds with negative curvature] *Leipzig Ber. Verhandl. Sächs. Akad. Wiss.* **91** (1939) 261–304.

[Hör85] L. Hörmander, On the Nash-Moser implicit function theorem, *Ann. Acad. Sci. Fenn.* Ser. A I Math. **10** (1985) 255–259.

[Hör90] L. Hörmander, The Nash-Moser implicit function theorem and paradifferential operators, in *Analysis, et cetera*, Academic Press, Boston, MA,

1990, pp. 429–449.

[HuaCL00] Q. Huang, F. Cong, and Y. Li, Persistence of hyperbolic invariant tori for Hamiltonian systems, *J. Differential Equations* **164** (2000), no. 2, 355–379.

[HubI04] J. Hubbard and Yu.S. Ilyashenko, A proof of Kolmogorov's theorem, *Discrete Contin. Dyn. Syst.* **10** (2004), nos. 1–2, 367–385.

[Ig98] P. Iglesias, Les origines du calcul symplectique chez Lagrange (French) [The origins of symplectic calculus in Lagrange's work] *Enseign. Math.* (2) **44** (1998), no. 3–4, 257–277.

[IzC66] F.M. Izrailev and B.V. Chirikov, Statistical properties of a non-linear string, *Dokl. Akad. Nauk SSSR* **166** (1) (1966) 57–59 (Russian) [English translation: *Soviet Physics Dokl.* **11** (1) (1966) 30–32].

[IzC68] F.M. Izrailev and B.V. Chirikov, Stochasticity of the simplest dynamical model with divided phase space (Russian), [Preprint, Novosibirsk Institute for Nuclear Physics of the USSR Academy of Sciences, no. 191, 1968].

[Jack91] E.A. Jackson, *Perspectives of Nonlinear Dynamics*, Vol. 1, corrected edition, Cambridge University Press, Cambridge, UK, 1991.

[Jaco42] C.G.J. Jacobi, Sur l'éliminations des noeuds dans le problème des trois corps (French) [On the elimination of the nodes in the three body problem], *Astron. Nachr.* **XX** (1842) 81–102.

[Jaco43] C.G.J. Jacobi, *Vorlesungen über Dynamik, gehalten an der Universität Königsberg im Wintersemester 1842–1843* (German) [Lectures on Dynamics, held at the University of Koenigsberg during winter semester 1842–1843], G. Reimer, Berlin, 1866 [available online]; [English translation: *Jacobi's Lectures on Dynamics* (2nd revised ed.) Hindustan Book Agency, New Delhi, 2009.]

[Je03] J.H. Jeans, On the vibrations set up in molecules by collisions, *Phil. Mag.* **6** (1903) 279–286.

[Je05] J.H. Jeans, On the partition of energy between matter and aether, *Phil. Mag.* **10** (1905) 91–98.

[Joh91] P. Johnson, *Modern Times: the world from the twenties to the nineties* (Revised Ed.), Harper Collins, New York, 1991.

[JoséS98] J.V. José and E.J. Saletan, *Classical Dynamics, a contemporary approach*, Cambridge University Press, Cambridge, UK, 1998.

[Jost68] R. Jost, Winkel und Wirkungsvariable für allgemeine mechanische Systeme (German) [Action and angle variables for general mechanical systems], *Helvetica Physica Acta* **41** (1968) 965–968.

[KalL08a] V. Kaloshin and M. Levi, An example of Arnold diffusion for near-integrable Hamiltonians, *Bull. Amer. Math. Soc.* (N.S.) **45** (3) (2008) 409–427.

[KalL08b] V. Kaloshin and M. Levi, Geometry of Arnold diffusion, *SIAM Rev.* **50** (4) (2008) 702–720.

[KalZ13] V. Kaloshin and K. Zhang, A strong form of Arnold diffusion for two and a half degrees of freedom, 2013 preprint (to appear).

[Kan48] L.V. Kantorovich, Functional analysis and applied mathematics, *Uspekhi Math. Nauk (N.S.)* **3** no. 6 (28) (1948) 89–185. (Russian) [English translation by C.D. Benster, Nat. Bur. Standards, Report no. 1509, US Dept. Commerce, NBS, Los Angeles, CA, March 7, 1952.]

[KapP03] T. Kappeler and J. Pöschel, *KdV & KAM* (Ergebnisse der Mathematik und ihrer Grenzgebiete 3, Folge **45**) Springer-Verlag, Berlin, 2003.

[KatH95] A. Katok and B. Hasselblatt, *Introduction to the Modern Theory of Dynamical Systems*, Cambridge University Press, Cambridge, UK, 1995.

[KhS86] K.M. Khanin and Ya.G. Sinai, Renormalization group method and Kolmogorov-Arnold-Moser theory, in *Nonlinear Phenomena in Plasma Physics and Hydrodynamics* (R.Z. Sagdeev, Ed.), Mir, Moscow, 1986, pp. 31–64.

[Kli72] M. Kline, *Mathematical Thought from Ancient to Modern Times*, Oxford U. Press, New York, 1972.

[Klo73] O. Klopp (Ed.), *Correspondenz von Leibniz mit Caroline* (German title, French text) [Leibniz' Correspondence with Caroline], Georg Olms Verlag, Hildesheim, New York, 1973 [reprint of the original edition published by Klindworth, Hannover, 1884].

[Kna87] A. Knauf, Ergodic and topological properties of Coulombic periodic potentials. *Comm. Math. Phys.* **110** (1987) no. 1, 89–112.

[Kni99] O. Knill, Weakly mixing invariant tori of Hamiltonian systems, *Comm. Math. Phys.* **204** (1) (1999) 85–88.

[Koč73] P.Ya. Kočina-Polubarinova (Ed.), Briefe von Karl Weierstrass an Sofie Kowalewskaya 1871–1891 (German title, German and Russian text) [Letters from Karl Weierstrass to Sophie Kovalevskaya 1871–1891] Izdat. Nauka, Moscow, 1973.

[Koe59] A. Koestler, *The Sleepwalkers: A History of Man's Changing Vision of the Universe*, Macmillan, New York, 1959.

[Koe60] A. Koestler, *The Watershed: A Biography of Johannes Kepler*, Anchor Books, Garden City, New York, 1960.

[Kol53] A.N. Kolmogorov, On dynamical systems with an integral invariant on the torus (Russian), *Dokl. Akad. Nauk SSSR* (NS) **93** (1953), 763–766.

[Kol54] A.N. Kolmogorov, On the preservation of conditionally periodic motions under a small change in Hamilton's function (Russian), *Dokl. Akad. Nauk SSSR (new series)* **98** (1954), 527–530. [English translation: Los Alamos Scientific Laboratory translation LA–TR–71–67, by H. Dahlby; first published in *Stochastic Behavior in Classical and Quantum Hamiltonian Systems* (Lecture Notes in Phys. Vol. 93; G. Casati and J. Ford, Eds.), Springer-Verlag, Berlin Heidelberg, 1979, pp. 51–56; reprinted in *Chaos* (B. Hao, Ed.), World Scientific Publishers, Singapore, 1984, pp. 81–86; also reprinted in Appendix A of this book.]

[Kol57] A.N. Kolmogorov, The general theory of dynamical systems and classical mechanics (Russian), *Proceedings of the International Congress of Mathematicians* (Amsterdam, 1954), Vol. 1, pp. 315–333, North Holland, Amsterdam, 1957. [Traduction française: *Sém. Janet*, 1957–1958, no. 6, Fac. Sci., Paris, 1958.] [English translation: *International Mathematical Congress in Amsterdam, 1954 (Plenary Lectures)*, pp. 187–208, Fizmatgiz, Moscow 1961; also reprinted as Appendix D in R.H. Abraham, *Foundations of Me-*

chanics, pp. 263–279, Benjamin, Reading, Mass., 1967; and as Appendix in R.H. Abraham and J.E. Marsden, *Foundations of Mechanics*, 2nd Ed., pp. 741–757, Benjamin/Cummings, Reading, Mass., 1978.]

[KolS00] A.N. Kolmogorov and A.N. Shiryaev, *Kolmogorov in Perspective* [translated from Russian by H.H. McFaden], (Hist. Math. **20**), Amer. Math. Soc., Providence, RI, 2000.

[Koo31] B.O. Koopman, Hamiltonian systems and linear transformations in Hilbert space, *Proc. Nat. Acad. Sci. USA* **17** (1931) 315–318.

[Kow88–90] S. Kowalevski, Mémoire sur un cas particulier du problème de la rotation d'un corps pesant autour d'un point fixe, où l'intégration s'effectue à l'aide des fonctions ultraelliptiques du temps (French) [Memoir on a particular case of the problem of rotation of a heavy body around a fixed point, where integration is effected with the help of ultraelliptic functions of time] (Prix Bordin de l'Académie des Sciences, Paris, 1888) in *Memoires presentés par divers savants étrangers à l'Académie des Sciences de l'Institut de France*, Tome 31, Paris, 1890, pp. 1–62; Sur le problème de la rotation d'un corps solide autour d'un point fixe. (French) [On the problem of rotation of a solid body around a fixed point] Acta Math. **12** (1889) 177–232.

[Kra00] H. Kragh, Max Planck: the reluctant revolutionary, *Physics World* **13** (2000) 31–35.

[Kran89] S.G. Krantz, Fractal Geometry, *Math. Intelligencer* **11** (4) (1989) 12–16; Steven G. Krantz replies *ibid.* p. 19.

[KranP02] S.G. Krantz and H.R. Parks, *The Implicit Function Theorem: History, Theory, and Applications*, Birkhäuser, Boston, 2002.

[Kub76] I. Kubo, Perturbed billiard systems. I. The ergodicity of the motion of a particle in a compound central field, *Nagoya Math. J.* **61** (1976) 1–57.

[Kuh62] T.S. Kuhn, *The Structure of Scientific Revolutions*, Univ. of Chicago Pr., Chicago, 1962.

[Kuk88a] S.B. Kuksin, Perturbation theory of conditionally periodic solutions of infinite-dimensional Hamiltonian systems and its applications to the Korteweg–de Vries equation (Russian) *Mat. Sb.* (N.S.) **136** (178) (1988), no. 3, 396–412, 431. [English translation: *Math. USSR-Sb.* **64** (1989), no. 2, 397–413.]

[Kuk88b] S.B. Kuksin, Perturbation of conditionally periodic solutions of infinite-dimensional Hamiltonian systems (Russian) *Izv. Akad. Nauk SSSR Ser. Mat.* **52** (1988), no. 1, 41–63, 240. [English translation: *Math. USSR-Izv.* **32** (1989), no. 1, 39–62.]

[Kuk93] S.B. Kuksin, *Nearly Integrable Infinite-Dimensional Hamiltonian Systems*, Springer-Verlag, Berlin, 1993.

[Kuk98] S.B. Kuksin, Elements of a qualitative theory of Hamiltonian PDEs, in *Proceedings of the International Congress of Mathematicians* (Berlin, 1998), Vol. II, *Doc. Math.* 1998, Extra Vol. II, pp. 819–829 (electronic).

[Kuk00] S.B. Kuksin, *Analysis of Hamiltonian PDEs* (Oxford Lecture Series in Mathematics and its Applications **19**), Oxford University Press, Oxford, UK, 2000.

[Kuk04] S.B. Kuksin, Fifteen years of KAM for PDE, in *Geometry, Topology, and*

Mathematical Physics, Amer. Math. Soc. Transl. Ser. 2, 212, Amer. Math. Soc., Providence, RI, 2004, pp. 237–258.

[Kuz07] Y.A. Kuznetsov, Conjugate maps, *Scholarpedia* 2(12):5420 (2007).

[LanL75] L.D. Landau and E.M. Lifshitz, *The Classical Theory of Fields*, Pergamon Press, New York, 1975.

[LanL76] L.D. Landau and E.M. Lifshitz, *Mechanics*, Pergamon Press, New York, 1976.

[Lap14] P.-S. de Laplace, *Essai philosophique sur les probabilités* (French) [Philosophical essay on probabilities], seconde édition revue et augmentée, Paris, Courcier, 1814.

[Las89] J. Laskar, A numerical experiment on the chaotic behaviour of the Solar System, *Nature* **338** (1989) 237–238.

[Las90] J. Laskar, The chaotic motion of the Solar System. A numerical estimate of the size of the chaotic zones, *Icarus* **88** (1990) 266–291.

[Las92] J. Laskar, La stabilité du système solaire (French) [The stability of the solar system], in *Chaos et déterminisme* (A. Dahan Dalmedico, J.-L. Chabert, et K. Chemla, Eds.), Editions du Seuil, Paris, 1992, pp. 170–211.

[Las94] J. Laskar, Large scale chaos and marginal stability in the solar system, in *Proceedings of the XIIth Internat. Congress Math. Phys. (Paris 1994)* Int. Press., Cambridge, UK, 1994.

[Las03] J. Laskar, Chaos in the solar system, *Ann. Henri Poincaré* **4**, Suppl. 2 (2003) S693–S705.

[Las13] J. Laskar, Stability of the solar system, to appear in *Celestial Mechanics* (A. Celletti, Ed.) a volume of the *Encyclopedia of Life Support Systems* (EOLSS), Developed under the Auspices of the UNESCO, Eolss Publishers, Oxford, UK, [http://www.eolss.net] (2013).

[LasG09] J. Laskar and M. Gastineau, Existence of collisional trajectories of Mercury, Mars and Venus with the Earth, *Nature*, Vol. 459, 11 June (2009) 817–819.

[LasR95] J. Laskar and P. Robutel, Stability of the planetary three-body problem, I. Expansion of the planetary Hamiltonian, *Celestial Mech. Dynam. Astronom.* **62** (3) (1995) 193–217.

[Lax02] P.D. Lax, Jürgen Moser, 1928–1999, *Ergod. Th. Dynam. Sys.* **22** (2002), 1337–1342.

[Laz72] V.F. Lazutkin, Existence of a continuum of closed invariant curves for a convex billiard (Russian), *Uspehi Mat. Nauk* **27** (1972), no. 3 (165), 201–202.

[Laz73] V.F. Lazutkin, Existence of caustics for the billiard problem in a convex domain (Russian), *Izv. Akad. Nauk SSSR* Ser. Mat. **37** (1973), 186–216.

[Laz74] V.F. Lazutkin, Concerning a theorem of Moser on invariant curves (Russian), *Problems in the dynamic theory of propagation of seismic waves*, Leningrad Steklov Mathem. Instit. No. 14 (1974), 109–120.

[Lea97] L. Leau, Etude sur les équations fonctionelles à une ou plusieures variables (French) [Study of functional equations in one or several variables] *Ann. Fac. Sci. Toulouse* **11** (1897).

[LevM01] M. Levi and J. Moser, A Lagrangian proof of the invariant curve theorem for twist mappings. In *Smooth Ergodic Theory and its Applications*

(Seattle, WA, 1999), 733–746, Proc. Sympos. Pure Math., **69** (A. Katok et al., Eds.), Amer. Math. Soc., Providence, RI, 2001.

[LiY75] T.Y. Li and J.A. Yorke, Period three implies chaos, *Amer. Math. Monthly* **82** (10) (1975) 985–992.

[LicL92] A.J. Lichtenberg and M.A. Lieberman, *Regular and Chaotic Dynamics*, 2nd ed. (Applied Math. Sciences Vol. 38), Springer-Verlag, New York, 1992.

[Lin01] D. Lindley, *Boltzmann's Atom: The Great Debate that Launched a Revolution in Physics*, The Free Press, New York, 2001.

[LindsP09] S. Lindström and E. Palmgren, Introduction: the three foundational programmes, in *Logicism, Intuitionism, and Formalism, what has become of them?* (Synthese Library Vol. 341, S. Lindström, E. Palmgren, K. Segerberg, V. Stoltenberg-Hansen, Eds.), Springer-Verlag, Dordrecht, 2009.

[LindsPSS09] S. Lindström, E. Palmgren, K. Segerberg, V. Stoltenberg-Hansen (Eds.), *Logicism, Intuitionism, and Formalism, what has become of them?* (Synthese Library Vol. 341), Springer-Verlag, Dordrecht, 2009.

[Lio82] P.-L. Lions, *Generalized Solutions of Hamilton-Jacobi Equations*, Pitman (Advanced Publishing Program), Boston, Mass., 1982.

[Liou35] J. Liouville, Mémoire sur l'intégration d'une classe de fonctions transcendantes (French) [Memoir on the integration of a class of transcendental functions], *J. Reine Angew. Math.* **13** (1835) 93–118.

[Liou55] J. Liouville, Note sur l'intégration des équations differentielles de la dynamique, présentée au Bureau des Longitudes le 29 juin 1853 (French) [Note on the integration of the differential equations of dynamics, presented at the Bureau of Longitudes June 29, 1853], *Journal de Mathématiques Pures et Appliquées* **20** (1855) 137–138.

[Lit59a] J.E. Littlewood, On the equilateral configuration in the restricted problem of three bodies, *Proc. London Math. Soc.* (3) **9** (1959), 342–372.

[Lit59b] J.E. Littlewood, The Lagrange configuration in celestial mechanics, *Proc. London Math. Soc.* (3) **9** (1959), 525–543.

[Loc90] P. Lochak, Effective speed of Arnol'd's diffusion and small denominators, *Phys. Lett. A* **143** (1–2) (1990) 39–42.

[Loc92] P. Lochak, Canonical perturbation theory: an approach based on joint approximations (Russian) *Uspekhi Mat. Nauk* **47** (288) (1992) 59–140; [English translation in *Russian Math. Surveys* **47** (6) (1992) 57–133.]

[Loc93] P. Lochak, Hamiltonian perturbation theory: periodic orbits, resonances and intermittency, *Nonlinearity* **6** (6) (1993) 885–904.

[Loc97] P. Lochak, Supplement to "Arnold diffusion: a compendium of remarks and questions," unpublished manuscript dated November 1, 1997 (5 pp.).

[Loc99] P. Lochak, Arnold diffusion; a compendium of remarks and questions, in *Hamiltonian Systems with Three or More Degrees of Freedom* (S'Agaró, 1995), NATO Adv. Sci. Inst. Ser. C, Math. Phys. Sci., Vol. 533, Kluwer Acad. Publ., Dordrecht, 1999, pp. 168–183.

[LocM05] P. Lochak and J.-P. Marco, Diffusion times and stability exponents for nearly integrable analytic systems, *Cent. Eur. J. Math.* **3** (3) (2005) 342–397 (electronic).

[LocM88] P. Lochak and C. Meunier, *Multiphase Averaging for Classical Sys-*

tems with Applications to Adiabatic Theorems [translated from French by H.S. Dumas], Springer-Verlag, New York, 1988.

[LocN92] P. Lochak and A.I. Neĭshtadt, Estimates of stability time for nearly integrable systems with a quasiconvex Hamiltonian, *Chaos* **2** (4) (1992) 495–499.

[Lor63] E.N. Lorenz, Deterministic nonperiodic flow, *J. Atmospheric Sci.* **20** (1963) 130–141.

[Lui96] S.H. Lui, An interview with Vladimir Arnold, *Hong Kong Math. Soc. Newsletter* 1996, no. 2, 2–8; reprinted in *Notices Amer. Math. Soc.* **44** (4) (1997) 432–438.

[Mac93] R.S. MacKay, Renormalisation in area-preserving maps, in *Advanced Series in Nonlinear Dynamics*, Vol. 6, World Scientific, River Edge, NJ, 1993.

[MacM87a] R.S. MacKay and J.D. Meiss, Survey of Hamiltonian Dynamics, in *Hamiltonian Dynamical Systems, a Reprint Selection*, R.S. MacKay and J.D. Meiss, Eds., IOP Publishing, Bristol, UK, 1987.

[MacM87b] R.S. MacKay and J.D. Meiss, Eds., *Hamiltonian Dynamical Systems, a Reprint Selection*, IOP Publishing, Bristol, UK, 1987.

[MacMP84] R.S. MacKay, J.D. Meiss, and I.C. Percival, Transport in Hamiltonian systems, *Phys. D* **13** (1–2) (1984) 55–81.

[MacP85] R.S. MacKay and I.C. Percival, Converse KAM: theory and practice, *Commun. Math. Phys.* **98** (1985) 469–512.

[Mack78] G.W. Mackey, Harmonic analysis as the exploitation of symmetry— a historical survey, *Rice University Studies* **64** (2,3) Spring-Summer 1978, 73–228 [Reprinted in *Bull. Am. Math. Soc.* **3** (1) July 1980, 543–698.]

[Macke92] M.C. Mackey, *Time's Arrow: The Origins of Thermodynamic Behavior*, Springer-Verlag, New York, 1992.

[Man75–89] B.B. Mandelbrot, *Les objets fractals : forme, hasard et dimension*, Flammarion, Paris, 1975; 3me éd. révisé 1989 (French) [English translation: *The Fractal Geometry of Nature*, W.H. Freeman, San Francisco, 1977; revised edition 1983].

[Man89a] B.B. Mandelbrot, Chaos, Bourbaki, and Poincaré, *Math. Intelligencer* **11** (3) (1989) 10–12.

[Man89b] B.B. Mandelbrot, Some "facts" that evaporate upon examination, *Math. Intelligencer* **11** (4) (1989) 17–19.

[Mañé87] R. Mañé, *Ergodic Theory and Differentiable Dynamics*, Springer-Verlag, Berlin, 1987 [English translation of Portuguese text *Introdução à Teoria Ergódica*, 1983].

[Mar96] J.-P. Marco, Transition le long des chaînes de tores invariants pour des systèmes hamiltoniens analytiques, *Ann. Inst. Henri Poincaré (Physique théorique)*, **64** (2) 1996, 205–252.

[Mar11] J.-P. Marco, Lectures on Arnold diffusion, preprint 2011.

[Mar13] J.-P. Marco, Generic hyperbolic properties of classical systems on the torus $\mathbb{T}^2$, preprint 2013 (to appear).

[MarG13] J.-P. Marco and M. Gidea, Generic hyperbolic properties of nearly integrable systems on $\mathbb{A}^{3n}$, preprint 2013 (to appear).

[MarS03] J.-P. Marco and D. Sauzin, Stability and instability for Gevrey quasi-convex near-integrable Hamiltonian systems, *Publ. Math. Inst. Hautes Etudes Sci.* No. 96 (2003) 199–275.

[MarkM70] L. Markus and K.R. Meyer, Generic Hamiltonian systems are not ergodic, in *Qualitative Methods of the Theory of Nonlinear Oscillations*, Tome 2 (Proceedings of the 5th International Conference on Nonlinear Oscillations, Kiev, Aug. 25–Sept. 4, 1969) Institute of Mathematics, Ukrainian Academy of Sciences, Kiev, 1970, pp. 311–332.

[MarkM74] L. Markus and K.R. Meyer, Generic Hamiltonian dynamical systems are neither integrable nor ergodic, *Memoirs Amer. Math. Soc.* No. 144, Amer. Math. Soc., Providence, RI, 1974.

[Marm01] S. Marmi, *An Introduction to Small Divisors Problems*, unpublished manuscript dated September 27, 2000 (91 pp.) [available online].

[MarmY02] S. Marmi and J.-C. Yoccoz, Some open problems related to small divisors, in *Dynamical Systems and Small Divisors* (Lectures given at the C.I.M.E. Summer School, held in Cetraro, Italy, June 13–20, 1998, L.H. Eliasson, S.B. Kuksin, S. Marmi and J.-C. Yoccoz, Eds.) pp. 175–191, Lecture Notes in Math. Vol. 1784, Springer-Verlag, Berlin, 2002.

[MarsR94] J.E. Marsden and T.S. Ratiu, *Introduction to Mechanics and Symmetry* (Texts in Applied Mathematics, Vol. 17), Springer-Verlag, New York, 1994 (2nd Ed. 1999).

[MartDS98] M. Martelli, M. Dang, and T. Seph, Defining chaos, *Mathematics Magazine* **71** (2) (1998) 112–122.

[Mas02] M. Mashaal, *Bourbaki: une société secrète de mathématiciens*, Éditions pour la science, Paris, 2002 [English translation: *Bourbaki: a secret society of mathematicians*, American Math. Soc., Providence, RI, 2006.]

[Mat82] J.N. Mather, Existence of quasiperiodic orbits for twist homeomorphisms of the annulus, *Topology* **21** (4) (1982) 457–467.

[Mat84] J.N. Mather, Nonexistence of invariant circles, *Ergodic Theory Dynam. Systems* **4** (1984) no. 2, 301–309.

[Mat89] J.N. Mather, Minimal measures, *Comment. Math. Helv.* **64** (1989), no. 3, 375–394.

[Mat91] J.N. Mather, Action minimizing invariant measures for positive definite Lagrangian systems, *Math. Z.* **207** (1991), no. 2, 169–207.

[Mat93] J.N. Mather, Variational construction of connecting orbits, *Ann. Inst. Fourier* (Grenoble) **43** (5) (1993) 1349–1386.

[Mat03] J.N. Mather, Arnol'd diffusion I, announcement of results, *Sovrem. Mat. Fundam. Napravl.* **2** (2003), 116–130 (electronic, Russian) [English translation in *J. Math. Sci.* (NY) **124** (2004) no. 5, 5275–5289].

[Mat06] J.N. Mather, Arnol'd diffusion II, preprint 2006, 160 pp.

[MatMNR00] J.N. Mather, H.P. McKean, L. Nirenberg, and P.H. Rabinowitz, Jürgen K. Moser (1928–1999), *Notices of the AMS* **47** (11) (2000) 1392–1405.

[Mei92] J.D. Meiss, Symplectic maps, variational principles, and transport, *Rev. Mod. Phys.* **64** (1992) 795–848.

[Mei07a] J.D. Meiss, Dynamical systems, *Scholarpedia* 2(2):1629 (2007).

[Mei07b] J.D. Meiss, Hamiltonian systems, *Scholarpedia* 2(8):1943 (2007).

[Mei07c] J.D. Meiss, *Differential Dynamical Systems*, Society for Industrial and Applied Mathematics, Philadelphia, 2007.

[Meln63] V.K. Melnikov, On the stability of a center for time-periodic perturbations (Russian) *Trudy Moskov. Mat. Obsc.* **12** (1963) 3–52 [English translation: *Trans. Moscow Math. Soc.* **12** (1963) 3–56].

[Meln65] V.K. Melnikov, On certain cases of conservation of almost periodic motions with a small change of the Hamiltonian function (Russian) *Dokl. Akad. Nauk SSSR* **165** (1965) 1245–1248.

[MeloS93] W. de Melo and S. van Strien, *One-Dimensional Dynamics*, Springer-Verlag, New York, 1993.

[MeyHO09] K.R. Meyer, G.R. Hall, and D. Offin, *Introduction to Hamiltonian Dynamical Systems and the N-Body Problem*, 2nd Ed. (Applied Mathematical Sciences, Vol. 90), Springer-Verlag, New York, 2009.

[Mic95] L. Michelotti, *Intermediate Classical Dynamics with Applications to Beam Physics*, John Wiley & Sons, New York, 1995.

[Mill96] A.I. Miller, *Insights of Genius: Imagery and Creativity in Science and Art*, Springer-Verlag, New York, 1996.

[Miln99] J. Milnor, *Dynamics in One Complex Variable*, Princeton Univ. Press, Princeton, NJ, 1999 [second edition 2000; third edition 2006].

[Min35] H. Mineur, Sur les systèmes mécaniques admettant n intégrales premières uniformes et l'extension à ces systèmes de la méthode de quantification de Sommerfeld (French) [On mechanical systems admitting n uniform first integrals and the extension to these systems of Sommerfeld's quantization method] *C.R. Acad. Sci. Paris* **200** (1935) 1571–1573.

[Min36] H. Mineur, Réduction des systèmes mécaniques à n degrés de liberté admettant n intégrales premières uniformes en involution aux systèmes à variables séparées (French) [Reduction of n-degree-of-freedom mechanical systems admitting n uniform first integrals in involution to systems with separated variables] *J. Math. Pures Appl.* **15** (1936) 385–389.

[Mit12] G. Mittag-Leffler, Zur Biographie von Weierstrass, *Act. Math.* **35** (1912) 29–65.

[Moe96] R. Moeckel, Transition tori in the five-body problem, *J. Differential Equations* **129** (1996), no. 2, 290–314.

[Mof90] H.K. Moffat, KAM-theory, *Bull. London Math. Soc.* **22** (1990) 71–73.

[Mol68] A. Molchanov, The resonant structure of the solar system, *Icarus* **8** (2) (1968) 203–215.

[MorG95a] A. Morbidelli and A. Giorgilli, Superexponential stability of KAM tori, *J. Statist. Phys.* **78** (5–6) (1995) 1607–1617.

[MorG95b] A. Morbidelli and A. Giorgilli, On a connection between KAM and Nekhoroshev's theorems, *Physica D* **86** (3) (1995) 514–516.

[Mos55] J.K. Moser, Stabilitätsverhalten kanonischer differentialgleichungssysteme (German) [Stability behavior of canonical systems of differential equations] *Nachr. Akad. Wiss. Göttingen, Math. Phys. Kl.* II (1955), 87–120.

[Mos61] J.K. Moser, A new technique for the construction of solutions of nonlinear differential equations, *Proc. Nat. Acad. Sci. U.S.A.* **47** (1961) 1824–1831.

[Mos62] J.K. Moser, On invariant curves of area-preserving mappings of an annulus, *Nachr. Akad. Wiss. Göttingen Math.-Phys. Kl. II* (1962) 1–20.

[Mos65] J.K. Moser, Combination tones for Duffing's equation, *Comm. Pure Appl. Math.* **18** (1965) 167–181.

[Mos66] J.K. Moser, A rapidly convergent iteration method and non-linear partial differential equations, I, *Ann. Scuola Norm. Sup.* Pisa (3) **20** (1966) 265–315; II, *ibid.*, 499–535.

[Mos67] J.K. Moser, Convergent series expansions for quasi-periodic motions, *Math. Ann.* **169** (1967) 136–176.

[Mos68] J.K. Moser, Lectures on Hamiltonian systems, *Mem. Am. Math. Soc.* **81** (1968) 1–60.

[Mos70] J.K. Moser, On the construction of almost periodic solutions for ordinary differential equations, *Proc. Internat. Conf. on Functional Analysis and Related Topics* (Tokyo, 1969), Univ. of Tokyo Press, Tokyo, 1970, pp. 60–67.

[Mos73] J.K. Moser, Stable and random motions in dynamical systems, with special emphasis on celestial mechanics, *Ann. Math. Studies 77*, Princeton University Press, Princeton, NJ, 1973.

[Mos75] J.K. Moser, Ist das Sonnensystem stabil?, *Neue Zürcher Zeitung*, May 14, 1975 (German) [English translation: Is the Solar System Stable? in *Math. Intelligencer* **1** (2) (1978) 65–71].

[Mos86] J.K. Moser, Recent developments in the theory of Hamiltonian systems, *SIAM Review* **28** (4) (1986) 459–485.

[Mos98] J.K. Moser, Dynamical systems—past and present, *Doc. Math.* Extra Vol. I (1998) 381–402.

[Mos99a] J.K. Moser, Recollections. Concerning the early development of KAM theory, in *The Arnoldfest: Proceedings of a Conference in Honour of V.I. Arnold for his Sixtieth Birthday (Toronto, ON, 1997)*, Fields Institute Communications 24, American Mathematical Society, Providence, RI, 1999, pp. 19–21.

[Mos99b] J.K. Moser, Old and new applications of KAM theory, in *Hamiltonian Systems with Three or More Degrees of Freedom (S'Agaró, 1995)*, NATO Adv. Sci. Inst. Ser. C Math. Phys. Sci., Vol. 533, Kluwer Acad. Publ., Dordrecht, 1999, pp. 184–192.

[Mos01] J.K. Moser, Remark on the paper "On invariant curves of area-preserving mappings of an annulus," *Regul. Chaotic Dyn.* **6** (2001) no. 3, 337–338.

[MosZ05] J.K. Moser and E.J. Zehnder, *Notes on Dynamical Systems (Courant lecture notes in mathematics 12)*, American Mathematical Society, Providence, RI, 2005.

[MouhV11] C. Mouhot and C. Villani, On Landau damping, *Acta Math.* **207**, no. 1 (2011) 29–201.

[Moul14] F.R. Moulton, *An Introduction to Celestial Mechanics* (2nd revised Ed.), Macmillan, New York, 1914.

[Mur91] J. Murdock, *Perturbations, Theory and Methods*, John Wiley & Sons, New York, 1991.

[Nab99] P. Nabonnand (Ed.), *La correspondance entre Henri Poincaré et Gösta*

Mittag-Leffler (French) [The Correspondence between Henri Poincaré and Gösta Mittag-Leffler] Birkhäuser Verlag, Basel, 1999.

[Nas98] S. Nasar, *A Beautiful Mind*, Simon and Schuster, New York, 1998.

[Nash56] J. Nash, The imbedding problem for Riemannian manifolds, *Annals of Math.*, **63** (1956), 20–63.

[NassP12] M. Nassiri and E. Pujals, Robust transitivity in Hamiltonian dynamics, *Ann. Sci. Ecole Norm. Sup. (4)* **45**, no. 2 (2012) 191–239.

[NassP13] M. Nassiri and E. Pujals, Robust transitivity in Hamiltonian dynamics II, preprint 2013.

[Nei81] A.I. Neishtadt, Estimates in the Kolmogorov theorem on conservation of conditionally periodic motions (Russian), *Prikl. Matem. Mekhan. SSSR* **45**, no. 6 (1981), 1016–1025.

[Nek71] N.N. Nekhoroshev, Behavior of Hamiltonian systems close to integrable (Russian), *Fun. Anal. Pril.* **5** (4) (1971), 82–84). [English translation: *Funct. Anal.* **5** (1971), 338–339.]

[Nek73] N.N. Nekhoroshev, Stable lower estimates for smooth mappings and for gradients of smooth functions (Russian) *Mat. Sbornik* **90** (132) (1973), 425–467. [English translation: *Math. USSR Sbornik* **19** (3) (1973), 425–467.]

[Nek77] N.N. Nekhoroshev, An exponential estimate of the time of stability of nearly integrable Hamiltonian systems (Russian), *Usp. Mat. Nauk. SSSR* **32** (6) (1977), 5–66. [English translation: *Russian Math. Surveys* **32** (6) (1977), 1–65.]

[Nek79] N.N. Nekhoroshev, An exponential estimate of the time of stability of nearly integrable Hamiltonian systems II (Russian), in *Tr. Sem. Petrows.* **5** (1979), 5–62. [English translation in *Topics in Modern Mathematics, Petrovskii Seminar No. 5* (O.A. Oleinik, Ed.), Consultants Bureau, London, 1980.]

[New99] I. Newton, *The Principia, Mathematical Principals of Natural Philosophy* (a new translation by I.B. Cohen and A. Whitman, assisted by J. Budenz; preceded by *A Guide to Newton's Principia*, by I.B. Cohen) Univ. of Calif. Press, Berkeley, 1999.

[Nie96] L. Niederman, Stability over exponentially long times in the planetary problem, *Nonlinearity* **9** (6) (1996) 1703–1751.

[Nie04] L. Niederman, Exponential stability for small perturbations of steep integrable Hamiltonian systems, *Ergodic Theory Dynam. Sys.* **24** (2) (2004) 593–608.

[Nie06] L. Niederman, Hamiltonian stability and subanalytic geometry, *Ann. Inst. Fourier* (Grenoble) **56** (2006) no. 3, 795–813.

[Nie07] L. Niederman, Prevalence of exponential stability among nearly integrable Hamiltonian systems, *Ergodic Theory Dynam. Sys.* **27** (3) (2007) 905–928.

[Neu29] J. von Neumann, Beweis des Ergodensatzes und des H-Theorems in der neuen Mechanik (German) [Proof of the ergodic hypothesis and of the H-theorem in the new mechanics], *Zeitschrift für Physik* **57** (1929) 30–70.

[Neu32] J. von Neumann, Proof of the quasi-ergodic hypothesis, *Proc. Natl. Acad. Sci. U.S.A.* **18** (1932) 70–82 .

[OxU41] J.C. Oxtoby and S.M. Ulam, Measure-preserving homeomorphisms and metrical transitivity, *Ann. of Math.* (2) **42** (1941) 874–920.

[Pal69] J. Palis, On Morse-Smale dynamical systems, *Topology* **8** (1969) 385–404.

[PalM82] J. Palis and W. de Melo, *Geometric Theory of Dynamical Systems, An Introduction*, Springer-Verlag, New York, 1982.

[Par82] I.O. Parasyuk, Conservation of quasiperiodic motions of reversible multifrequency systems (Russian, with English summary) *Dokl. Akad. Nauk Ukrain. SSR Ser.* A (9) **85** (1982) 19–22.

[Par84a] I.O. Parasyuk, Non-Poisson commutative symmetries and multidimensional invariant tori of Hamiltonian systems (Russian, English summary) *Dokl. Akad. Nauk Ukrain. SSR Ser.* A 1984, no. 10, 14–17.

[Par84b] I.O. Parasyuk, Preservation of multidimensional invariant tori of Hamiltonian systems (Russian) *Ukrain. Mat. Zh.* (4) **36** (1984) 467–473.

[Pat87] A. Patrascioiu, The ergodic hypothesis, a complicated problem in mathematics and physics, in *Los Alamos Science* No. 15, Special Issue: Stanislaw Ulam 1909–1984 (1987) 263–279.

[Perc79] I.C. Percival, A variational principle for invariant tori of fixed frequency, *J. Phys. A* **12** (1979) no. 3, L57–L60.

[Pére92] R. Pérez-Marco, Solution complète au problème de Siegel de linéarisation d'une application holomorphe au voisinage d'un point fixe (d'après J.-C. Yoccoz) (French) [Complete solution to Siegel's problem of linearizing a holomorphic mapping in the neighborhood of a fixed point (following J.-C. Yoccoz)], *Séminaire Bourbaki*, 44me année, N° 753, 1991–92 (31 pp.).

[PerrW94] A.D. Perry and S. Wiggins, KAM tori are very sticky: Rigorous lower bounds on the time to move away from an invariant Lagrangian torus with linear flow, *Physica D* **71** (1994) 102-121.

[Pet83] K.E. Petersen, *Ergodic Theory*, Cambridge U. Press, Cambridge, UK, 1983.

[Pete93] I. Peterson, *Newton's Clock: Chaos in the Solar System*, W.H. Freeman and Co., New York, 1993.

[PettiCCFC05] M. Pettini, L. Casetti, M. Cerruti-Sola, R. Franzosi, and E. G. D. Cohen, Weak and strong chaos in Fermi-Pasta-Ulam models and beyond, *Chaos* **15** (1) 015106 (2005) 13 pp.

[Pf17] G.A. Pfeiffer, On the conformal mapping of curvilinear angles. The functional equation $\phi[f(x)] = a_1\phi(x)$, *Trans. Amer. Math. Soc.* **18** (2) (1917) 185–198.

[Pi09] G. Pinzari, On the Kolmogorov set for many-body problems, PhD thesis, Università degli Studi Roma Tre, April 2009 [available online].

[Poi79] J.H. Poincaré, Sur les propriétés des fonctions définies par les équations aux différences partielles (French) [On the properties of functions defined by partial difference equations] (Première thèse de doctorat, présentée à la Faculté des Sciences de Paris, 1er août, 1879); Reprinted in Œuvres de Henri Poincaré, Tome 1, Gauthier-Villars, Paris, 1928, pp. IL–CXXXII.

[Poi86] J.H. Poincaré, Sur les intégrales irrégulières des équations linéaires (French) [On irregular integrals of linear equations], *Acta Math.* **8** (1886); Reprinted in Œuvres de Henri Poincaré Tome 1, pp. 290–332.

[Poi90] J.H. Poincaré, Sur le problème des trois corps et les équations de la dynamique (French) [On the three body problem and the equations of dynam-
ics], *Acta Math.* **13** (1890), 1–270; Reprinted in *Œuvres de Henri Poincaré*
Tome 7, pp. 262–479.

[Poi92–99] J.H. Poincaré, *Les Méthodes nouvelles de la mécanique céleste* **I**, **II**,
III, Gauthier-Villars, Paris, 1892, 1893, 1899. [English translation: New
Methods of Celestial Mechanics, Vols. 1, 2, 3 (D. Goroff, Ed.) American
Institute of Physics, Woodbury, NY, 1993.]

[Poi95] J.H. Poincaré, Analysis situs (Latin title, French text), *J. Ecole Polytechnique* **1** (1895) 1–121. Reprinted in *Œuvres de Henri Poincaré* Tome 6,
pp. 193–288.

[Poi08] J.H. Poincaré, *Science et méthode*, Edition Ernest Flammarion, Paris,
1908. [English translation: *Science and Method*, Dover Publications Inc.,
New York, 2003.]

[Poi12] J.H. Poincaré, Sur un théorème de géométrie (French) [On a theorem of
geometry], *Rend. del. Circ. Math. Palermo* **33** (1912) 375–407.

[Poi15–56] J.H. Poincaré, *Œuvres de Henri Poincaré* (French) [The Works of
Henri Poincaré] 11 tomes, Gauthier-Villars, Paris, 1915–1956.

[PonB05] A. Ponno and D. Bambusi, Korteweg-de Vries equation and energy
sharing in Fermi-Pasta-Ulam, *Chaos* **15** (1) 015107 (2005) 5 pp.

[Pop04] G. Popov, KAM theorem for Gevrey Hamiltonians, *Ergodic Theory Dynam. Sys.* **24** (2004) 1753–1786.

[Por98] N.A. Porter, *Physicists in Conflict*, Institute of Physics Publishing, Bristol, UK, 1998.

[Pös80] J. Pöschel, Über invariante Tori in differenzierbaren Hamiltonschen
Systemen (German) [On invariant tori in differentiable Hamiltonian sys-
tems] Diplomarbeit, Rheinische Friedrich-Wilhelms-Universität, Bonn, 1979.
Beiträge zur Differentialgeometrie, 3. Bonner Mathematische Schriften, 120.
Universität Bonn, Mathematisches Institut, Bonn, 1980. 103 pp.

[Pös82] J. Pöschel, Integrability of Hamiltonian systems on Cantor sets, *Comm.
Pure Appl. Math.* **35** (5) (1982), 653–696.

[Pös90] J. Pöschel, Small divisors with spatial structure in infinite-dimensional
Hamiltonian systems, *Comm. Math. Phys.* **127** (2) (1990), 351–393.

[Pös93] J. Pöschel, Nekhoroshev estimates for quasi-convex Hamiltonian systems,
Math. Z. **213** (2) (1993) 187–216.

[Pös96] J. Pöschel, A KAM theorem for some nonlinear partial differential equations, *Ann. Scuola Norm. Sup. Pisa* Cl. Sci. (4), **23** (1): 119–148 (1996).

[Pös99] J. Pöschel, On Nekhoroshev's estimate at an elliptic equilibrium, *Internat. Math. Res. Notices* no. 4 (1999) 203–215.

[Pös01] J. Pöschel, A lecture on the classical KAM theorem, in *Smooth Ergodic
Theory and its Applications (Seattle, WA, 1999), Proc. Sympos. Pure Math.*
69, Amer. Math. Soc., Providence, RI, 2001, pp. 707–732.

[Pös11] J. Pöschel, KAM à la R, *Regul. Chaotic Dyn.* **16**, nos. 1–2 (2011) 17–23.

[Pya69] A.S. Pyartli, Diophantine approximations on submanifolds of Euclidean
space (Russian), *Funkcional. Anal. i Prilozen.* **3** (1969), 59–62 [English
translation: *Functional Anal. Appl.* **3** (1969), 303–306].

[Rin06] B.W. Rink, Proof of Nishida's conjecture on anharmonic lattices, *Comm. Math. Phys.* **261** (2006), 613–627.

[Rin09] B.W. Rink, Fermi Pasta Ulam systems (FPU): mathematical aspects, *Scholarpedia*, 4 (12): 9217 (2009).

[Rob95] P. Robutel, Stability of the planetary three-body problem II, KAM theory and existence of quasiperiodic motions, *Celestial Mech. Dynam. Astronom.* **62** (3) (1995) 219–261.

[Rota97] G.-C. Rota, *Indiscrete Thoughts*, Birkhäuser, Boston, MA, 1997.

[RouN09] C. Rouvas-Nicolis and G. Nicolis, Butterfly effect, *Scholarpedia* 4(5):1720 (2009).

[Rud86] W. Rudin, *Real and Complex Analysis* (3rd Ed.), McGraw-Hill, New York, 1986.

[Rue91] D. Ruelle, *Chance and Chaos*, Princeton U. Press, Princeton, NJ, 1991.

[Rüss76] H. Rüssmann, On optimal estimates for the solutions of linear difference equations on the circle, *Celestial Mech.* **14** (1) (1976) 33–37.

[Rüss89] H. Rüssmann, Nondegeneracy in the perturbation theory of integrable dynamical systems, in *Number Theory and Dynamical Systems*, (York, 1987; M.M. Dodson and J.A.G. Vickers, Eds.), London Math. Soc. Lecture Note Ser., Vol. 134, Cambridge Univ. Press, Cambridge, 1989, pp. 5–18; reprinted in *Stochastics, Algebra and Analysis in Classical and Quantum Dynamics* (Marseille, 1988; S. Albeverio, P. Blanchard, and D. Testard, Eds.), Math. Appl., 59, Kluwer Acad. Publ., Dordrecht, 1990, pp. 211–223.

[Rüss01] H. Rüssmann, Invariant tori in non-degenerate nearly integrable Hamiltonian systems, *Regul. Chaotic Dyn.* **6** (2) (2001), 119–204.

[Rüss10] H. Rüssmann, KAM-iteration with nearly infinitely small steps in dynamical systems of polynomial character, *Discrete Contin. Dyn. Syst.*, Ser. S, **3**, no. 4 (2010) 683–718.

[Sa04] D.A. Salamon, The Kolmogorov-Arnold-Moser theorem, *Math. Phys. Electron. J.* **10**, Paper 3 (2004), 37 pp. (electronic).

[SaZ89] D.A. Salamon and E. Zehnder, KAM theory in configuration space, *Comment. Math. Helv.* **64** (1) (1989) 84–132.

[SanVM07] J.A. Sanders, F. Verhulst, and J. Murdock, *Averaging Methods in Nonlinear Dynamical Systems*, 2nd Ed. (Applied Mathematical Sciences, Vol. 59), Springer-Verlag, New York, 2007.

[Sard00] Z. Sardar, *Thomas Kuhn and the Science Wars*, Icon Books, Cambridge, U.K., 2000.

[Sart52] G. Sarton, *A Guide to the History of Science*, Chronica Botanica, Waltham, MA, 1952.

[Sche79] J. Scheurle, Über die Konstruktion invarianter Tori, welche von einer stationären Grundlösung eines reversiblen dynamischen Systems abzweigen (German, with English summary) [On the construction of invariant tori which bifurcate from a stationary solution of a reversible dynamical system], in *Constructive Methods for Nonlinear Boundary Value Problems and Nonlinear Oscillations* (Proc. Conf., Math. Res. Inst., Oberwolfach, 1978), pp. 134–144, Internat. Ser. Numer. Math., 48, Birkhäuser, Basel-Boston, Mass., 1979.

[Schr71] E. Schröder, Über iterierte Functionen (German) [On iterated functions], *Math. Ann.* **3** (1871) 296–322.

[Schw60] J.T. Schwartz, On Nash's implicit functional theorem, *Comm. Pure Appl. Math.* **13** (1960) 509–530.

[Ser72] F. Sergeraert, Un théorème de fonctions implicites sur certains espaces de Fréchet et quelques applications (French) [A theorem on implicit functions on certain Fréchet spaces and some applications], *Ann. Sci. École Norm. Sup.* **5** (1972) 599–660.

[Sev91] M.B. Sevryuk, Lower-dimensional tori in reversible systems, *Chaos* **1** (1991) no. 2, 160–167.

[Sev95] M.B. Sevryuk, KAM-stable Hamiltonians, *J. Dynam. Control Sys.* **1** (1995) 351–366.

[Sev96] M.B. Sevryuk, Invariant tori of Hamiltonian systems nondegenerate in the sense of Rüssmann, *Dokl. Math.* **53** (1996) 69–72.

[Sev98] M.B. Sevryuk, The finite-dimensional reversible KAM theory, *Phys. D* **112** (1998) nos. 1–2, 132–147.

[Sev03] M.B. Sevryuk, Classical KAM theory at the dawn of the twenty-first century (dedicated to Vladimir Igorevich Arnold on the occasion of his 65th birthday), *Mosc. Math. J.* **3** (3) (2003) 1113–1144, 1201–1202.

[Sev13] M.B. Sevryuk, Classical Hamiltonian perturbation theory, to appear in *Celestial Mechanics* (A. Celletti, Ed.) a volume of the *Encyclopedia of Life Support Systems* (EOLSS), Developed under the Auspices of the UNESCO, Eolss Publishers, Oxford, UK, [http://www.eolss.net] (2013).

[Sie42] C.L. Siegel, Iteration of analytic functions, *Ann. of Math.* (2) **43** (1942) 607–612.

[Sie54] C.L. Siegel, Über die Existenz einer Normalform analytischer Hamiltonscher Differentialgleichungen in der Nähe einer Gleichgewichtslösung (German) [On the existence of a normal form for analytic Hamiltonian differential equations in the neighborhood of an equilibrium solution], *Math. Ann.* **128** (1954) 144–170.

[Sie56] C.L. Siegel, *Vorlesungen über Himmelsmechanik* (German) [Lectures on Celestial Mechanics], Springer-Verlag, Berlin, 1956.

[SieM71] C.L. Siegel and J.K. Moser, *Lectures on Celestial Mechanics*, Springer-Verlag, Berlin, 1971.

[Sim72] G.F. Simmons, *Differential Equations with Applications and Historical Notes*, McGraw-Hill, New York, 1972.

[Sin63] Ya. G. Sinai, On the foundations of the ergodic hypothesis for a dynamical system of statistical mechanics, *Dokl. Akad. Nauk SSSR* **153** (1963) 1261–1264 (Russian) [English translation: *Soviet Math. Dokl.* **4** (1963) 1818–1822.]

[Sin70] Ya. G. Sinai, Dynamical systems with elastic reflections. Ergodic properties of dispersing billiards, *Uspehi Mat. Nauk* **25** no. 2 (152), (1970) 141–192 (Russian) [English translation: *Russian Math. Surveys* **25** (1970) 137–189.]

[SinC87] Ya. G. Sinai and N.I. Chernov, Ergodic properties of some systems of two-dimensional disks and three-dimensional balls (Russian), *Uspekhi Mat. Nauk* **42** no. 3 (255) (1987) 153–174.

[Skl93] L. Sklar, *Physics and Chance: Philosophical Issues in the Foundations of Statistical Mechanics*, Cambridge Univ. Press, Cambridge, UK, 1993.

[Skl09] L. Sklar, Philosophy of statistical mechanics, *The Stanford Encyclopedia of Philosophy* (Summer 2009 Edition), E.N. Zalta (Ed.), URL = http://plato.stanford.edu/archives/fall2008/entries/statphys-statmech/.

[Sma65] S. Smale, Diffeomorphisms with many periodic points, in *Differential and Combinatorial Topology* (A Symposium in Honor of Marston Morse), pp. 63–80, Princeton Univ. Press, Princeton, NJ, 1965.

[Sma80] S. Smale, On the problem of reviving the ergodic hypothesis of Boltzmann and Birkhoff, in *Nonlinear Dynamics* (Proceedings of the International Conference on Nonlinear Dynamics held by the New York Academy of Sciences, December 17–21, 1979; R.H.G. Helleman, Ed.) Annals of the New York Academy of Sciences, Vol. 357, New York Academy of Sciences, New York, 1980, pp. 260–266.

[Ster57] S. Sternberg, Local contractions and a theorem of Poincaré, *Amer. J. Math.* **79** (1957) 809–824.

[Ster58] S. Sternberg, On the structure of local homeomorphisms of euclidean n-space, II, *Amer. J. Math.* **80** (1958) 623–631.

[Ster59] S. Sternberg, The structure of local homeomorphisms, III, *Amer. J. Math.* **81** (1959) 578–604.

[Stew89] I. Stewart, *Does God Play Dice? The Mathematics of Chaos*, Basil Blackwell Ltd, Oxford, UK, 1989.

[Stew95] I. Stewart, Bye-bye Bourbaki: paradigm shifts in mathematics, *Mathematical Gazette* **79** (486) (1995) 496–498.

[Sul01] D. Sullivan, Reminiscences of Michel Herman's first great theorem, *SMF Gazette* **88** avril (2001) 90–93 [available online].

[Su07] K.F. Sundman, Recherches sur le problème des trois corps (French) [Research on the three body problem], *Acta Soc. Scien. Fennicae* **34** (6) (1907) 1–43.

[Su09] K.F. Sundman, Nouvelles recherches sur le problème des trois corps (French) [New research on the three body problem], *Acta Soc. Scien. Fennicae* **35** (9) (1909) 1–27.

[Su12] K.F. Sundman, Mémoire sur le problème des trois corps (French) [Memoir on the three body problem], *Acta Math.* **36** (1912) 105–179.

[SusW88] G.J. Sussman and J. Wisdom, Numerical evidence that the motion of Pluto is chaotic, *Science* **241** (1988) 433–437.

[SusW92] G.J. Sussman and J. Wisdom, Chaotic evolution of the solar system, *Science* **257** (1992) 56–62.

[Szá96] D. Szász, Boltzmann's ergodic hypothesis, a conjecture for centuries? *Studia Sci. Math. Hungaricae* **31** (1996) 299–322; reprinted in *Encyclopaedia Math. Sci.*, **101**, Springer-Verlag, Berlin, 2000, pp. 421–448.

[Szá08] D. Szász, Some challenges in the theory of (semi)-dispersing billiards, *Nonlinearity* **21**, no. 10 (2008) T187–T193.

[Tak08] L.A. Takhtajan, *Quantum Mechanics for Mathematicians* (Graduate Studies in Mathematics Vol. 95), American Mathematical Society, Providence, R.I., 2008.

[Tay96] M.E. Taylor, *Partial Differential Equations, Basic Theory*, Springer-Verlag, New York, 1996.

[Tou99] S. Toulmin, The idol of stability, in *The Tanner Lectures on Human Values*, Vol. 20 (G.B. Peterson, Ed.), University of Utah Press, Salt Lake City, 1999, pp. 327–354 [available online].

[TouG61] S. Toulmin and J. Goodfield, *The Fabric of the Heavens, the development of astronomy and dynamics*, Harper & Brothers, New York, 1961.

[Tre04] D. Treschev, Evolution of slow variables in a priori unstable Hamiltonian systems, *Nonlinearity* **17** (5) (2004) 1803–1841.

[TreZ10] D. Treschev and O. Zubelevich, *Introduction to the Perturbation Theory of Hamiltonian Systems*, Springer Monographs in Mathematics, Springer-Verlag, Berlin, 2010.

[Tru84] C. Truesdell, The computer: ruin of science and threat to mankind, in *An Idiot's Fugitive Essays on Science*, Springer-Verlag, New York, 1984, pp. 594–631.

[TuR98] D. Turaev and V. Rom-Kedar, Elliptic islands appearing in near-ergodic flows, *Nonlinearity* **11**, no. 3 (1998) 575–600.

[Tz04] S.I. Tzenov, *Contemporary Accelerator Physics*, World Scientific, Singapore, 2004.

[Uf07] J. Uffink, Compendium to the foundations of classical statistical physics, in *Philosophy of Physics* (J. Butterfield and J. Earman, Eds.) North-Holland, Amsterdam, 2007, 923–1074.

[Uf08] J. Uffink, Boltzmann's Work in Statistical Physics, *The Stanford Encyclopedia of Philosophy* (Winter 2008 Edition), E.N. Zalta, Ed., URL = http://plato.stanford.edu/archives/win2008/entries/statphys-Boltzmann/

[Veks44] V.I. Veksler, A new method of the acceleration of relativistic particles (Russian), *Dokl. Akad. Nauk SSSR* **43**: 346 (1944).

[Ver12] F. Verhulst, *Henri Poincaré, Impatient Genius*, Springer Science & Business Media, New York, 2012.

[Vit07] P.M.B. Vitanyi, Andrey Nikolaevich Kolmogorov, *Scholarpedia* 2(2):2798 (2007).

[Wal82] P. Walters, *An Introduction to Ergodic Theory*, Springer-Verlag, New York, 1982.

[Wan91] Q.-D. Wang, The global solution of the n-body problem, *Celestial Mech. Dynam. Astron.* **50** (1991) 73–88.

[Wat02] P. Watson, *The Modern Mind : an intellectual history of the 20th century*, Harper Collins, New York, 2002.

[Wat05] P. Watson, *Ideas: A History of Thought and Invention, from Fire to Freud*, Harper Perennial, New York, 2005.

[Way96] C.E. Wayne, An introduction to KAM theory, in *Dynamical Systems and Probabilistic Methods in Partial Differential Equations*, Lectures in Appl. Math. Vol. 31, Amer. Math. Soc., Providence, RI, 1996, pp. 3–29.

[Wein92] S. Weinberg, *Dreams of a Final Theory*, Pantheon Books, New York, 1992.

[Weis97] T.P. Weisert, *The Genesis of Simulation in Dynamics: Pursuing the Fermi-Pasta-Ulam Problem*, Springer, New York, 1997.

[Wes80] R.S. Westfall, *Never at Rest, a Biography of Isaac Newton*, Cambridge University Press, UK, 1980.

[Wes93] R.S. Westfall, *The Life of Isaac Newton*, Cambridge University Press, UK, 1993.

[Wey46] H. Weyl, *Classical Groups*, Princeton Univ. Press, Princeton, NJ, 1946.

[White99] M. White, *Isaac Newton: The Last Sorcerer*, Helix Books, New York, 1999.

[Whitn34] H. Whitney, Differentiable functions defined in closed sets, I, *Trans. Amer. Math. Soc.* **36** (2) (1934) 369–387.

[Wie93] H. Wiedemann, *Particle Accelerator Physics, basic principles and linear beam dynamics*, Springer-Verlag, Heidelberg-Berlin, 1993.

[Wie95] H. Wiedemann, *Particle Accelerator Physics II, nonlinear and higher-order beam dynamics*, Springer-Verlag, Heidelberg-Berlin, 1995.

[Wig92] S. Wiggins, *Chaotic Transport in Dynamical Systems*, Springer-Verlag, New York, 1992.

[Wig94] S. Wiggins, *Normally Hyperbolic Invariant Manifolds in Dynamical Systems*, Springer-Verlag, New York, 1994.

[Wig90–03] S. Wiggins, *Introduction to Applied Dynamical Systems and Chaos*, Springer-Verlag, New York, 1990 [2nd edition 2003].

[Wil00] K. Wille, *The Physics of Particle Accelerators, An Introduction*, Oxford Univ. Press, Oxford, UK, 2000.

[Wils01] E.J.N. Wilson, *Introduction to Particle Accelerators*, Oxford Univ. Press, New York, 2001.

[Xia93] Z. Xia, Arnold diffusion in the elliptic restricted three-body problem, *J. Dynam. Differential Equations* **5** (1993), no. 2, 219–240.

[Xia94] Z. Xia, Arnold diffusion and oscillatory solutions in the planar three-body problem, *J. Differential Equations* **110** (1994), no. 2, 289–321.

[Ya02] B.H. Yandell, *The Honors Class, Hilbert's problems and their solvers*, A.K. Peters, Ltd., Natick, Mass., 2002.

[Yo88] J.-C. Yoccoz, Linéarisation des germes de difféomorphismes holomorphes de $(C, 0)$ (French; English summary) [Linearization of germs of holomorphic diffeomorphisms of $(C, 0)$] *C.R. Acad. Sci. Paris Sér. I Math.* **306**, N° 1 (1988) 55–58.

[Yo92] J.-C. Yoccoz, Travaux de Herman sur les tores invariants (French) [Herman's work on invariant tori] *Séminaire Bourbaki*, 1991/92, Exp. 754, *Astérisque* N° 206 (1992) 311–344.

[Yo94] J.-C. Yoccoz, Interview in *Pour la science*, N° 203, septembre 1994, p. 24.

[Yo95a] J.-C. Yoccoz, Recent developments in dynamics, in *Proceedings of the International Congress of Mathematicians (Zürich, 1994)*, Vol. 1, Birkhäuser, Basel, 1995, pp. 246–265.

[Yo95b] J.-C. Yoccoz, Théorème de Siegel, nombres de Bruno et polynômes quadratiques (French) [Siegel's theorem, Bruno numbers, and quadratic polynomials] *Astérisque* N° 231 (1995) 3–88.

[Yo02] J.-C. Yoccoz, Analytic linearization of circle diffeomorphisms, in *Dynamical Systems and Small Divisors* (Lectures at the C.I.M.E. Summer School, held in Cetraro, Italy, June 13–20, 1998; L.H. Eliasson, S.B. Kuksin,

S. Marmi and J.-C. Yoccoz, Eds.) pp. 125–174, Lecture Notes in Math. Vol. 1784, Springer-Verlag, Berlin, 2002.

[Zab05] N.J. Zabusky, Fermi-Pasta-Ulam, solitons and the fabric of nonlinear and computational science: History, synergetics, and visiometrics, *Chaos* **15** (1) 015102 (2005) 16 pp.

[ZabK65] N.J. Zabusky and M.D. Kruskal, Interactions of solitons in a collisionless plasma and the recurrence of initial states, *Phys. Rev. Lett.* **15** (1965) 240–243.

[Zas05] G.M. Zaslavsky, Long way from the FPU-problem to chaos, *Chaos* **15** (1) 015103 (2005) 10 pp.

[Zdr87] S. Zdravkovska, Conversation with Vladimir Igorevich Arnol'd, *Math. Intelligencer* **9** (4) (1987) 28–32.

[ZdrD93–07] S. Zdravkovska and P.L. Duren (Eds.), *Golden Years of Moscow Mathematics*, Hist. Math. **6**, Amer. Math. Soc., Providence, RI, and London Math. Soc., London, 1993 (2nd Edition 2007).

[Ze74] E. Zehnder, An implicit function theorem for small divisor problems, *Bull. Amer. Math. Soc.* **80** (1) (1974) 174–179.

[Ze75–76] E. Zehnder, Generalized implicit function theorems with applications to some small divisor problems, I, *Comm. Pure Appl. Math.* **28** (1) (1975) 91–140; *ibid.* II, **29** (1) (1976) 49–111.

[Ze76] E. Zehnder, Moser's implicit function theorem in the framework of analytic smoothing, *Math. Ann.* **219** (2) (1976) 105–121.

[Zh11] K. Zhang, Speed of Arnold diffusion for analytic Hamiltonian systems, *Inventiones math.* **186** (2011) no. 2, 255–290.

Index

*Page numbers in italics indicate glossary entries in Appendix F.

Printed in the United States
By Bookmasters